The Lighting Management Handbook

THE LIGHTING MANAGEMENT HANDBOOK

By Craig DiLouie

Published by
THE FAIRMONT PRESS, INC.
700 Indian Trail
Lilburn, GA 30247

Library of Congress Cataloging-in-Publication Data

DiLouie, Craig, 1967-
The lighting management handbook / by Craig DiLouie.
p. cm.
Includes bibliographical references and index.
ISBN 0-88173-169-2
1. Electric lighting--Handbooks, manuals, etc. 2. Electric power-Conservation--Handbooks, manuals, etc. 3. Interior lighting--Handbooks, manuals, etc. 4. Office buildings--Lighting--Equipment and supplies--Maintenance and repair--Handbooks, manuals, etc. I. Title.
TK4175.D55 1994 621.32--dc20 93-31081
CIP

The Lighting Management Handbook / by Craig DiLouie.

Published by The Fairmont Press, Inc.
700 Indian Trail
Lilburn, GA 30247

Printed in the United States of America

10 9 8 7 6 5 4 3 2 1

ISBN 0-88173-169-2 FP

ISBN 0-13-147406-5 PH

While every effort is made to provide dependable information, the publisher, authors, and editors cannot be held responsible for any errors or omissions.

Distributed by PTR Prentice Hall
Prentice-Hall, Inc.
A Paramount Communications Company
Englewood Cliffs, NJ 07632

Prentice-Hall International (UK) Limited, London
Prentice-Hall of Australia Pty. Limited, Sydney
Prentice-Hall Canada Inc., Toronto
Prentice-Hall Hispanoamericana, S.A., Mexico
Prentice-Hall of India Private Limited, New Delhi
Prentice-Hall of Japan, Inc., Tokyo
Simon & Schuster Asia Pte. Ltd., Singapore
Editora Prentice-Hall do Brasil, Ltda., Rio de Janeiro

DEDICATION

This book is dedicated to Richard Reisser, my grandfather; to Eileen DiLouie, my mother; and to the fond memory of Ronald Lieberman, teacher and mentor … for giving me the kind of belief and inspiration that makes great things possible.

CONTENTS

PREFACE xiii

INTRODUCTION xv

The Mission of Planned Lighting Management
Effective Illumination
Efficient Illumination
Retrofitting

ACKNOWLEDGMENTS xix

CHAPTER 1: FUNDAMENTALS AND TECHNOLOGY: GETTING STARTED

Lighting fundamentals 1
- Light and vision 1
- The lighting system 2
- The mission of the lighting system 3
- Quantity of light 4
  - Luminous flux
  - Illuminance
- Quality of light 5
  - Luminance
  - Color rendition
- Lighting system efficiency 7
  - Input watts
  - Relative efficiency

Fundamentals of fluorescent lighting 8
- Composition 9
  - Traditional lamp composition
  - Traditional ballast composition
- Theory of operation 11
- Lighting circuits 12
  - Preheat circuit
  - Instant start circuit
  - Rapid start circuit
- Lamp identification 14
  - Rapid start (40 watts or less) and preheat lamps
  - High output rapid start lamps
  - Very high output rapid start lamps
  - Instant start lamps
  - Note on specialty lamps
- Lamp and ballast life 16
  - Lamp life
  - Ballast life

Lamp efficiency: It's the law....17
Ballast efficiency: It's the law....18
A quick note on "standard"....18
Lighting upgrade equipment....18
Upgrade lamp types....19
Example: Energy-saving T12 lamps – F40T12/ES lamp
Example: T10 lamps – F40T10 lamp
Example: T8 lamps – F32T8 lamp
Upgrade ballast types....24
Example: Energy-efficient ("premium") magnetic ballasts
Example: Cathode cut-out ("hybrid") ballasts
Example: Electronic ballasts
Other upgrade equipment types....26
Rapid start fluorescent system comparisons, 2-lamp fixture....27
Rapid start fluorescent system comparisons, 3-lamp fixture....29
Rapid start fluorescent system comparisons, 4-lamp fixture....30
Questions to ask when choosing equipment....32
Does the area need it?
Is the equipment compatible?
How old is the existing system?
Can light level be sacrificed in certain areas?
What is the track record of the equipment's manufacturer?

CHAPTER 2: THE BUILDING SURVEY: SCOUTING FOR OPPORTUNITY
Getting started....37
Establish a plan....37
Prioritize benefits
Gather education, training and resources
Prioritize use of labor
Collect building, lighting system, financial and occupant information
Determine required light levels per room and address quality issues
Research financing opportunities
Establish lighting management options
Determine light levels provided by new equipment options
Conduct an economic analysis of the options
Complete an environmental benefits assessment
Conduct a full economic analysis
Present the proposal to the decisionmaker
Conduct a trial installation
Full installation
Maintain the system
Monitor results
Market environmental commitment
Collect information....40
Collect general information....40
Collect occupant information....40
Area-by-area facility survey....42
Equipment....42
Light meter
Mechanical counter
Access equipment

Using a light meter 43
Completing the area-by-area survey 44
Survey Worksheet 45
Completing the Survey Worksheet 47

CHAPTER 3: EFFECTIVE ILLUMINATION: MEETING THE "PEOPLE PRIORITY"

People need light ... and the right lighting 51
Value of increased human performance 52
Specifying quantity of light 53
Time 53
Size 54
Contrast 54
Specifying light level 54
Light Level Worksheet #1 57
Specifying lumen output 60
Coefficient of utilization 62
Fixture efficiency
Room proportions
Room surface reflectances
Procedure to determine coefficient of utilization
Non-recoverable light loss factors 65
Ballast factor
Ambient fixture temperature (fluorescent systems)
Supply voltage variation
Optical factor
Fixture surface depreciation
Lamp position (HID systems)
Recoverable light loss factors 68
Lamp burnouts
Lamp lumen depreciation
Fixture dirt depreciation
Room surface dirt depreciation
Light Level Worksheet #2 76
Complete Light Level Worksheet #2 78
Completing Light Level Worksheet #2
Completing Light Level Worksheet #3 80
Light Level Worksheet #3 81
Quality of Light 82
Luminance 82
Discomfort glare
Disability glare
Direct glare
Reflected glare
Visual comfort probability (VCP)
Color and quality of light 87
Correlated color temperature
Color rendering
Let the application determine color quality
Completing the Lighting Quality Worksheet 92
Lighting Quality Worksheet 93

CHAPTER 4: RETROFIT ECONOMICS: FINDING THE BOTTOM LINE
Make the system efficient 97
How utilities charge for service 98
Energy use charge 98
Demand charge 99
Demand charge and the ratchet clause
Power factor charge 99
Fuel adjustment charge 100
Other charges 100
Reduce the electric bill 100
Demand for power 103
Unit power density (UPD) 103
Energy consumption 103
Upgrade the lighting system for energy savings 104
Reduce light levels 104
Increase light levels 104
Maintain light levels 105
Decrease hours of use 105
Proper maintenance 105
Choosing the option 105
Measuring energy savings 107
Payback and ROI formula
Complete the Retrofit Economics Worksheet 108
Retrofit Economics Worksheet 109
Retrofit Economics Worksheet, Page 1
Retrofit Economics Worksheet, Page 2
Retrofit Economics Worksheet, Page 3
Retrofit Economics Worksheet, Page 4
Retrofit Economics Worksheet, Page 5
Retrofit Economics Worksheet, Page 6
Calculating net HVAC savings 119
Determining actual savings 119
Measuring air cooling energy savings 121
Measuring additional heat required after retrofit 122
Freezing new cooling requirements 124

CHAPTER 5: MAINTENANCE: MAKING THE SYSTEM WORK FOR THE LONG HAUL
The mission of planned lighting maintenance 127
Anatomy of poor maintenance 127
Anatomy of planned maintenance 128
Maintenance in principle
Maintenance in application
Benefits of planned lighting maintenance 128
Consistent proper light level 129
Effective use of labor 129
Effective use of resources 129
Effective use of energy 130
Depreciation cushion
Reduce depreciation cushion with maintenance

Plan ahead
Group relamping 131
The lighting system at end of life 131
Cost of lamp lumen depreciation
Cost of burnouts
Group relamping versus spot relamping 137
Cleaning 138
The lighting system at end of life 139
Effects of dirt depreciation
Maintenance operations and methods 141
Methods
Use the right cleaning method
Use the right equipment
Managing a maintenance program 143
Survey existing conditions 143
Determine maintenance interval 143
Complete Maintenance Worksheet #1
Maintenance Worksheet #1 145
Choosing the most economical option
Maintenance Worksheet #2 149
Completing Maintenance Worksheet #2
Determine energy management opportunity 153
Establish a permanent maintenance policy 153
Fluorescent lamp and ballast disposal 154
Fluorescent lamps 154
Fluorescent ballasts 156
How to identify PCB ballasts
TSCA regulations for leaking ballasts
CERCLA regulations for non-leaking ballasts
State guidelines
Disposal options

CHAPTER 6: RETROFIT FINANCING: PUTTING MONEY TO WORK
How to pay for your lighting upgrade 163
Overview of utility demand side management (DSM) programs 164
What is DSM? 164
Why utilities are promoting DSM 166
Least-cost planning
Environmental costing and the Clean Air Act
Financial incentive mechanisms set up by states
Current progress and future outlook for DSM 169
Types of utility-sponsored DSM programs 170
Rebates 170
Technology-based
Energy reduction
Direct assistance 192
Low-interest loans 192
Getting involved 192
Be prepared to stay energy-efficient permanently 192
Manufacturer rebate 193

Third-party financing 193
Conventional financing....193
Lease-purchase....194
Lease-purchase with guaranteed savings 195
Shared savings....196
Rolling financing 198

CHAPTER 7: GREEN ENGINEERING: ASSESSING ENVIRONMENTAL IMPACT
Lighting upgrades bring green benefits 201
Energy's relationship with the environment....202
Carbon dioxide (CO_2) and global warming.... 202
Sulphur dioxide (SO_2) and acid rain....204
Nitrogen oxides (NO_x) and ozone pollution (smog) 204
Corporations can help the environment by helping themselves 205
Marketing green benefits....206
Three out of four Americans are environmentalists....206
Selling green benefits to senior management 207
Federal Green Lights program....208
Measuring deferred emissions of air pollution 221
Environmental Assessment Worksheet #1 222
Calculating conserved resources 223
Environmental Assessment Worksheet #2 224
An innovative idea: using an avoided environmental cost (AEC) factor....227
Costs incorporated into AEC 228
Cost of increasing energy demand
Cost of decrease in agriculture
Calculating AEC

CHAPTER 8: GETTING HELP: ASSOCIATIONS, TRADE GROUPS, PROFESSIONALS, GOVERNMENT, PUBLICATIONS
Getting help 233
Professional and trade associations....234
Federal government resources 235
Publications....236

CHAPTER 9: SUCCESS STORIES: QUALITY LIGHTING UPGRADES
Cutting through red tape at the University of Oklahoma....239
Using a move to re-think lighting strategy at Knorr Brake 240
Teamwork at Augusta College....240
Customized upgrades at multiple facilities at Bank of America 241
Acton schools score high in efficiency 242
Elmbrook school district makes the grade: starts with a pilot project.... 242
Oak Park Hospital cures lighting ills.... 243
Lighting consultant worth his weight in gold, client says.... 244

APPENDIX I: GENERAL TECHNICAL INFORMATION....247

APPENDIX II: WORKSHEETS....255

APPENDIX III: PRODUCT GUIDES 275

Preface

This book is written for professionals responsible for managing and maintaining their organization's lighting system. These include energy managers, facility engineers and facility managers. Energy management professionals from electric utilities, lighting management companies, energy service companies and other organizations as well as students can also benefit from this book.

This book is designed to both educate and train you toward effective management and upgrade of an existing lighting system. Although designed for professionals, this book can also serve as a textbook for students of lighting desiring to learn more about improving and managing lighting systems.

First, we will discuss the mission of planned lighting management. Then we will review the basics of lighting, fluorescent lighting and typical fluorescent upgrades. (For a comprehensive review of lighting technologies and their application, consult John Fetters' book, *Lighting Retrofits: Solutions that Work*, the companion book in this two-volume set.) The material will also enhance your education about demand side management trends, the value of quality lighting, the value of maintenance, marketing "green" benefits and other important information.

At the same time, the book trains you to use this information to implement a lighting management program, going step by step — chapter by chapter — from the building survey on through economic analysis, maintenance and assessing environmental impact.

By the time you are finished with this book, you will be able to investigate your own lighting upgrade opportunities or speak in an educated manner with a competent lighting professional. Further, this book can serve as basic training material for future managers at your company.

Introduction

THE MISSION OF PLANNED LIGHTING MANAGEMENT

The discipline of planned lighting management requires taking action to ensure that the lighting system consistently provides the most effective illumination at the lowest operating and maintenance cost. Reaching this goal involves providing *quality lighting* to occupants, and *retrofitting* — installing energy-efficient equipment in existing lighting systems — to reduce energy consumption. The sum of these products and practices, implemented via a lighting management program, is known as a lighting upgrade.

Because a lighting upgrade is a multifaceted operation, and can pose a significant initial cost, planning is essential. Proper planning and careful execution is the best way to avoid costly mistakes. This way, the right equipment will be chosen, the right resources gathered and the right actions taken.

EFFECTIVE ILLUMINATION

The recent gold rush to save lighting energy has raised concerns about how system performance relates to human performance. Clearly, light is for people, and the right light level and quality are necessary for people to work and play at their best. As salaries, wages and benefits allocated by a company to its personnel can consume as much as 75% of its operating budget, an even modest boost in productivity can produce profits far greater than typical energy savings. A successful lighting management program makes this original purpose of the lighting system its first priority.

Therefore, the overall system — which includes the lighting system interacting with the workplace environment — must yield the highest quality and recommended quantity of light on a consistent basis. This is achieved through proper specification of light levels according to the demands of the tasks being performed and to occupant needs. It is also achieved through installation of specialized "people-sensitive" equipment, adjustments in the work environment where possible, innovative design and planned maintenance.

EFFICIENT ILLUMINATION

The best quantity and quality of light must be provided at the lowest operating and maintenance cost. This is achieved through installation of energy-efficient equipment, again through adjustments in the work environment where possible, and through planned maintenance.

Retrofitting

Since 1983, companies have spent more money modernizing their facilities than building new ones. According to the National Lighting Bureau, almost two out of three nonresidential buildings in the United States are at least 25 years old. While the structures of these buildings are sound, the building systems they were originally equipped with have been rendered obsolete by advancing technology. Building system "retrofit," therefore, has become an energy buzzword for the 1980s and '90s.

Most fluorescent lighting systems installed during the 1960s and '70s were of the easily recognizable workhorse package: a 2'x4' fixture, covered with a plastic lens, housing four F40 cool white lamps powered by two "standard" (traditional) magnetic ballasts. The fixtures are roughly box-shaped and their interiors coated with white enamel paint. Most lighting systems at that time were designed to provide light levels considered excessive by today's standards because energy was relatively cheap and the prevailing philosophy of lighting design was "more light, better sight." And many systems did not benefit from planned maintenance unless the facility employed a lighting maintenance company.

The change started in the 1970s. Two energy crunches made most companies re-evaluate their use of energy as energy prices skyrocketed. Many decided that one way to conserve energy was to remove lamps and ballasts from lighting fixtures. While delamping reduced energy use, it also resulted in a corresponding drop in light levels. The original purpose of the lighting system — to consistently provide recommended illumination to occupants — was seconded to the need to cut energy costs. Hence, worker productivity and safety were often placed at risk.

As President Ronald Reagan told America that energy conservation meant being "too hot in the summer, too cold in the winter," a different philosophy had already emerged in the market. The lighting industry produced a new generation of lighting technologies. Some equipment types were simply updated versions of old ones designed to provide better performance, better light quality and/or greater operating flexibility. Other equipment, such as occupancy sensors, reflectors, fixtures and other products, presented innovative new approaches to lighting spaces for less energy. This equipment in turn has undergone numerous improvements and has become increasingly reliable over the years.

Overall, new lighting technology represented the new philosophy of energy efficiency — and the lighting retrofit was born.

As with nearly any new technology, advanced lighting equipment costs more than the traditional workhorse package. When installing energy-efficient equipment, a tradeoff takes place — a higher sum is paid in advance to install the equipment so that the long-term benefit of lower operating costs can be reaped. Fortunately, it is most often an uneven trade in the lighting manager's favor. Depending on electric rates and type of upgrade, the operating cost savings typically recover the initial cost in three years or less.

GET STARTED

This book is designed to help you plan your lighting upgrade, from choosing equipment to assessing its environmental impact. As noted in the Preface, John Fetters supplies a comprehensive review of the latest upgrade technologies in the companion book of this two-volume set. This set is designed so that each book complements the other. Our focus here is how to implement a lighting upgrade.

So we do not start off blind, however, in Chapter 1 we will review the fundamentals of fluorescent lighting. Then we will review new fluorescent lamps and ballasts on the market and compare systems and equipment considerations. This information will give us a starting point so that we can then begin exploring planned lighting management.

Let us begin.

Acknowledgments

Special thanks to the following lighting professionals for their valuable assistance in completing this book. All are noted industry experts and consulted on this book to help ensure its technical accuracy and comprehensiveness.

Norm Blake
Manager of Specification Markets and Legislative Affairs
Osram Sylvania, Inc.
Sylvania Lighting Center
Danvers, MA 01923

C. Charles Occhino, CLMC
Vice President
Aetna Corp.
175 Brookline Street
Cambridge, MA 02139

Thomas S. Madonia, P.E.
Consultant
670 West Hazelwood Avenue
Rahway, NJ 07065

Source: National Lighting Bureau

CHAPTER 1

Fundamentals and Technology: Getting Started

LIGHTING FUNDAMENTALS

LIGHT AND VISION

Light is radiant energy which travels in waves composed of vibrating electric and magnetic fields. Light waves have both frequency and length; the ranges of frequency and wavelength differentiate light from other forms of radiant energy such as heat and radio waves. These properties are expressed on the electromagnetic spectrum.

Certain light waves comprise a portion of the spectrum called the visible light spectrum. Visible light is capable of exciting the eye's retina, producing a visual sensation which we call vision. The process of seeing, therefore, requires at least one functioning eye and visible light.

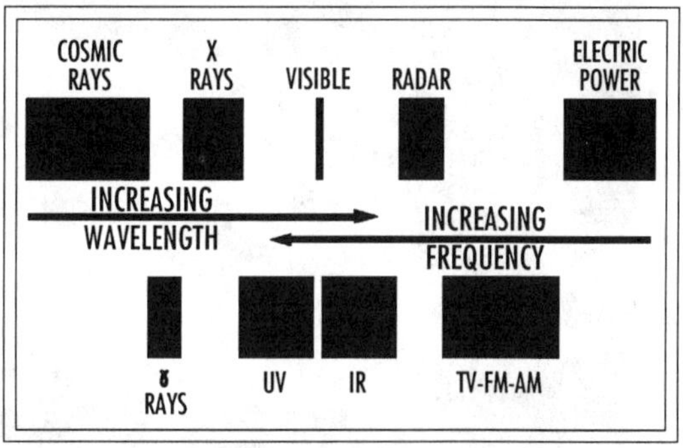

Figure 1-1. The electromagnetic spectrum. Visible light occupies a small portion along with ultraviolet light, infrared (heat) and other forms of radiated energy. *Source: Osram Sylvania, Inc.*

THE LIGHTING SYSTEM

Visible light can be natural or artificial. Artificial light is typically produced by electric-powered lighting systems. Nearly all of these systems generate light by sending electrical current either through an element (which heats until it glows) or through gases. In either case, the desired result is that atoms become excited until they emit radiant energy. The incandescent light bulb is an example of the first method. Fluorescent and high intensity discharge (HID) systems (mercury, high pressure sodium, metal halide and low pressure sodium) rely on gaseous discharge.

All lighting systems contain a lamp, which is the physical light source. In gaseous discharge systems, the lamp must be controlled by a ballast, a device which starts the lamp and/or regulates the electrical current. This combination is part of a luminaire, commonly called a light fixture, which is the lamp, ballast and other equipment designed to distribute the light. This equipment may include housing

Figure 1-2. The three basic light sources: high intensity discharge (left), incandescent (center) and fluorescent (right). Gaseous discharge sources (high intensity discharge and fluorescent) require a ballast to provide startup voltage and regulate the current after startup. *Source: Philips Lighting Company.*

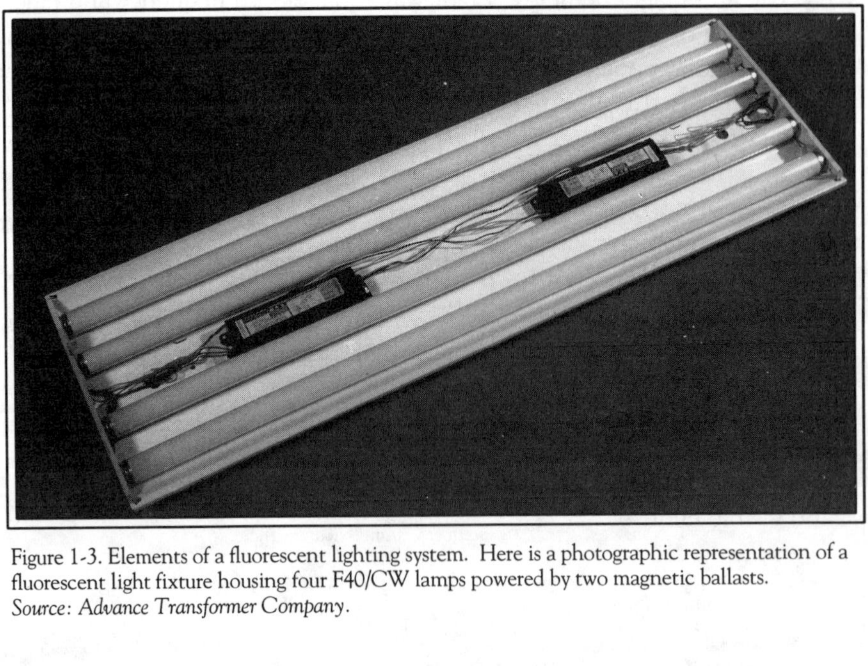

Figure 1-3. Elements of a fluorescent lighting system. Here is a photographic representation of a fluorescent light fixture housing four F40/CW lamps powered by two magnetic ballasts. *Source: Advance Transformer Company.*

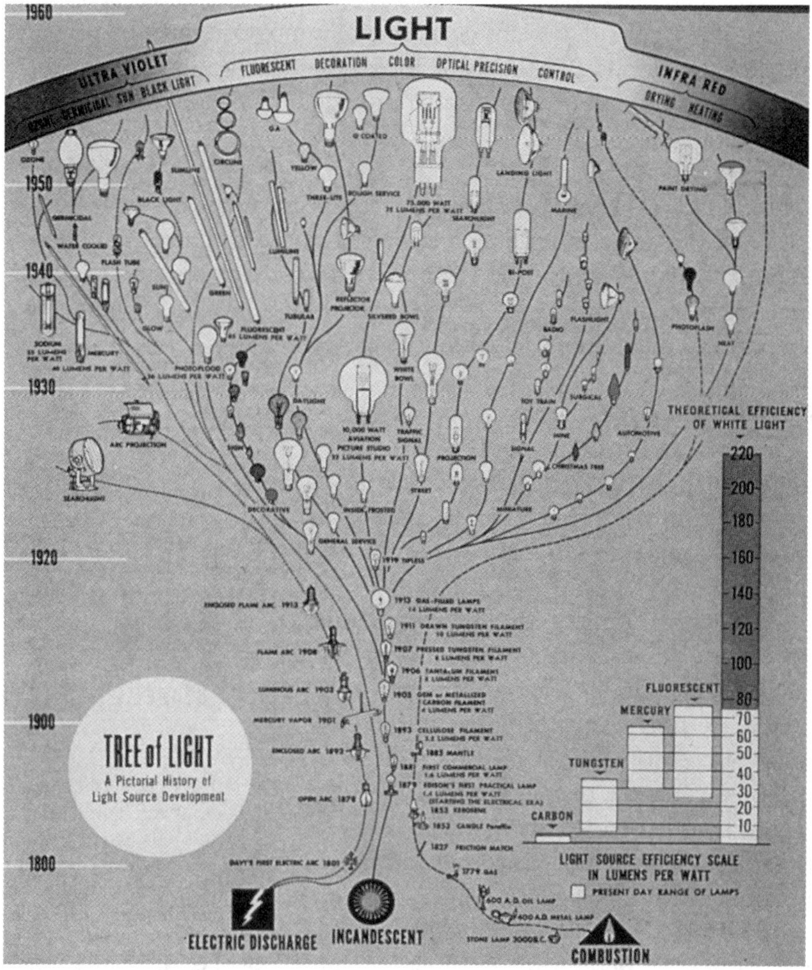

Figure 1-4. The "TREE OF LIGHT" showing the evolution of light sources.
Source: Philips Lighting Company.

for the lamp and shielding material such as a plastic lens which serves the additional purpose of diffusing (softening or scattering) the light or distributing it in a controlled pattern.

THE MISSION OF THE LIGHTING SYSTEM

The lighting system's mission, of course, is to provide illumination so that people can perform activities (called tasks) in areas lacking natural light.

The mere existence of light, however, is no guarantee of successfully performing a task. The right *quantity* of illumination must be provided for people to work and play safely at their best. At the same time, the right *quality* of illumination must also be provided for people to efficiently and safely perform visual tasks.

QUANTITY OF LIGHT

Quantity of light in an area is viewed as the relationship between the amount of light leaving the lamp and the amount of light reaching the area where people need it. This is called luminous flux (lumen or light output) and illuminance (light level), respectively.

Luminous flux

Luminous flux, commonly referred to as lumen or light output, is the quantity of light leaving the lamp. It is measured in lumens. Lamps are rated in both initial and mean lumens.

A catalog rating of initial lumens shows how many lumens are produced once the lamp has stabilized. Fluorescent and HID lamps are considered stabilized after 100 hours of operation.

Mean lumens reveals the average light output over lamp life, after continued operation depreciates the lamp's performance. For fluorescent lamps, mean lumens are typically determined at 40% of rated life.

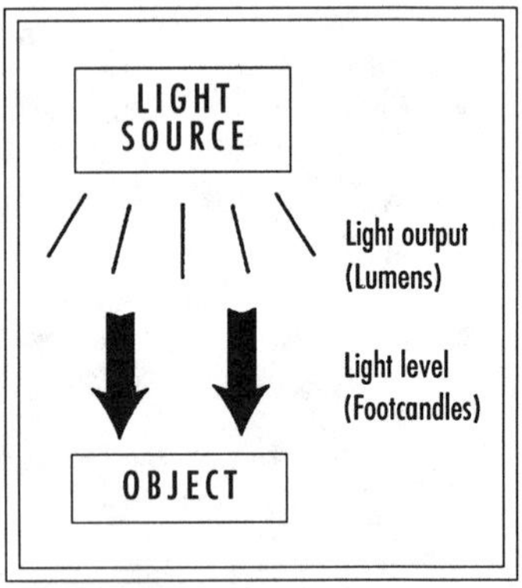

Figure 1-5. Relationship between light output and light level. Once the light is reflected from the object, it is perceived as brightness. *Source: The interNational Association of Lighting Management Companies.*

An F32T8 fluorescent lamp, for example, may be rated in a manufacturer's catalog at 2,900 initial lumens and 2,600 mean lumens.

Besides this depreciation, other factors affect lumen output, including variations in supply voltage and the lamp's interaction with the ballast.

The term "lumen output" is also used to describe the amount of light leaving the entire fixture. Factors such as the characteristics of the fixture, fixture surface depreciation (scratching and other damage to fixture surfaces), and dirt and dust accumulation on the lamp and fixture affect fixture lumen output.

Illuminance

Illuminance, commonly referred to as light level, is the amount of light measured on the workplane or task in the lighted area. In a typical office, the workplane would be considered an invisible plane resting on top of the

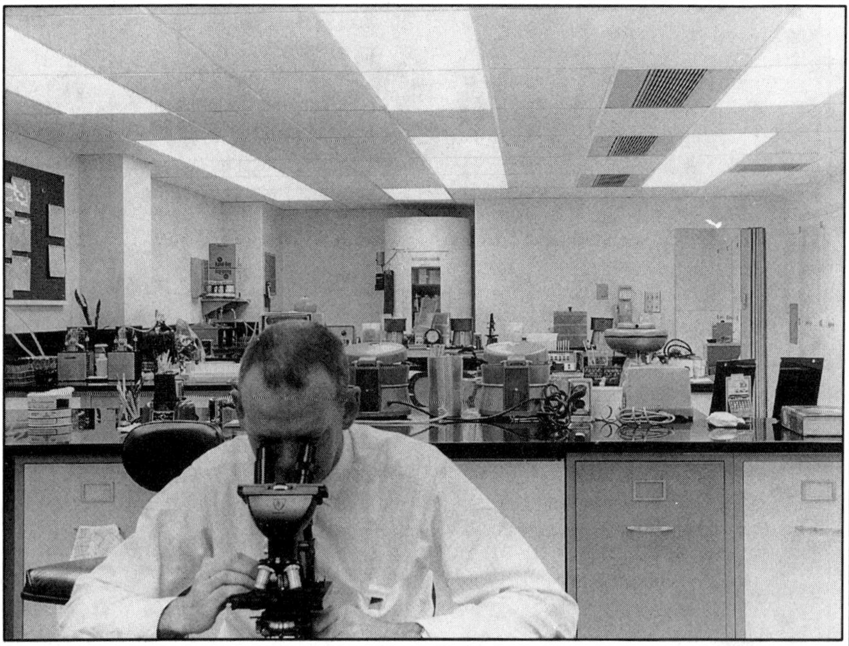

Figure 1-6. Insufficient quantity of light on a task can result in the task being performed below top efficiency and in costly errors. In addition, eyestrain can result in headaches and fatigue. *Source: National Lighting Bureau.*

desks, where most of the important visual tasks are performed. Numerous factors interfere with the quantity of lumens reaching this area, including room surface reflectances, room size and proportions, dirt and dust accumulation, and the characteristics of the fixture itself.

Light level is measured in footcandles (or lux [metric]) with a light meter. The relationship between lumen output and light level is manifested as the footcandle, because a footcandle represents one lumen of light density per square foot (while one lux is one lumen per square meter). Like lumens, footcandles can be determined as initial and maintained quantities. Initial footcandles tells the light level after new lamps are installed. Maintained footcandles is light level after all light loss factors (such as dirt, dust and lamp lumen depreciation) are considered for a given period of time.

QUALITY OF LIGHT

Quality of light describes how light leaving the lamp interacts with the work environment — room surfaces such as desk tops, computer screens and the like — and directly with human eyes. Just as people need the right amount of light to perform tasks productively, so do they need

Figure 1-7. In many applications, ensuring quality lighting is as important as ensuring proper quantity of light. In this industrial application, for example, proper attention must be given to color quality, contrast and brightness in addition to light levels.
Source: National Lighting Bureau.

good lighting quality to perform tasks comfortably. Ensuring proper lighting quality primarily involves controlling luminance and color rendition.

Luminance

Luminance, or photometric brightness, is the light leaving a lamp or an object's surface after it has been struck with light from a lamp. Brightness is the light we can actually see and that allows us to perceive objects. Uncontrolled brightness can cause glare. Direct glare can result from looking at the light source itself. Reflected glare can result from light reflecting from the task or other surfaces and objects in the room.

Color rendition

Besides visual comfort, lighting quality also describes the quality of the light itself. Lamps vary in their ability to show the color of objects — how most people expect them to appear. Color rendering is the ability of lamps to show or render colors accurately and show color shade variations more clearly. The color-rendering ability of a lamp can be judged using its Color Rendering Index (CRI) rating (on a scale of 0 to 100). A traditional F40 cool white (CW) fluorescent lamp has a CRI rating of 62, considered "good." A trichromatic (tri-phosphor) lamp, by contrast, holds a CRI rating of 70 to 85, considered "excellent."

LIGHTING SYSTEM EFFICIENCY

The lighting system is part of an electrical system — electricity is converted to light using the lamp or lamp-ballast system as a medium. The amount of electric power a particular lighting system requires to operate at any given moment is expressed as input watts. Once the lumen output of a system and its input watts are determined, its relative efficiency can be determined.

Input watts

A watt is a unit of measurement of required electric power. Rated input wattage for any electrical equipment is the amount of power it requires to operate at any given instant of time.

When lamp manufacturers determine lumen output for a particular fluorescent lamp, they operate the lamp on a special "reference ballast" specified by American National Standards Institute (ANSI) standards for that lamp type. The reference ballast is a laboratory device. It simulates the performance of a "perfect" ballast (a ballast that can transfer source

Figure 1-8. Ballasts are tested in a laboratory to determine input wattage and performance of lamp-ballast systems. *Source: Advance Transformer Company.*

power to the lamp with minimal power loss). The resulting power draw of the lamp is its nominal wattage. No commercial ballast is perfect, however. When a fluorescent lamp and ballast operate together as a system under actual field conditions, some efficiency is lost. The lamp is typically driven at a lower power, so it will not actually produce all of its rated initial lumens, and will not require its full input wattage. Ballast and fixture manufacturers publish input wattage ratings for lamps and ballasts operating together as a system.

When consulting a ballast manufacturer's literature, up to three input wattage ratings may be published. The ANSI number is the result of a manufacturer test conducted using an ANSI-specified procedure (typically called the "bench test" because the lamps and ballasts are operated bare on a bench). The next one or two ratings are manufacturer's ratings, the result of tests conducted in actual open and/or enclosed fixtures. The manufacturer's tests are generally considered more realistic simulations; and if that manufacturer's product is chosen, this number can be used to help complete a simple payback analysis. When comparing the relative efficiencies of ballasts, however, the ANSI number should be used because it reflects results from a common test procedure.

Relative efficiency

Two methods are used to compare the relative efficiency of lighting systems. The more common method uses the existing system as the standard. Compare lumen output and input wattage of proposed retrofit systems to this standard to determine if the proposed system is more efficient and therefore possibly a good economic opportunity.

The second method provides a purely relative comparison if there is no starting reference. It is based on comparing efficacy, an expression of lumen output produced per unit of power consumed (lumens per watt). Efficacy, which can be thought of as miles-per-gallon figures apply to cars, can be a good tool for comparison. It can show, for example, that one system produces *somewhat less* lumen output than another, but consumes *much less* energy, which in some cases is an acceptable tradeoff.

FUNDAMENTALS OF FLUORESCENT LIGHTING

For brevity and simplicity, fluorescent lighting will be the primary focus for the rest of this chapter. The true scope of this book is not to showcase new equipment types so much as how to specify, install and maintain them to make the lighting system effective at the lowest operating and maintenance cost. Before we can improve a fluorescent system on paper, we must understand how this system traditionally operates. This fundamentals section introduces the reader to traditional fluorescent lighting. In the next section, we will examine upgrade options.

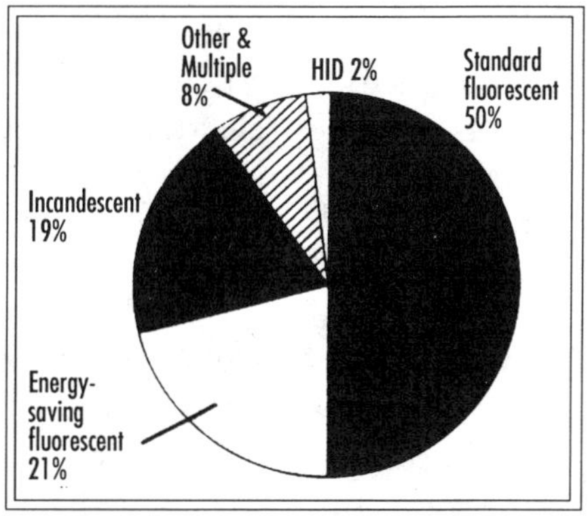

Figure 1-9. Predominant lighting equipment in commercial buildings — 1986. "Predominant lighting equipment" is defined as the type of equipment that lighted the greatest fraction of the building's floorspace. The figure indicates that standard fluorescent lighting represents the majority of lighting equipment in use. *Source: Energy Information Administration, 1988.*

COMPOSITION

All fluorescent lighting systems contain a lamp, which is the actual light-producing component. No fluorescent lamp can operate without a ballast. Named for the lead, iron or sand ballast used to stabilize ships in choppy ocean waters, the fluorescent ballast stabilizes the lamp's power supply — transforming supply voltage to the proper voltage to start the lamp, then regulating its operation by limiting the flow of current through the lamp.

Traditional lamp composition

A typical fluorescent lamp is a glass tube capped at each end by a base. Depending on the circuit, the base is tipped by a single pin, two pins or a

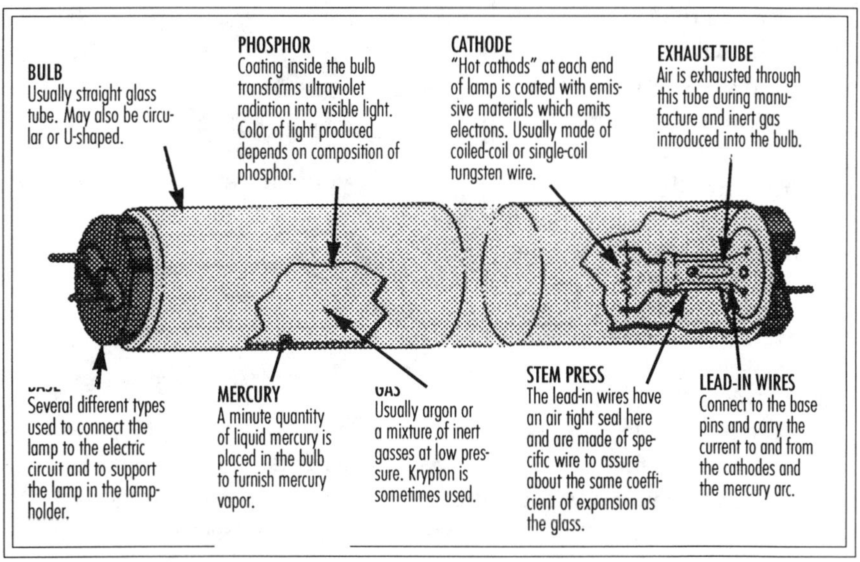

Figure 1-10. Composition of a traditional fluorescent lamp. *Source: Osram Sylvania, Inc.*

recessed double contact. The base supports the lamp inside the fixture and allows electricity to flow into the lamp. At the ends of the lamp's interior are its cathodes (electrodes), which are distant cousins of the filament windings inside an incandescent light bulb. The lamp is filled with gases and a small amount of mercury. The glass bulb is coated with phosphor on the inside and in most cases, silicon on the outside.

Traditional ballast composition

Magnetic ballasts, also called "core-and-coil" ballasts, contain a steel or iron core, wire coils and insulating material housed in a steel case. The core consists of stacked steel laminations. Insulated wire, made of copper, is coiled around the core. Together, these components act both as a transformer and a current-limiting choke. As the heart of an electrical device, they produce heat. Insulation material, which can be asphalt, conducts this heat away from the coils, because over time heat breaks down the insulation around the coils which in turn causes the ballast to fail.

Additionally, the ballast is likely to include a thermal protection device and/or a capacitor.

Thermal protection device. Under mandate of the 1984 U.S. National Electrical Code by Underwriters Laboratories Inc., all ballasts installed indoors from then on in the United States must be designated "Class P," meaning they contain a thermal protection device. If, due to abnormal conditions, the ballast begins to overheat, the thermal switch disconnects the ballast from the power source. When the ballast cools sufficiently, it is reconnected to the power source and restarts the lamps. If the cause of the overheating is not corrected, the ballast will continue to cycle the lamp on and off.

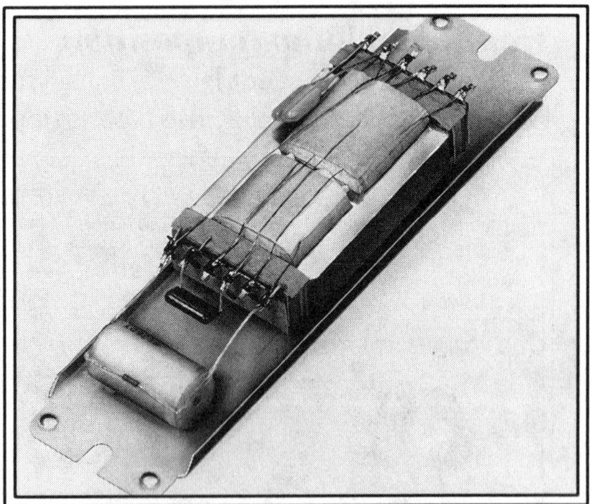

Figure 1-11. Interior of a magnetic ballast.
Source: Advance Transformer Company.

Capacitor. In high power factor ballasts (as opposed to low, or "normal," power factor ballasts), a capacitor corrects power factor. Corrected power factor does not significantly change input watts, but high power factor ballasts require less line current. This allows more fixtures to be installed on each branch circuit and avoids possible charges set by some

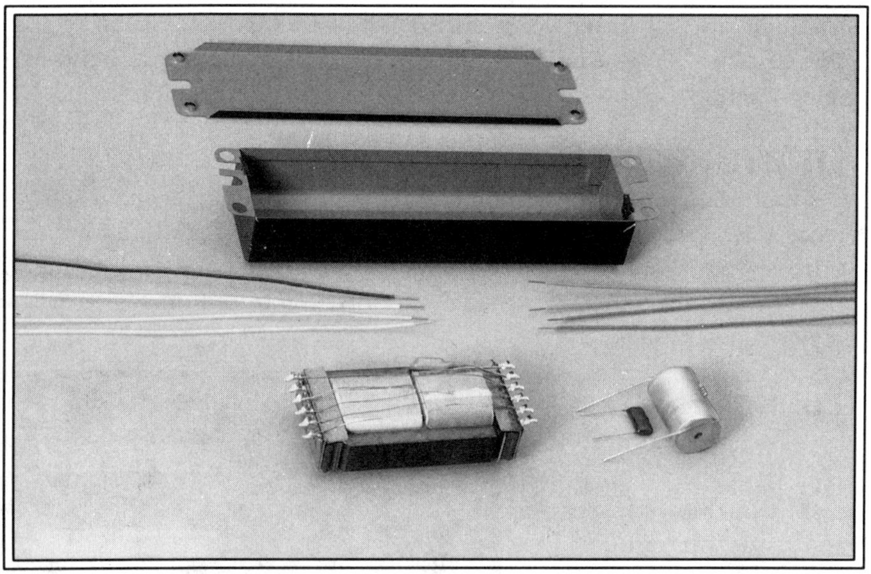

Figure 1-12. Components of a magnetic ballast.
Source: Advance Transformer Company.

utilities. For these reasons, low power factor ballasts should be avoided unless no high power factor ballast is available for the lamp in use.

THEORY OF OPERATION

When the wall switch is flicked into the "on" position, the ballast receives a message, in the form of electrical input via the closed circuit to start the lamp. It responds by providing the right voltage to activate the lamp cathodes. Materials on the heated cathodes emit electrons which conduct an arc of electrical current between the cathodes. As these electrons collide with electrons of atoms of gaseous mercury, the collisions produce ultraviolet energy. This energy is then converted to visible light by the phosphor coating on the inside of the glass tube. This process is called "fluorescence."

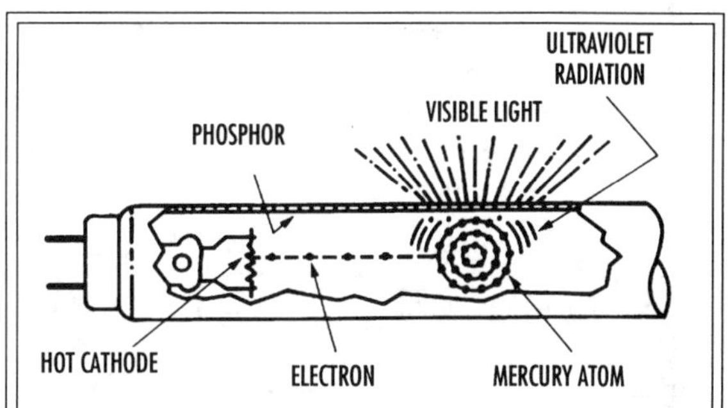

Figure 1-13. Fluorescent lamp operation.
Source: Osram Sylvania, Inc.

Once in operation, lamps exhibit negative electrical resistance; as current increases, the resistance to its flow decreases. If not controlled, the lamp would fail seconds after receiving the first current. Once the lamps are started, the ballast regulates the current to prevent immediate failure and operates the lamp at its design wattage.

LIGHTING CIRCUITS

Lighting systems operate on a preheat, instant start or rapid start circuit. Each circuit has its own mode of operation for starting lamps and other special characteristics, and can be electromagnetic or electronic. In this section, we will review magnetic only; electronic operation will be reviewed later in this chapter.

Preheat circuit

The preheat system, the first commercially available fluorescent system, was introduced in 1938 at the World's Fair in New York City and San Francisco. In a preheat system, the ballast needs the help of a starter to preheat the cathodes. Once sufficiently heated, the starter opens so that the ballast can supply startup voltage to the lamp. Two to three seconds of such warm-up time are

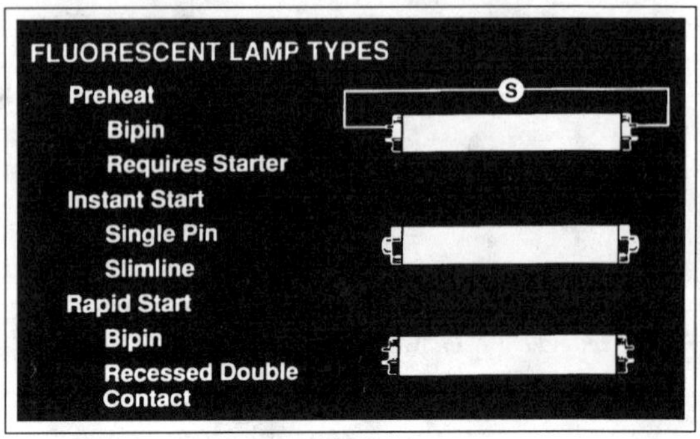

Figure 1-14. Basic fluorescent lamp types.
Source: MagneTek, Lighting and Electric Products.

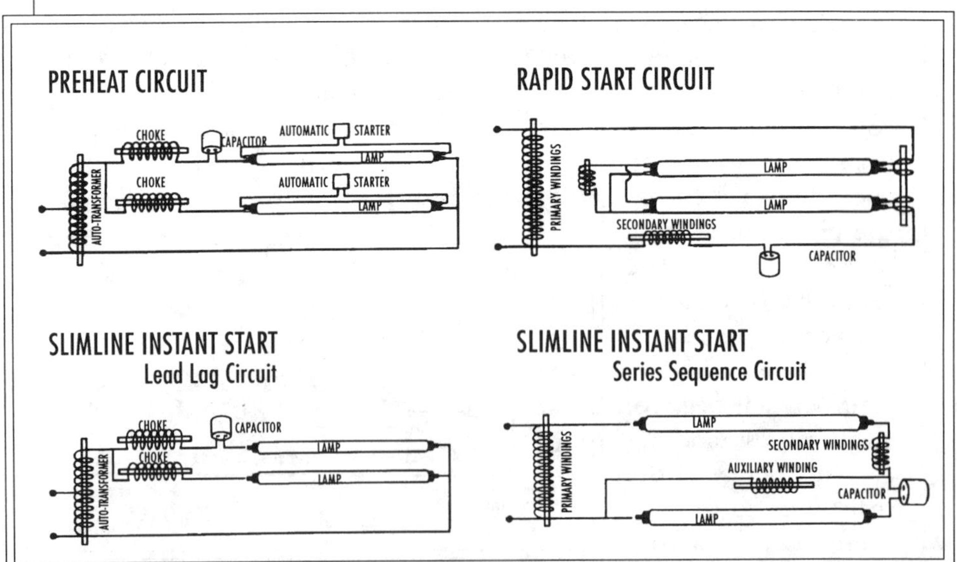

Figure 1-15. Wiring diagrams for preheat, rapid start and instant start circuits.
Source: Advance Transformer Company.

required before the lamp starts. Today, most preheat lamps are small tubular fluorescents consuming four to 20 watts. Preheat lamps are capped with a two-pin (bi-pin) base.

Up to two lamps can be powered by a single ballast. Ballasts operate preheat lamps on a parallel circuit. This means that if a ballast is powering two lamps and one lamp fails, its companion will continue to light because they are operated independently.

Instant start circuit

Instant start systems were introduced in 1944 to reduce the warm-up time required before the lamps started. They are also known as "slimline" systems because the original instant start lamps were thinner in diameter than the preheat lamps used at the time. These lamps are easily identified by a single-pin base capping each end of the tube. In operation, the ballast provides a higher starting voltage, heating the cathodes quickly enough so that the lamps start immediately. Although they do not need a starter, many instant start systems require a larger ballast, which should be avoided in noise-sensitive areas such as conference rooms because of its relatively noisy hum. In addition, higher starting voltage can shorten lamp life.

Up to two lamps can be powered by a single ballast. Instant start ballasts can be either lead-lag or series sequence ballasts. Lead-lag ballasts start and operate the lamps on a parallel circuit. Series sequence ballasts start the lamps in sequence and operate them in series. The latter means that if the first lighted lamp on the circuit fails, its companion lamp will fail to light. If the second lamp fails, its companion will be "stuck" in a starting mode and light only dimly. Series sequence ballasts for instant start systems, despite presenting a disadvantage if one lamp fails, are about one-third smaller than lead-lag ballasts, and less expensive.

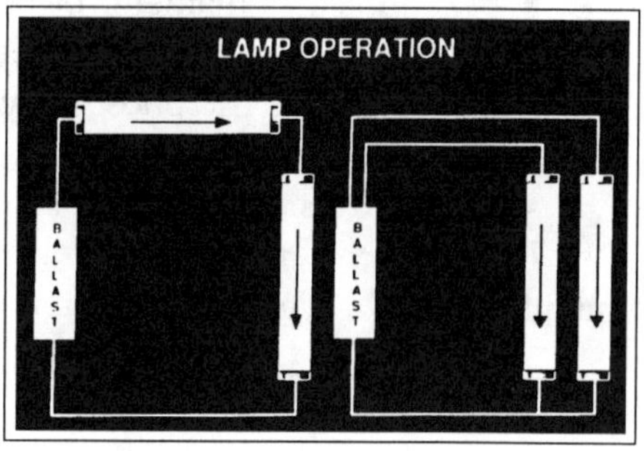

Figure 1-16. Lamp operating circuits. The left diagram shows series sequence operation; the right diagram shows parallel operation. *Source: MagnaTek, Lighting and Electronic Products.*

Rapid start circuits

Rapid start systems were introduced in 1952 and are the most popular fluorescent lighting systems for commercial applications. They do not need a starter nor a larger ballast, as the lamps start at a lower voltage than instant start lamps because the ballast constantly heats the cathodes. The lamps light almost immediately. They are capped with bi-pin bases.

Up to four lamps can be powered by a single ballast. The smaller ballast is less expensive and makes less noise than most instant start ballasts. It ignites the lamps on a series sequence circuit.

High light output rapid start systems. High output (HO) and very high output (VHO) lamps (also called 1500-milliamp lamps or Power Groove lamps) are rapid start lamps which operate at higher electrical currents (typically 800 and 1500 milliamps, respectively) to produce higher lumen output. They feature a recessed double-contact base. VHO lamps possess a lower efficacy than standard rapid start and HO lamps.

LAMP IDENTIFICATION

All lamps are identified by a designation code which reveals basic information about their physical dimensions and operating characteristics. Note that manufacturer's codes may differ slightly from one to the next. The best method to become familiar with lamp codes is to review catalogs published by the various manufacturers.

Rapid start (40 watts or less) and preheat lamps

For these systems, let us look at an F30T12/CW/RS lamp as an example.

- "F" indicates that this lamp is a fluorescent.
- "30" shows rated wattage.
- "T" means the lamp is tubular.
- "12" indicates diameter of the lamp in eighths of an inch — this lamp is 1-1/2 inches in diameter.
- "CW" reveals the phosphor color and appearance of the lamp; this lamp is a cool white lamp.
- "RS" means this is a rapid start lamp.

Exception: The code for traditional 40-watt rapid start lamps reads simply F40/CW. Product evolution in the past two decades led to the sim-

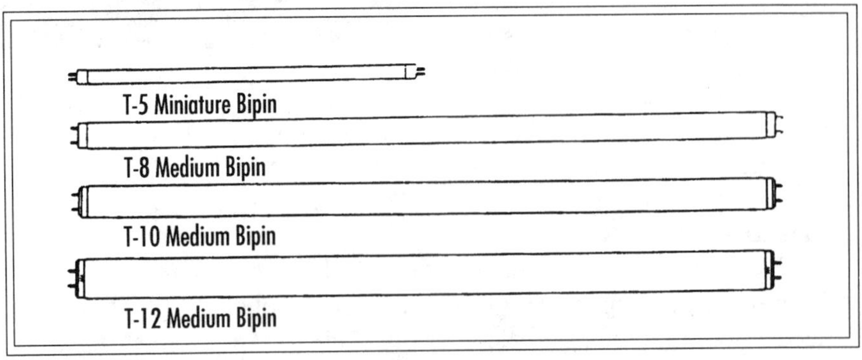

Figure 1-17. Fluorescent lamps carry a "T" in their designation code which indicates they are tubular; the number after the "T" indicates diameter in eighths of an inch.
Source: Philips Lighting Company.

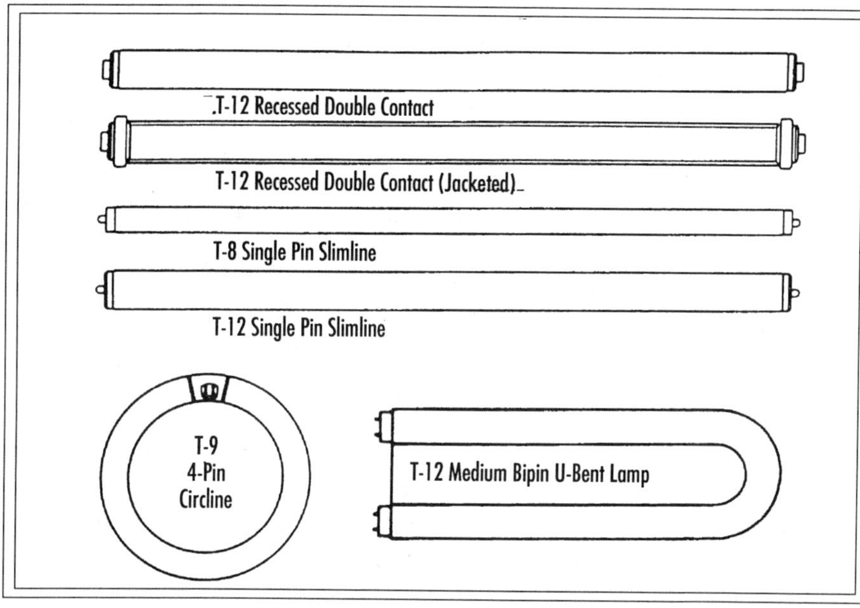

Figure 1-18. Fluorescent lamp shapes. *Source: Philips Lighting Company.*

plification of this code; in addition, the "40" has lost its original meaning. For example, a lamp may be designated as "F40T12/ES," but "ES" indicates that this is an "energy-saving" lamp consuming only 34 watts. Actual manufacturer suffixes may be "SS" for SuperSaver, "EW" for Econ-o-Watt, "WM" for Watt-Miser or others.

Preheat lamps: Preheat lamps do not have "RS" at the end of the code series.

High output rapid start lamps

We will examine an F48T12/WW/HO as an example.

- "F" indicates that this lamp is a fluorescent.

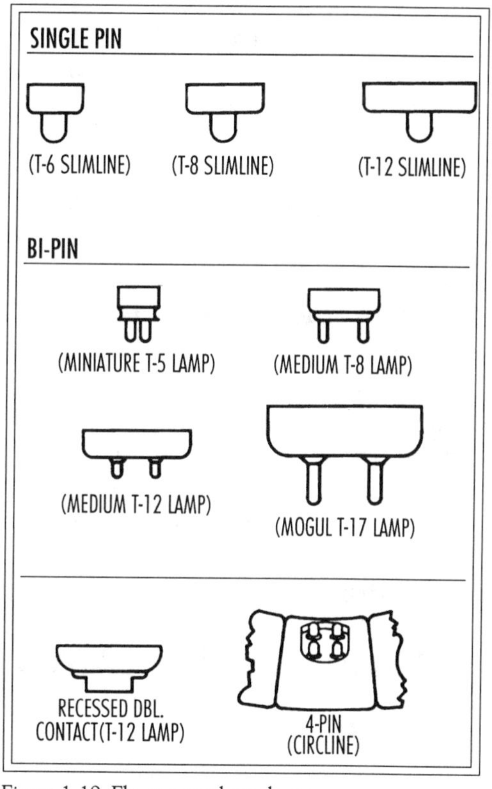

Figure 1-19. Fluorescent lamp bases.
Source: Osram Sylvania, Inc.

- "48" shows the nominal length of the lamp in inches.
- "T" means the lamp is tubular.
- "12" indicates diameter of the lamp in eighths of an inch — this lamp is 1-1/2 inches in diameter.
- "WW" reveals this lamp is a warm white lamp.
- "HO" means this is a high output lamp operating on 800 milliamps.

Very high output rapid start lamps

Let us review the code for an F72T12/CW/VHO as an example.

- "F" indicates that this lamp is a fluorescent.
- "72" describes the nominal length of the lamp in inches.
- "T" means the lamp is tubular.
- "12" indicates diameter of the lamp in eighths of an inch — this lamp is 1-1/2 inches in diameter.
- "CW" reveals this lamp is a cool white lamp.
- "VHO" means this is a very high output lamp operating on 1500 milliamps.

Note: Instead of "VHO," it may read "1500" or "Power Groove."

Instant start lamps

We will look at an instant start F96T12/WWX as an example.

- "F" means this is a fluorescent lamp.
- "96" describes the nominal length of the lamp in inches.
- "T" means tubular.
- "12" is the diameter of the lamp in eighths of an inch.
- "WWX" reveals this is a deluxe warm white lamp.

Note on specialty lamps

"FC" instead of "F" indicates that the lamp is circular. "FB" instead of "F" describes the lamp as bent or U-shaped. The suffix "U" can also indicate a U-shaped lamp, followed by a "/" and a number. The number reveals the distance between the lamp's legs in inches.

Again, consult manufacturers' catalogs for a more detailed view of lamp codes. No lamp should be ordered unless its characteristics and application are thoroughly understood, as well as its compatibility with all elements of the new upgraded system including the ballast and fixture.

LAMP AND BALLAST LIFE

Lamp life

Lamp manufacturers publish rated life for their lamps in their literature. Rated lamp life is an average. If a lamp has a rated life of 20,000 hours, then at 20,000 hours of operation one-half of a large group of lamps are expected to fail. Fluorescent lamps are typically rated at 12,000 to 20,000 hours. Factors affecting lamp life include supply voltage variations and "hours per start" — the length of the period the lamps are left on before being shut off and then restarted. Fluorescent lamps are typically rated by manufacturers at three hours per start (three hours on, 20 minutes off). Lamps generally will last longer if operated at longer hours per start. For example, F40/CW rapid start lamps — rated at 20,000 hours at three hours per start — can have an average rated life of 40,000 hours if operated continuously after starting.

Most fluorescent lamps operate at their peak efficiency at 77 degrees Fahrenheit (25 degrees Celsius). Temperatures above or below 77 degrees will reduce lumen output. Temperatures within fixtures are typically hotter than 77 degrees; in typical operation in a fixture, therefore, the actual light delivered from a fluorescent lamp may be lower than the catalog rating.

Ballast life

According to the Certified Ballast Manufacturers Association, traditional ballast life ranges from 12 to 15 years (approximately 75,000 hours of normal use). High ambient temperatures can reduce this range, while some cooler temperatures can as much as double it.

New ballast types such as the electronic ballast, due to their construction, are estimated to last up to 25 years. These ballast types are reviewed later in this chapter.

LAMP EFFICIENCY: IT'S THE LAW

On October 24, 1992, President George Bush signed the National Energy Policy Act into law. In regards to lighting, the new energy law mandates minimum efficiency standards for certain electric lamps. Eighteen to 36 months after enactment, lamp manufacturers will no longer be allowed to produce many popular fluorescent and incandescent lamp types which do not meet the new efficiency standards.

One example is the traditional F40/CW, the most widely used fluorescent lamp prior to the law's passage. Energy-saving equivalents (F40T12/ES cool white) can continue to be manufactured. Other affected fluorescent lamps include certain colors in F40 two-foot U-shaped, eight-foot instant start lamps and HO lamps. Incandescent types which are affected include many popular reflector lamp types.

In addition, many states have implemented or are considering implementing energy efficiency standards for lamps. State standards, however, cannot be more stringent than national standards for the lamps covered by the Act.

BALLAST EFFICIENCY: IT'S THE LAW

The Federal Ballast Energy Law (Public Law 100-357) was enacted in 1988 as part of the National Appliance Energy Conservation Amendments (NAECA '88). This law sets minimum ballast efficacy standards for four major fluorescent ballast types which represent some 85% of all installed fluorescent ballasts. Traditionally used magnetic ballasts do not comply and are no longer manufactured; they are now being replaced in the fixture by energy-efficient magnetic ballasts or even-more-efficient alternatives. As of 1991, ballasts submitted for testing by the Department of Energy and complying with the provisions of NAECA '88 carry an Ⓔ symbol on their labels. Ballasts not covered under the Act include dimming ballasts and ballasts used in areas where temperatures reach 0 degrees Fahrenheit or lower.

A QUICK NOTE ON "STANDARD"

These two landmarks of energy legislation essentially eliminated the standard lamp and ballast and created new standards. For example, most standard magnetic ballast types were eliminated from manufacture; energy-efficient "premium" magnetic ballasts are now considered the new standard as the most popularly used ballast type. During contact with the lighting community, however, one may often hear the words "standard lamps and ballasts" referring to the old standards. This kind of language is commonly employed during retrofit situations, when the lighting professional refers to an older system as "standard." In this book, these lamps and ballasts are referred to as "traditional."

LIGHTING UPGRADE EQUIPMENT

Dramatic advances have been made in the construction and performance of fixtures, lamps, ballasts and shielding material. New lighting systems can reduce energy consumption significantly while producing comparable light levels. Other devices, such as reflectors and occupancy sensors, have also been developed to control the system and enhance its performance.

Note that each type of new equipment, while offering numerous advantages, have individual disadvantages which should be given serious consideration during the selection process. Consult with the supplier as to exact specifications for its products, such as the level of maintained lumen output throughout lamp life (lumen maintenance), ballast factor and other specifications. The reader may also find it beneficial to retain an outside contractor or consultant; experienced independent professionals can help sort through manufacturer claims and determine advantages and disadvantages of various equipment types.

Here are lists of lamp, ballast and other equipment types:

UPGRADE LAMP TYPES

T8 lamps	Compact fluorescent twin- and quad-tube
T10 lamps	U-Bent fluorescent lamps
T12 high lumen	HID metal halide
T12 High Output	HID high pressure sodium
T12 Very High Output	HID low pressure sodium
	Halogen incandescent lamps (general service, reflectorized, tubular)

Example: Energy-saving T12 lamps ... F40T12/ES lamp

An F40T12/ES is basically a traditional fluorescent lamp with krypton gas added to the argon gas inside the bulb, assisting operation. The result: The lamp consumes six fewer watts (15% energy savings) than a traditional F40/CW lamp, making it rated a 34-watt lamp. It is distinguishable from other lamps by the nickname "ES," which stands for "energy-saving." Actual manufacturer code letters include "SS," "EW," "WM" or others which designate the lamp as an energy-saving product.

No special ballast is required, although increased savings can occur if operated by an energy-efficient magnetic, cathode cut-out or electronic ballast. Actual input wattage savings are seven watts per lamp when energy-saving lamps are installed with energy-efficient ballasts.

Figure 1-20. Operating an inefficient lighting system results in waste. In the next series of illustrations, we will briefly examine common lighting efficiency problems and the new technologies and techniques used to improve efficiency.
Source: U.S. Environmental Protection Agency, Green Lights Program.

Figure 1-21. A typical problem producing waste is operating the lighting system when an area is unoccupied. Controls such as occupancy sensors and energy management systems can generate energy savings in these situations.

Source: U.S. Environmental Protection Agency, Green Lights Program.

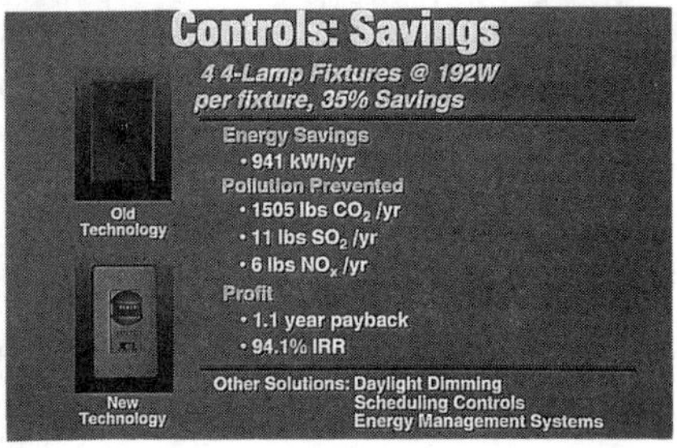

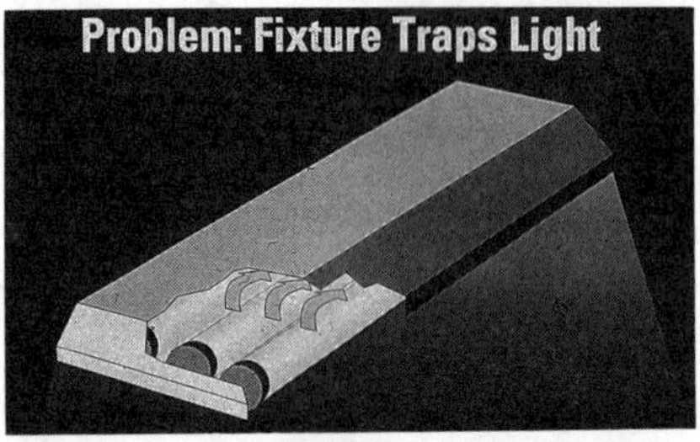

Figure 1-22. A typical problem with traditional light fixtures is that light is trapped above the lamps and does not reach the task area. Reflectors deliver more of this light to the workplane, presenting the opportunity to turn the excess light into energy savings by delamping.

Source: U.S. Environmental Protection Agency, Green Lights Program.

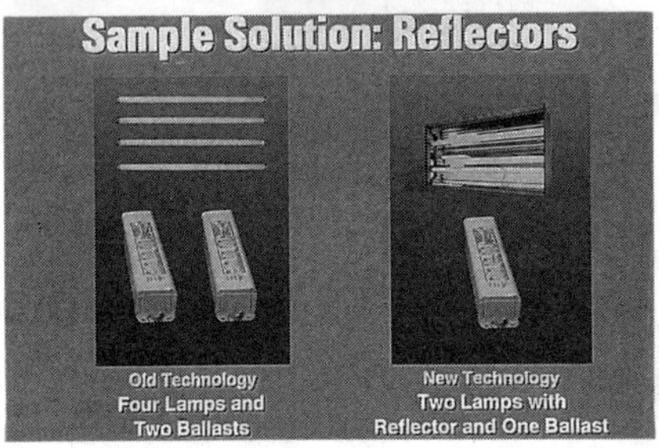

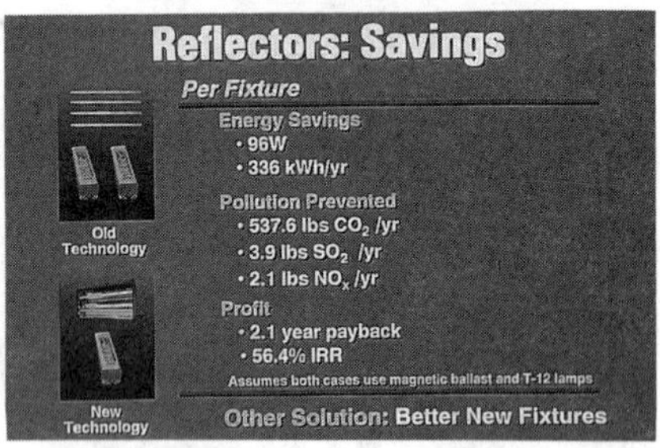

However, an energy-saving lamp also sacrifices 10-20% of its lumen output (a trichromatic version is available, however, which produces higher lumen output besides better color rendition). Additionally, these lamps cannot be dimmed as easily as traditional F40/CWs.

Overall, the F40T12/ES can be a good lamp choice if the area is presently overlighted and/or if a scheduled lighting maintenance program is planned.

Figure 1-23. Traditional ballasts are inefficient relative to commercially available alternatives. Electronic ballasts, for example, can generate high energy savings and yield greater performance.

Source: U.S. Environmental Protection Agency, Green Lights Program.

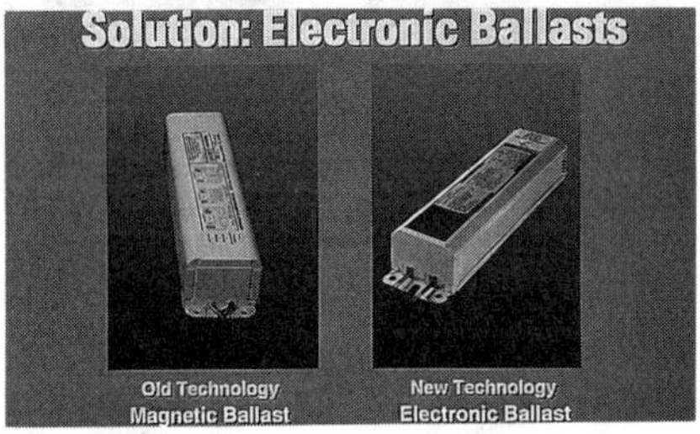

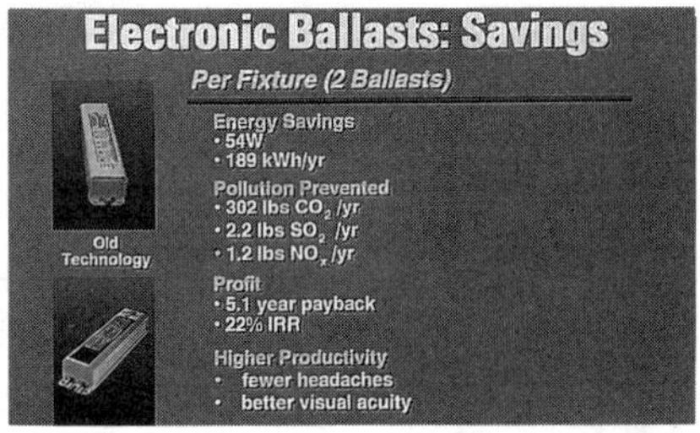

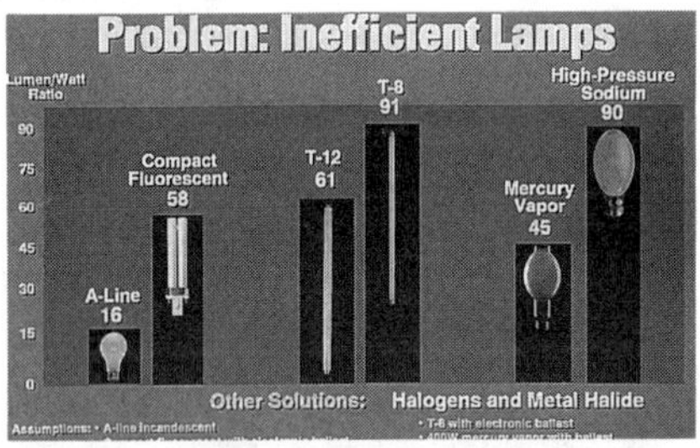

Figure 1-24. Traditional lamps are also inefficient relative to today's alternatives. A variety of lamps are available to produce energy savings and possibly better color rendition.

Source: U.S. Environmental Protection Agency, Green Lights Program.

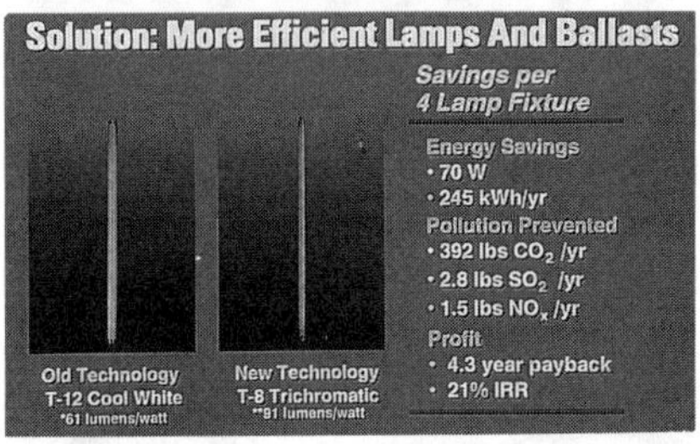

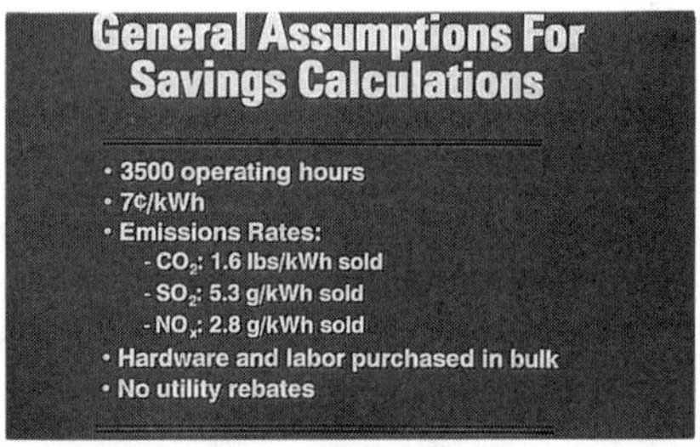

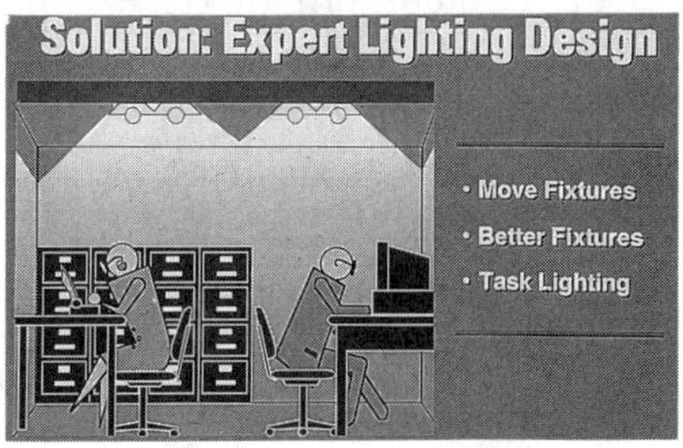

Figure 1-25. The design of traditional lighting systems has resulted in overlighted and inefficiently lighted spaces, presenting the opportunity to redesign the system to reflect today's lighting recommendations and standards of energy efficiency.
Source: U.S. Environmental Protection Agency, Green Lights Program.

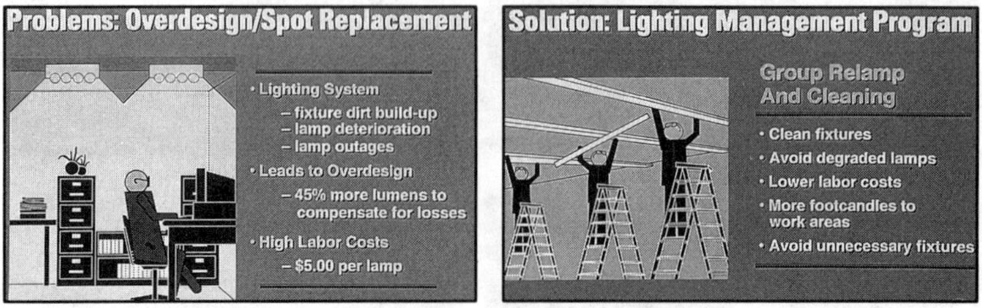

Figure 1-26. A planned lighting maintenance program optimizes the use of the lighting system and broadens retrofit opportunities (see Chapter 5).
Source: U.S. Environmental Protection Agency, Green Lights Program.

Example: T10 lamps ... F40T10 lamp

The F40T10 lamp is designed for use in rapid start circuits and is fitted with a medium bi-pin base for easy replacement of existing fluorescent lamps.

The F40T10 lamp's benefits come from two distinctive features. As opposed to one coat of halophosphate phosphor on the glass tube to produce visible light, the F40T10 is braced by a second phosphor coat — a blend of red, blue and green phosphors called rare earths. These three primary colors label this type of lamp a "triphosphor lamp" or "trichromatic lamp." Triphosphor technology allows the lamp to produce more lumens per watt.

And as its code name "T10" tells us, this lamp's bulb diameter is 1-1/4 inches wide. Thin diameter allows more light that is normally trapped in the "house side" of the fixture (above the lamps), or absorbed by the lamps themselves, to be reflected down onto the workplane.

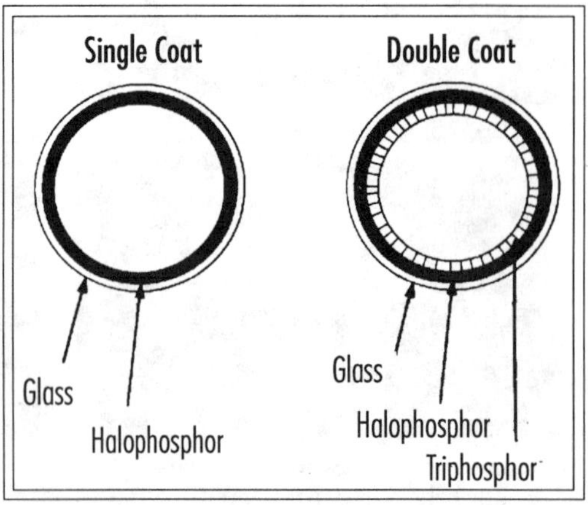

Figure 1-27. Fluorescent dual-coat technology.
Source: Osram Sylvania, Inc.

Because of these two features, light level can be expected to increase 21% after replacement of traditional F40/CWs. The F40T10 lamp is known for its high lumen output at about 3,700 lumens, and efficacy as high as 92 lumens per watt (when used with electronic ballasts). By increasing lumen output, an opportunity is presented to delamp.

In addition, the presence of the three rare earth phosphors substantially improves the lamp's color rendering ability; the F40T10 has a Color Rendering Index (CRI) rating of 80 or higher.

However, F40T10 lamps are currently only available as four-foot straight lamps. No "U"-shaped or circular lamps are now available. And although the F40T10 is compatible with any standard rapid start ballast, some cathode cut-out ballasts will not start this lamp over the full range of primary voltage (according to the California Energy Commission, 1990).

Example: T8 lamps ... F32T8 lamp

F32T8 (also sometimes called FO32T8) lamps are designed for rapid start or instant start circuits, capped with medium bi-pin bases to allow easy replacement of traditional fluorescent lamps. A ballast changeout is also required. The T8 family of lamps also includes two-foot (17-watt), three-foot (25-watt), five-foot (40-watt) and eight-foot (59-watt) straight tubes, and one-, 1.5- and two-foot U-shaped lamps.

Unlike the F40T12/ES, the F32T8 does not use a krypton-argon gas mix for starting, but instead relies purely on argon. Like the F40T10, its two distinctive features are triphosphor technology and reduced diameter.

The triphosphor second coat increases the lamp's ability to convert ultraviolet energy into visible light, and increases the lamp's color rendering ability. F32T8 lamps produce 2,900-3,000 lumens and hold a medium to high CRI rating of 70 to 85. They are available in four color temperatures: 3000 Kelvin, 3500 Kelvin, 4100 Kelvin and 5100 Kelvin.

As its name "T8" tells us, the lamp is even thinner than the T10 — being about one inch in diameter. Although the lamp is one-third thinner than the T12, the T8 is fitted with a medium bi-pin base to allow replacement of any equivalent-length rapid start fluorescent lamp. Again, thinner bulb diameter allows more lumen output which is normally diffused within the fixture to be delivered to the task.

F32T8-electronic ballast systems can reduce energy consumption by some 12-40% while producing comparable light levels as a traditional four-lamp system. On "low-wattage" electronic ballasts, approximately 13% of lumen output is sacrificed for even greater energy savings which can reach 43%.

Besides high efficiency and color rendering, another benefit of this lamp's construction is that because it draws less wattage, it runs cooler than 40-watt lamps, producing less heat inside the fixture. The cooler bulb wall temperature, combined with trisphosphor technology, improves both lumen output and lumen maintenance (the amount of lumens produced by the lamp over its life, as lumen output drops gradually as the lamp ages).

Because F32T8 lamps operate on a reduced 265-milliamp current as opposed to 430 milliamps for standard lamps, they can only operate on a compatible special magnetic ballast or an electronic ballast. Instant-start-mode electronic ballasts can decrease rated lamp life by about 25% to 15,000 hours at three hours per start. Life at 12 hours per start or more, however, is equal to the rated life for a rapid start electronic ballast.

UPGRADE BALLAST TYPES

Energy-efficient magnetic
Cathode cut-out (also called heater cut-out or "hybrid" ballasts)
Electronic
Dimming electronic
Step-dimming electronic
Fluorescent high output magnetic and electronic
HID electronic

Example: Energy-efficient ("premium") magnetic ballasts

The transformer component of a traditional rapid start magnetic ballast is a core of stacked steel laminations, around which insulated copper wire is wound. In an energy-efficient magnetic ballast, the copper wire is replaced by larger, higher-quality copper wire reinforced by ferrous materials. The result is that a system powered by an energy-efficient ballast requires 10% less power than a traditional ballast. Less electrical input generally means cooler operation, which extends typical life to about 25 years.

Example: Cathode cut-out ("hybrid") ballasts

A cathode cut-out ballast is essentially an energy-efficient rapid start magnetic ballast with a built-in feature that furthers efficiency. Rapid start lamp cathodes are constantly heated, which reduces start-up time to about one second. The cathode cut-out ballast's feature is an electronic switch which stops the ballast from heating the cathodes after lamp startup, resulting in an additional two watts of power savings per lamp. This combination of magnetic and electronic components labels this ballast a "hybrid" ballast. Systems powered by cathode

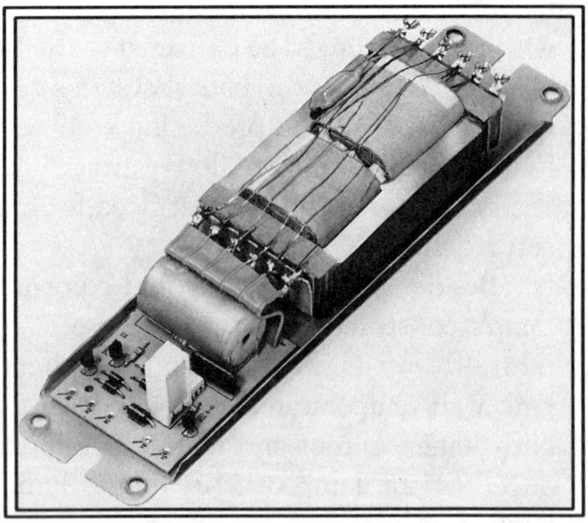

Figure 1-28. Interior of a "hybrid" ballast.
Source: Advance Transformer Company.

cut-out ballasts generally produce 8-15% fewer lumens while consuming 17-32% less energy, although new full-lumen-output versions are now available. Cathode cut-out ballasts are designed to last 25 years.

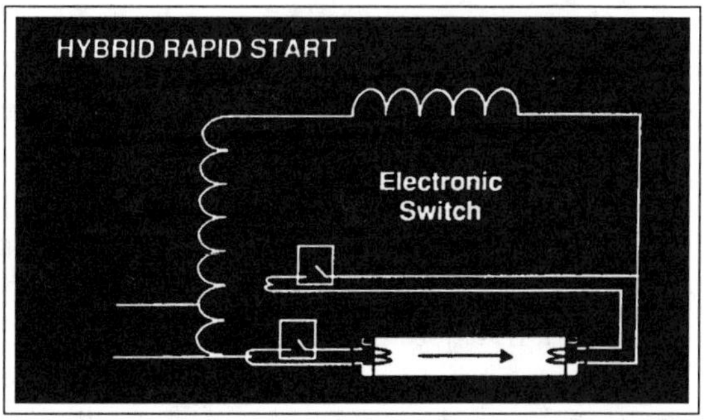

Figure 1-29. Hybrid rapid start circuit.
Source: MagneTek, Lighting and Electronic Products.

Example: Electronic ballasts

Inside some electronic ballasts, there is no core, coil and asphalt at all, but instead a series of electronic components housed in resin. One section of components converts the alternating current (AC) line voltage to direct current (DC) voltage. Another section inverts the DC voltage to high-frequency AC voltage. The third section provides the proper voltage to the lamp and acts as the current-regulating circuit. Most electronic ballasts operate at a higher frequency (25,000-60,000 Hertz or cycles per second compared to 60 Hertz for magnetic and cathode cut-out ballasts).

Electronic ballasts offer benefits which include energy efficiency, versatility, lower weight, less audible noise, reduced lamp flicker and long life.

Energy efficiency. High-frequency operation increases lamp efficiency by about 10%. Systems featuring electronic ballasts can achieve the highest energy savings of any lamp-ballast system (12-40%) while producing comparable light levels. Other electronic ballasts can increase energy savings at the expense of lumen output.

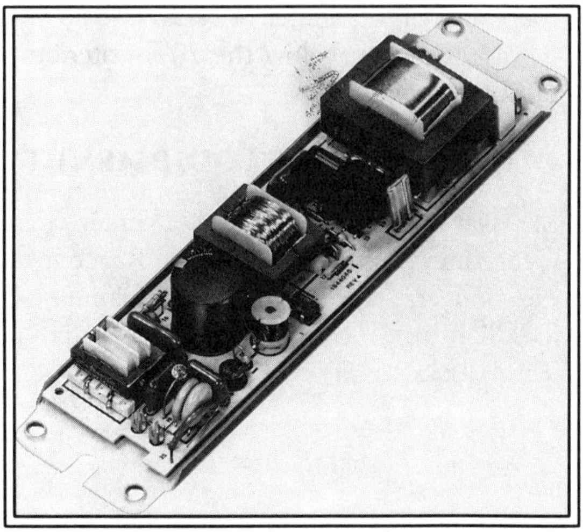

Figure 1-30. Interior of an electronic ballast. Note that electronic ballasts represent a product family with numerous manufactured lines presenting varying features, benefits and efficiencies. This electronic ballast does not rely solely on electronic components; it also contains a core and coils (magnetic components).
Source: Advance Transformer Company.

Versatility. Some electronic ballast models can operate up to four lamps on a single ballast (some building codes prohibit this practice, however). Some electronic ballast models have dimming capabilities, which allow them to be used in conjunction with a dimming system. A component of this system is a photocell, a device which measures the amount of external light (such as daylight) entering a room. As the sun rises or other light enters the room, the system dims the lamp accordingly; as clouds obscure the sun or the external light is removed, the system increases lumen output.

Lower weight. The absence of heavy components such as the core, coils and asphalt in most electronic ballast models makes the ballast lighter than magnetic and cathode cut-out ballasts.

Less audible noise. High frequency of operation reduces the hum sometimes associated with fluorescent lighting. This can increase the quality of lighting in a space.

Reduced flicker. A magnetic or cathode cut-out ballast operating at a frequency of 60 Hertz experiences 60 cycles of operation per second. This means the ballast turns the lamps off and on 120 times during each second. This causes lamp flicker that may not be perceived, but nonetheless can have a negative effect on some occupants. Most electronic ballast models operate at 25,000-60,000 Hertz, which virtually eliminates flicker.

Long life. Electronic ballasts, due to cooler operation, are projected to last 25 years.

The electronic ballast can be either rapid or instant start and power the lamps on a parallel or series circuit. Most new models generate harmonic distortion below the 20% criterion for eligibility in many electric utility rebate programs.

OTHER UPGRADE EQUIPMENT TYPES

Specular reflectors
Occupancy sensors
New louvers
Task lighting
Time clocks
Energy management systems
Reactive devices (current limiters)
Dimming controls
New fixtures

2-LAMP SYSTEM OPTIONS, FLUORESCENT RAPID START SYSTEMS

OPTION	LAMPS	BALLAST	RATED INITIAL LUMENS[1]	BALLAST FACTOR	LUMEN OUTPUT	INPUT WATTAGE	EFFICACY (LUMENS/WATT)	RATED LIFE (HOURS)[2]	ANNUAL ENERGY COST ($)[3]
#1	(2) F40T12	(1) TRADITIONAL MAG.[4]	3050	0.95	5795	96	60	20,000	$30.72
#2	(2) F40T12	(1) T12 EFF. MAG.	3050	0.95	5795	86	67	20,000	$27.52
#3	(2) F40T12	(1) T12 HYBRID	3050	0.81	4941	71	70	20,000	$22.72
#4	(2) F40T12	(1) FLO T12 HYBRID[5]	3050	0.95	5795	80	72	20,000	$25.60
#5	(2) F40T12	(1) T12 ELEC.	3050	0.88	5368	71	76	20,000	$22.72
#6	(2) F40T12 TRICHROMATIC	(1) T12 EFF. MAG.	3300	0.95	6270	86	73	20,000	$27.52
#7	(2) F40T12 TRICHROMATIC	(1) T12 HYBRID	3300	0.81	5346	71	75	20,000	$22.72
#8	(2) F40T12 TRICHROMATIC	(1) FLO T12 HYBRID	3300	0.95	6270	80	78	20,000	$25.60
#9	(2) F40T12 TRICHROMATIC	(1) T12 ELEC.	3300	0.88	5808	71	82	20,000	$22.72
#10	(2) F40T12/ES	(1) TRADITIONAL MAG.	2775	0.88	4884	80	61	20,000	$25.60
#11	(2) F40T12/ES	(1) T12 EFF. MAG.	2775	0.88	4718	72	66	20,000	$23.04
#12	(2) F40T12/ES	(1) T12 HYBRID	2775	0.77	4274	58	74	20,000	$18.56
#13	(2) F40T12/ES	(1) FLO T12 HYBRID	2775	0.88	4884	66	74	20,000	$21.12
#14	(2) F40T12/ES	(1) T12 ELEC.	2775	0.80	4440	59	75	20,000	$22.72
#15	(2) F40T12/ES TRICHROMATIC	(1) T12 EFF. MAG.	2925	0.88	5148	72	72	20,000	$23.04
#16	(2) F40T12/ES TRICHROMATIC	(1) T12 HYBRID	2925	0.77	4505	58	78	20,000	$18.56

[1] Initial lumen output, ballast factor and input wattage for this data and data on the next two pages are typical values. Actual values will vary by manufacturer of lamp and ballast, as well as fixture effects. Input wattage is based on results of ANSI-specified procedures called the "bench test."

[2] Rated life is based on three hours per start.

[3] Based on 4,000 operating hours per year and an energy unit price of $0.08 per kWh.

[4] No longer manufactured. For comparison purposes only.

[5] "FLO" represents full light output version of hybrid (cathode cutout) ballast.

2-LAMP SYSTEM OPTIONS, FLUORESCENT RAPID START SYSTEMS (CONTINUED)

OPTION	LAMPS	BALLAST	RATED INITIAL LUMENS	BALLAST FACTOR	LUMEN OUTPUT	INPUT WATTAGE	EFFICACY (LUMENS/WATT)	RATED LIFE (HOURS)	ANNUAL ENERGY COST ($)
#17	(2) F40T12/ES TRICHROMATIC	(1) FLO T12 HYBRID	2925	0.88	5148	66	78	20,000	$21.12
#18	(2) F40T12/ES TRICHROMATIC	(1) T12 ELEC.	2925	0.85	4973	59	84	20,000	$18.88
#19	(2) F32T8 TRICHROMATIC	(1) T8 MAG.	2900	0.95	5510	71	78	20,000	$22.72
#20	(2) F32T8 TRICHROMATIC	(1) T8 HYBRID	2900	0.87	5046	62	81	20,000	$19.84
#21	(2) F40T10 TRICHROMATIC	(1) T12 EFF. MAG.	3700	0.95	7030	86	82	24,000	$27.52
#22	(2) F40T10 TRICHROMATIC	(1) T12 ELEC.	3700	0.88	6512	71	92	24,000	$22.72

3-LAMP SYSTEM OPTIONS, FLUORESCENT RAPID START SYSTEMS

OPTION	LAMPS	BALLAST	RATED INITIAL LUMENS	BALLAST FACTOR	LUMEN OUTPUT	INPUT WATTAGE	EFFICACY (LUMENS/WATT)	RATED LIFE (HOURS)[1]	ANNUAL ENERGY COST ($)[2]
#1	(3) F40T12	(1) T12 ELEC.	3050	0.86	7869	110	72	20,000	$35.20
#2	(3) F40T12 TRICHROMATIC	(1) T12 ELEC.	3300	0.86	8514	110	77	20,000	$35.20
#3	(3) F40T12/ES	(1) T12 ELEC.	2775	0.80	6660	94	71	20,000	$30.08
#4	(3) F40T12/ES TRICHROMATIC	(1) T12 ELEC.	2925	0.86	7547	110	69	20,000	$35.20
#5	(3) F32T8 TRICHROMATIC	(1) T8 ELEC.	2900	0.93	8265	88	94	15,000[3]	$28.16
#6	(3) F40T10 TRICHROMATIC	(1) T12 ELEC.	3700	0.86	9546	110	87	24,000	$35.20

[1]Rated life is based on three hours per start.
[2]Based on 4,000 operating hours per year and an energy unit price of $0.08 per kWh.
[3]Instant start versus rapid start operating mode for less than 12 hours per start; for 12 hours per start or more, standard life (20,000 hours) applies.

4-LAMP SYSTEM OPTIONS, FLUORESCENT RAPID START SYSTEMS

OPTION	LAMPS	BALLAST	RATED INITIAL LUMENS	BALLAST FACTOR	LUMEN OUTPUT	INPUT WATTAGE	EFFICACY (LUMENS/WATT)	RATED LIFE (HOURS)[1]	ANNUAL ENERGY COST ($)[2]
#1	(4) F40T12	(2) TRADITIONAL MAG.[3]	3050	0.95	11590	192	60	20,000	$61.44
#2	(4) F40T12	(2) T12 EFF. MAG.	3050	0.95	11590	172	67	20,000	$55.04
#3	(4) F40T12	(2) T12 ELEC.	3050	0.88	10736	142	76	20,000	$45.44
#4	(4) F40T12	(2) T12 HYBRID	3050	0.81	9982	142	70	20,000	$45.44
#5	(4) F40T12	(2) FLO T12 HYBRID[6]	3050	0.95	11590	160	72	20,000	$51.20
#6	(4) F40T12 TRICHROMATIC	(2) T12 EFF. MAG.	3300	0.95	12540	172	73	20,000	$55.04
#7	(4) F40T12 TRICHROMATIC	(2) T12 ELEC.	3300	0.88	11616	142	72	20,000	$45.44
#8	(4) F40T12 TRICHROMATIC	(2) T12 HYBRID	3300	0.81	10692	142	75	20,000	$45.44
#9	(4) F40T12 TRICHROMATIC	(2) FLO T12 HYBRID	3300	0.95	12540	160	78	20,000	$51.20
#10	(4) F40T12/ES	(2) TRADITIONAL MAG.	5550	0.88	9768	160	61	20,000	$51.20
#11	(4) F40T12/ES	(2) T12 EFF. MAG.	2775	0.88	11100	144	77	20,000	$46.08
#12	(4) F40T12/ES	(2) T12 ELEC.	2775	0.85	9435	118	80	20,000	$37.76
#13	(4) F40T12/ES	(2) T12 HYBRID	2775	0.77	8547	116	74	20,000	$37.12
#14	(4) F40T12/ES	(2) FLO T12 HYBRID	2775	0.88	9768	132	74	20,000	$42.24
#15	(4) F40T12/ES TRICHROMATIC	(2) T12 EFF. MAG.	2925	0.88	10296	144	72	20,000	$46.08
#16	(4) F40T12/ES TRICHROMATIC	(2) T12 ELEC.	2925	0.85	9945	118	84	20,000	$37.76

[1]Rated life is based on three hours per start.
[2]Based on 4,000 operating hours per year and an energy unit price of $0.08 per kWh.
[3]No longer manufactured. For comparison purposes only.
[4]"FLO" represents full light output version of hybrid ballast.

4-LAMP SYSTEM OPTIONS, FLUORESCENT RAPID START SYSTEMS (CONTINUED)

OPTION	LAMPS	BALLAST	RATED INITIAL LUMENS	BALLAST FACTOR	LUMEN OUTPUT	INPUT WATTAGE	EFFICACY (LUMENS/WATT)	RATED LIFE (HOURS)	ANNUAL ENERGY COST ($)
#17	(4) F40T12/ES TRICHROMATIC	(2) T12 HYBRID	2925	0.77	9009	116	78	20,000	$37.12
#18	(4) F40T12/ES TRICHROMATIC	(2) FLO T12 HYBRID	2925	0.88	10296	132	78	20,000	$42.24
#19	(4) F32T8 TRICHROMATIC	(2) T8 MAG.	2900	0.95	11020	142	78	20,000	$45.44
#20	(4) F32T8 TRICHROMATIC	(2) T8 HYBRID	2900	0.87	10092	124	81	20,000	$39.68
#21	(4) F32T8 TRICHROMATIC	(1) T8 ELEC.[1]	2900	0.93	10788	112	96	15,000[2]	$35.84
#22	(4) F40T10 TRICHPOMATIC	(2) T12 EFF. MAG.	3700	0.95	14060	172	82	24,000	$55.04
#23	(4) F40T10 TRICHOMATIC	(2) T12 ELEC.	3700	0.88	13024	142	92	24,000	$45.44

[1]Building codes in some states do not allow four lamps to be operated by a single ballast.
[2]Instant start versus rapid start operating mode for less than 12 hours per start; for 12 hours per start or more use 20,000 hours or apply normal life escalations for longer burning cycles.

QUESTIONS TO ASK WHEN CHOOSING EQUIPMENT

The primary factors which affect choice of equipment include its efficiency (efficacy, input watts, operating cost), lumen output, rated life, color quality, total owning cost and initial cost. These characteristics must be matched to corporate priorities and the requirements of the application.

The Illuminating Engineering Society of North America's *Lighting Handbook* discusses the lighting needs of almost every application, from hospitals to schools to office buildings, and should be consulted in addition to other sources.

Here are a few basic questions which should be asked when choosing equipment:

Does the area need it?

New equipment should only be used if the installation warrants its use. For example, if responsible occupants can turn lights off themselves, automatic lighting controls may not be necessary. Electronic ballasts may not be appropriate if a cathode cut-out ballast can provide acceptable results and a much faster payback due to a lower initial cost.

Is the equipment compatible?

All equipment chosen must be compatible with the equipment from the old system and with each other. The equipment must also be compatible with the application.

How old is the existing system?

The age and condition of the existing system must be considered. The installation of entirely new fixtures may be warranted if the existing fixtures are in poor condition. In addition, many new fixtures are now designed to more effectively deliver light to the task and can come equipped with energy-efficient components.

Can light level be sacrificed in certain areas?

In areas that are found to be underlighted, installing systems that save energy but also give up light level is generally not desirable. In an overlighted area or a corridor where the primary visual task is orientation, however, a new system which saves energy and gives up light level may be appropriate.

What is the track record of the equipment's manufacturers?

Once preferences for type of equipment have been established, check the manufacturers' track records for reliability and warranties. Note if a manufacturer has been in existence for many years and holds a reputation for introducing reliable equipment and supporting it with a strong warranty.

Sources:

ABCs of Electronic Ballasts: A Handy Guide to State-of-the-Art Fluorescent Lamp Control. Rosemont, Illinois: Advance Transformer Company, October 1989.

Bleeker, Nicholas and Anthony Krayzckowski. *Fluorescent System Performance & Economic Cost Comparisons For New and Existing Installations.* Somerset, New Jersey: Philips Lighting Company.

Lighting Equipment Options for Modern Buildings. Bethesda, Maryland: National Electrical Contractors Association, 1989.

Lighting Upgrade Manual, First Edition. Washington, DC: U.S. Environmental Protection Agency Green Lights Program, November 1991.

Lindsey, Jack. *Applied Illumination Engineering.* Lilburn, Georgia: The Fairmont Press, 1991.

MagneTek, Lighting and Electronic Products, Paramus, New Jersey.

Profiting from Lighting Modernization. Washington, DC: National Lighting Bureau, 1987.

Retrofit Lighting Technologies: Cost-Effective Lighting for Commercial Applications. Palo Alto: Electric Power Research Institute, July 1991. EPRI CU-3040R.

Selecting a Light Source. Danvers, Massachusetts: Sylvania Lighting Division. Color Clinic Unit 3, Lighting Application Bulletin 0-369.

Further reading:

Babcock, Robert, PE, "Understanding Total Harmonic Distortion — And How to Tackle It," *MainLighter: The Lighting Management & Maintenance Magazine,* April 1992.

Blake, Norm, "National Energy Efficiency Act to Outlaw Standard Lamps," *MainLighter: The Lighting Management & Maintenance Magazine,* November 1991.

Blake, Norm, "Senate Energy Bill Passes by Wide Margin," *MainLighter: The Lighting Management & Maintenance Magazine,* April 1992.

Blake, Norm, "House Energy Bill Passed," *Lighting Management & Maintenance,* July 1992.

Blanton, Richard, "Energy-Efficient Outdoor Lighting — The Boons of the Boom," *MainLighter: The Lighting Management & Maintenance Magazine,* September 1991.

Commercial Lighting Efficiency Resource Book. Palo Alto, California: Electric Power Research Institute, September 1991. EPRI CU-7427.

Compact Fluorescent Lamps: Energy-Efficient Alternatives to Incandescent Lighting. Palo Alto, California: Electric Power Research Institute, July 1991. EPRI CU-2042R.

Fundamentals of Commercial and Industrial Lighting, Fifth Edition. Washington, DC: The Electrification Council, 1984.

Getting the Most from Your Lighting Dollar, Second Edition. Washington, DC: National Lighting Bureau.

Gould, Scott, "Lighting and Energy Conservation: A Review of Current Technologies," *Strategic Planning for Energy and the Environment*, Spring 1990.

Guide to Performance Evaluation of Efficient Lighting Products, First Edition. Troy, New York: Rensselaer Polytechnic Institute (Lighting Research Center), June 1991.

Hathaway, Alden, "Master/Satellite Systems — Innovative Design," *MainLighter: The Lighting Management & Maintenance Magazine*, March 1992.

IES Education Series, ED-100: Introductory Lighting. New York: Illuminating Engineering Society of North America, July 1988.

Jensen, Brad, "Occupancy Sensors: Application and Troubleshooting," *Lighting Management & Maintenance*, July 1992.

Jensen, Brad, "Occupancy Sensors: Their Effect on Lamp Life," *MainLighter: The Lighting Management & Maintenance Magazine*, June 1992.

Jones, Richard, "Parabolic Louvers Beat the VDT Glare Problem," *MainLighter: The Lighting Management & Maintenance Magazine*, February 1992.

Kaloudis, James D. "Applying Occupancy Sensors for Lighting Control," *Electric Design and Installation*, June 1991.

Kessel, J. "Performance of Retrofit Optical Reflectors," *Strategic Planning for Energy and the Environment*, Fall 1990.

Lovins, Amory and R. Sardinsky. *The State of the Art: Lighting*. Snowmass, Colorado: Rocky Mountain Institute, March 1988.

Luminaire Retrofit Performance. Palo Alto, California: Electric Power Research Institute, March 1987. EPRI EM-5094.

McGowan, Terry. "Energy-Efficient Lighting." In *Electricity: Efficient End-Use and New Generation Technologies, and Their Planning Implications*, Thomas Johansson, Birgit Bodlund, and Robert Williams, eds. Sweden: Lund University Press, 1989.

Occupancy Sensors: Positive On/Off Lighting Control. Palo Alto, California: Electric Power Research Institute, 1992. EPRI BR-100323.

Rea, Mark S., Ivan Pasini and Louise Jutras, "Lighting Performance Measured in a Commercial Building," *Lighting Design & Application*, January 1990.

Rosenfeld, Arthur H. and David Hafemeister, "Energy-Efficient Buildings," *Scientific American*, July 1990.

Rubenstein, Francis, "50 Percent Energy Savings with Automatic Lighting Controls," *Strategic Planning for Energy and the Environment*, Fall 1991.

Specifier Reports: Electronic Ballasts, National Lighting Product Information Program. Troy, New York: Rensselaer Polytechnic Institute (Lighting Research Center), 1991.

Specifier Reports: Power Reducers, National Lighting Product Information Program. Troy, New York: Rensselaer Polytechnic Institute (Lighting Research Center), 1992.

Specifier Reports: Specular Reflectors, National Lighting Product Information Program. Troy, New York: Rensselaer Polytechnic Institute (Lighting Research Center), 1992.

Specular Retrofit Reflectors: Improving the Efficiency of Fluorescent Fixtures. Palo Alto, California: Electric Power Research Institute, October 1991. EPRI CU-2046.

Technical Lighting Bulletins. North Carolina Alternative Energy Corporation, January 1991.

Technology Assessment: Energy-Efficient Commercial Lighting. Berkeley, California: Lawrence Berkeley Laboratory, Applied Science Division, March 1989. LBL-27032.

Thumann, Albert, PE, Editor. *Lighting Efficiency Applications*, Second Edition. Lilburn, Georgia: The Fairmont Press, 1992.

Verderber, R.R. and O. Morse, *Performance of Electronic Ballasts and Other New Lighting Equipment.* Palo Alto, California: Electric Power Research Institute, March 1986. EPRI EM-4510.

Zarlenga, James W., "Reflector Retrofits: What You Should Know About Installation, Maintenance, Warranties," *MainLighter: The Lighting Management & Maintenance Magazine*, April 1992.

Source: Colorado Lighting, Inc.

CHAPTER 2

The Building Survey: Scouting for Opportunity

GETTING STARTED

A lighting upgrade starts with a commitment and a vision. Commitment must come from senior management to support thorough research of a lighting upgrade. The vision is the benefits of quality lighting, fit to corporate priorities, which include efficient use of energy and increased human performance. Corporate commitment and a thorough grasp of the benefits will provide the momentum to start and finish the lighting upgrade.

ESTABLISH A PLAN

These action steps will see the upgrade through from start to finish:

1. *Prioritize benefits*. Benefits of a lighting upgrade potentially include increased human performance, lower operating costs, lower maintenance costs, higher retail sales, a safer work environment, greater security, night landscaping/beautification, proper atmosphere, and reduced air cooling requirements. The benefits should be prioritized to answer the simple question, "Why are we doing this?" and to help in choosing strategies and equipment.

2. *Gather education, training and resources*. Numerous organizations

offer educational courses, seminars, publications, trade shows and/or other activities to both educate and train people about lighting upgrades (see Chapter 8: "Getting Help: Associations, Trade Groups, Professionals, Government, Publications"). In addition, computer software, survey equipment and other materials can be obtained from these organizations and suppliers.

One resource which is important to successfully complete a lighting upgrade is the *IES Lighting Handbook* published by the Illuminating Engineering Society of North America (IESNA) in New York City.

3. *Prioritize use of labor.* Labor will be required to perform the survey and analysis, and then install and maintain the new lighting equipment. Either in-house personnel or an outside contractor(s) perform this work. Outside sources of expertise include lighting management companies, energy service companies, lighting designers, lighting suppliers/manufacturers, electrical contractors, consultants and the local electric utility. Some of these sources perform only one or a combination of these services, while others are full-service. The main advantage of utilizing a professional contractor is that company's experience and expertise in lighting upgrades. The benefit of having the upgrade performed by experienced companies, and the cost of equipment and personnel to perform those services in-house, must be weighed against the cost the company charges.

4. *Collect building, lighting system, financial and occupant information.* To better specify the new lighting system, information about the building, various economic factors and occupant needs should be collected. In addition, the lighting system should be physically inspected and surveyed by area to determine the condition of the existing system, its energy consumption, and whether or not it provides appropriate lighting quantity and quality to the area's occupants.

5. *Determine required light levels per room and address quality issues.* Review IESNA recommendations and prescribe light levels appropriate to the type of tasks performed. Also review opportunities to create a more "user-friendly" lighting system through, for example, installation of task lighting, high color quality lamps, or equipment which reduces glare.

6. *Research financing opportunities.* Research financing opportunities such as utility incentives, shared savings contracts, leasing options and other types of agreements. A rebate check from the local utility may change corporate equipment preferences and can substantially reduce the payback period.

7. *Establish lighting management options.* Establish several options featuring combinations of energy-efficient components and planned maintenance of the lighting system. In addition, other lighting management options can be considered, including the implementation of lighting

schedules and a campaign to encourage employees to turn off lights when not in use.

8. *Determine light levels provided by new equipment options.* Options which do not provide minimum recommended levels should be eliminated.

9. *Conduct an economic analysis of the options.* Calculate energy savings, simple paybacks and internal rates of return derived from their installation. Determine other economic benefits such as projected increases in productivity or sales. Incorporate the benefits of planned maintenance of the lighting system.

10. *Complete an environmental benefits assessment.* The environmental benefit of energy conservation should be calculated. Energy conservation is generally good for the natural environment; calculating the environmental benefit can result in valuable information which the marketing communications department can pass along to the public. This will position the upgrade project as even more valuable to senior management.

11. *Conduct a full economic analysis.* To conduct a full life-cycle costing analysis, the financial manager should research and analyze more-detailed financial information. This information includes cost of capital and financing agreements; inflation rates for labor, equipment and electricity; corporate tax rates; financing agreements; equipment depreciation schedules; and other factors used to determine the present and future value of an investment.

12. *Present the proposal to decisionmaker.* Present the complete proposal to the decisionmaker for review. All viable options should be presented, with focus on the most favored option.

13. *Conduct a trial installation.* A trial installation of new equipment should be completed and the results carefully monitored quantitatively (energy savings) and qualitatively (occupant response). Adjustments can be made to the upgrade, or another option can be considered and implemented, should the equipment's performance fail to meet expectations.

14. *Full installation.* Install new equipment. Old lamps, ballasts and other equipment should be properly disposed in accordance with Federal and state guidelines.

15. *Maintain the system.* The lighting system should be properly maintained to maximize its effectiveness over the long term and to further economize use of labor, resources and energy.

16. *Monitor results.* The results of the upgrade can be measured both quantitatively and qualitatively to assess value of the upgrade investment.

17. *Market environmental commitment.* As stated earlier, the marketing communications department can take the environmental benefits of the upgrade to the public.

COLLECT INFORMATION

Accurate and complete information can help the upgrading company choose the best lighting upgrade option, eliminate wasteful lighting practices, fully realize the benefits of lighting management, and provide an effective lighting system for its users. Getting this information requires performing a area-by-area survey of the lighting system. In addition, general building and occupant information should be gathered.

COLLECT GENERAL INFORMATION

Obtain a floorplan or reflected ceiling plan of the facility. If one is not available, create a floor-by-floor line drawing of the facility on graph paper. Sometimes, what is actually used is a combination of floorplan and drawing as the building may have been remodeled since the original floorplans were drawn.

Each area (room, hallway or group of rooms) should be labeled by number and a generic description of the space. The numbers can be assigned based on areas in a floor, by interior and exterior areas, by all areas or any other means that appears easiest to work with. Examples: Area #1: Parking Lot, Area #2: Lobby, and so on.

During the area-by-area survey conducted later, it can be of great help to put a sticker on room door hinges or other out-of-the-way place which carries the same identification.

COLLECT OCCUPANT INFORMATION

An excellent source of information to tap before starting a lighting upgrade is the experience of the occupants of the space that is lighted. An Occupant Questionnaire can be a simple and effective tool to collect information about how occupants regard their lighting system. Each group of questionnaires can be correlated to the room description and number in the layout.

The Occupant Questionnaire should be tailored to the characteristics of the facility and the type of work performed there. Questions put to a factory worker, for example, will usually differ from those asked of an office worker. Examples:

- Rate, on a scale of 1 to 5, how important you feel good lighting is for you to do your job, with 1 being "of minimal importance" and 5 being "very important"?

- Overall, how satisfied are you with the lighting in your area? Rate the lighting on a scale of 1 to 5, with 1 being "poor" and 5 being "excellent."

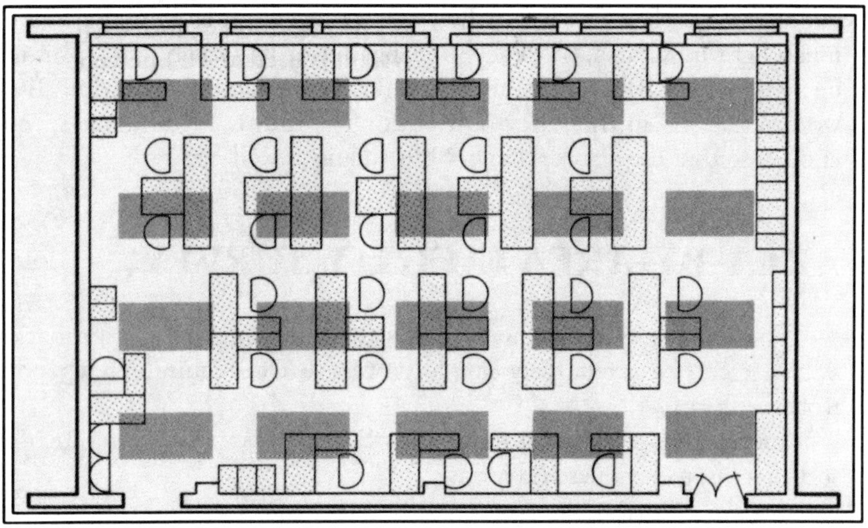

Figure 2-1. Room sketch indicating locations of workstations and light fixtures.
Source: National Lighting Bureau.

- Do you have trouble with light reflecting off your computer screen?
- Do the parking lot lights make you feel safe when you walk to or from your car after sunset?
- Do you feel that customers often return products because they were dissatisfied with the color they chose when they were here?
- Do customers complain about it being too dim or too bright in your area?
- Are you bothered by glare either from the lights themselves or reflecting off windows or other objects?
- Are there enough light switches available in your area?
- Do you often find that the lights are left on when the lavatory is unoccupied? (This question can continue for storage rooms, etc.)

The responses to the Occupant Questionnaire can provide insight into problems with the lighting system and opportunities to improve its effectiveness. Note that the responses will be written by people who know little about lighting; in fact, they rarely notice the lighting system unless it malfunctions. For example, when an employee writes that there is too much light in an area, he or she probably means it is too bright. This may be a problem of lighting quality (glare) instead of lighting quantity. Try to word questions so that they do not confuse the reader and yet specific enough so that the answers are not misleading as well.

AREA–BY–AREA FACILITY SURVEY

The Facility Survey usually yields the most important information. While it can be a relatively simple process, it can require extensive leg-work.

Successfully conducting the Facility Survey depends on using the right survey tools, techniques and forms.

EQUIPMENT

Before beginning the survey, the right equipment should be assembled and the surveyor should become acquainted with its proper use. This equipment can include an accurate light meter, a mechanical counter, access equipment and a notepad and forms. There are also situations which may warrant the use of a watt meter to measure actual fixture wattage.

Light meter

A high-quality light meter (photometer) should be obtained. This hand-held device measures the amount of light impacting the work-plane and room surfaces. Besides the ability to measure light level in

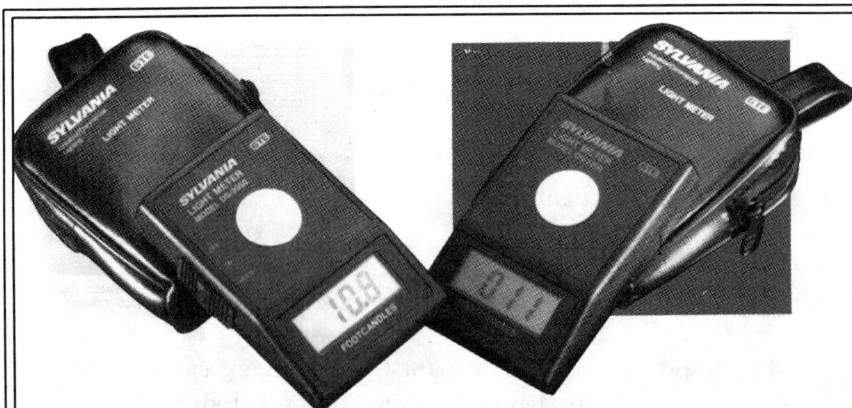

Figure 2-2. Light meters.
Source: Osram Sylvania, Inc.

footcandles, it can be used to determine the ability of the room's surfaces to reflect light onto the workplane.

Mechanical counter

Although counting fixtures appears to be a simple task, in a large area it can be frustrating when the surveyor's attention is distracted and he or she loses count. The mechanical counter is a hand-held device used to count fixtures. To count one fixture and record the number, simply press a button on the counter, which automatically advances the count by one.

Access equipment

Indoors, a step ladder should allow quick access to most fixtures. Outdoors and in industrial areas lighted by high-bay fixtures, aerial platforms may be required. Access equipment provides ability to measure energy consumption and determine the condition of the fixtures.

USING A LIGHT METER

The light meter is used to determine average light level in an area, measure horizontal and vertical illuminances impacting a task, and measure room surface reflectances.

Measure light level. The following steps comprise a simplified photometric survey procedure used to measure light level. Actual photometric surveys are generally more complicated. Be sure to follow all manufacturer instructions on the light meter's use, calibration and other care for the sensor.

1. *Turn lights on and allow time for warm-up.* After being turned on, fluorescent and HID lamps should be allowed to "warm up" for about 20 minutes. Incandescent lamps stabilize immediately and therefore require no warm-up time.

2. *Eliminate the effect of daylight.* To gain the light level produced only by the lighting system, daylight and other external light must be eliminated in the room. The measurement, therefore, should take place at night and/or with the blinds closed. If daylight cannot be prevented, take one measurement with the lights on and one with the lights off within seconds of each other; the difference in the readings is the light level contributed by the lighting system.

3. *Position the light meter.* Hold the light meter horizontally on the workplane (if the visual task is vertical, such as a bookshelf, the meter should be positioned vertically). Be careful not to shadow the meter or reflect light from your clothing, which will affect the measurement. When conducting a photometric survey, wear darker clothing so as not to reflect light onto the meter.

4. *Record light level.* Light level falling onto the primary room task(s) should be measured. Often, it is desirable to mark and record the location of the reading so that light level can be measured there again after the upgrade for comparison purposes.

Measure room surface reflectances. The reflectances of walls, ceiling and floors can be gained by performing this procedure:

1. *Record light level at the surface.* Hold the light meter flat against the surface with the sensor facing into the room. Again, be careful not to let shadows, or light reflected from clothing, interfere with the accuracy of the reading. Record light level. This is called the incident reading.

2. *Record light level reflected by the surface.* Hold the light meter about one to 1-1/2 feet away from the surface with the sensor parallel to and facing the surface. Record light level. This is called the reflected reading.

3. *Determine surface reflectance:*

$$\frac{\text{Reflected Reading}}{\text{Incident Reading}} = \text{Reflectance}$$

For example, if 100 footcandles are impacting a room surface and 75 footcandles are measured as reflected by that surface, the reflectance for that surface is 75 ÷ 100 = 0.75 or 75% reflectance.

COMPLETING THE AREA-BY-AREA SURVEY

An area-by-area survey can be performed by completing the Survey Worksheet on the following pages for each area on the floorplan. The worksheet is completed with an example; blank forms can be found in Appendix II. See pages following the Survey Worksheet for a line-by-line guide to successfully completing the form.

As each area is completed in turn, a sticker bearing the area's description and number or simply the number can be adhered near the doorhinges (or other discrete place) to match the area numbers on the floorplan. This can help with analysis, installation and maintenance.

Survey Worksheet: Area-by-Area Survey, Page 1

Area description: WORD PROCESSING DEPT. Area No.: 4

Room Length 120 ' x Room Width 60 ' = Room Area 7,200 ft^2

Daylight ____ is ✓ is not present.

Occupancy hours 6:00 (a.m.)/p.m. to 9:00 a.m./(p.m.) 5 days per week

room ceiling reflectance: 80 %

room walls reflectance: 50 %

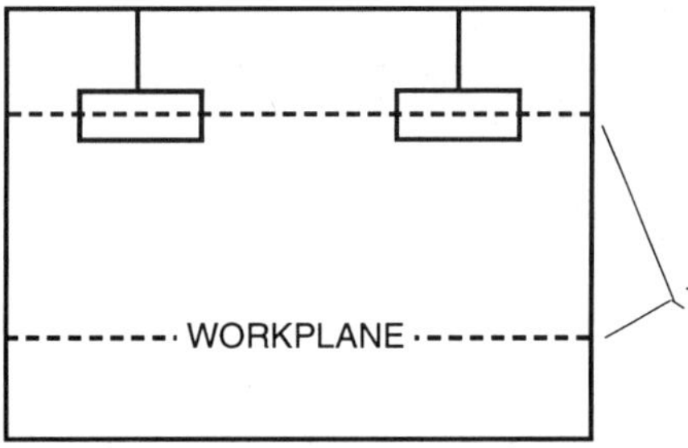

10 ' room cavity

	Fixture Type #1	Fixture Type #2	Fixture Type #3
1. Type of fixture	FLUORESCENT		
	RAPID START		
2. Nominal size (2' x 4', etc.)	2' x 4'		
3. Number of lamps per fixture	4		
4. Type of lamps	F40/CW		
5. Type of ballasts	MAGNETIC		
6. Type of shielding material	PLASTIC LENS		
7. Physical condition (excellent, good, fair, poor)	GOOD		

Survey Worksheet: Area-by-Area Survey, Page 2 Area No. 4

	Fixture Type #1		Fixture Type #2		Fixture Type #3		TOTAL
8. Degree of dirt accumulation (very clean, clean, medium, dirty, very dirty)	CLEAN						
9. Total number of burnouts	0						
10. Number of fixtures	90 (8'×10' SPACING)						
11. Input watts per fixture	192	x		x			
12. Total watts	17,280	=		=		=	17,280
13. Square footage						÷	7,200
14. Watts per square foot						=	2.4
	÷ 1,000		÷ 1,000		÷ 1,000		
15. Demand for power in kilowatts	17.28	=		=		=	
16. Annual operating hours	3,900	x		x		x	
17. Energy consumption in kWh	67,392	=		=		=	

VISUAL TASKS PERFORMED	LIGHT LEVEL ON TASK	RANGE OF AGE OF OCCUPANTS	LIGHTING QUALITY PROBLEMS
1. READING PEN-WRITTEN MATERIAL	60 Fc	40-60	1. OCCUPANTS COMPLAIN AREA LOOKS "DRAB"; MAY NOT BE RELATED TO LIGHTING
2. TYPING ON COMPUTER KEYBOARDS	60 Fc	40-60	
	Fc		2. ROWS B AND C COMPLAIN OF OVERHEAD GLARE
	Fc		

COMPLETING THE SURVEY WORKSHEET

The Survey Worksheet is essentially divided into three portions, identified below as Section 1, Section 2 and Section 3.

Section 1

First, write down this area's description and identification number. Determine the room's area, expressed in square feet, by multiplying the space's length (feet) by its width (feet).

Indicate if daylight is or is not available during occupancy hours; this will allow you to determine if equipment options such as a photocell/dimming system operating with dimmable ballasts are viable.

Now write in the hours of the day the room's lights are on. Total hours for the year will be later determined by multiplying the number of hours per day by the days per week the lights are on by the number of operating weeks in the year. Comparing this figure to the company's operating hours will assist in determining whether lighting is being wasted, and hence open an opportunity for the use of controls.

Next, on the illustration, use measuring tape to determine the height of the room cavity (in feet). The room cavity is the space between the bottom of the fixtures (typically the ceiling) and the workplane, the invisible horizontal plane that defines where the majority of important visual tasks are performed in the space. In office applications, the workplane is typically 2.5 feet above the floor — normal desk height.

Finally, measure the reflectances of the room's surfaces. This can be done using a light meter.

Section 2

1. Write down what type of fixtures are in the room.
2. Indicate what the nominal size of the fixture is, expressed as width (feet) x length (feet).

   1 x 1   2 x 2   3 x 3   4 x 4
   1 x 2   2 x 4
   1 x 3
   1 x 4
   1 x 8
3. Write down the number of lamps in each fixture.
4. Identify the type of lamps used.
5. Identify the type of ballasts used.
6. Identify the type of shielding material/lens used.
7. Rate the fixtures' general physical condition as excellent, good, fair or poor based on opinion.
8. Write down the total number of burnouts in all of the fixtures of

this type.

9. Record the average dirt accumulation on the fixtures, indicating whether the fixtures are, in general, very dirty, dirty, clean or very clean. This is based on opinion; a more sophisticated procedure is available in Chapter 3: "Effective Illumination: Meeting the 'People Priority.' "
10. Indicate the number of fixtures present in the area of this type.
11. Indicate the input wattage per fixture. Obtain from the manufacturer literature or through a field survey using a watt meter or energy monitoring equipment.
12. Multiply the number of fixtures by the input wattage per fixture to determine total watts for each fixture type. Add the figure in each column and write the sum in the "TOTAL" column.
13. Indicate in the "TOTAL" column the area of the space expressed in square feet.
14. Divide total watts by total area to determine unit power density, or watts per square foot.
15. Divide total watts by 1,000 to determine demand for power in kilowatts.
16. Indicate annual operating hours for this fixture type (remember to include weekends where applicable):

ANNUAL OPERATING HOURS = OPERATING HOURS PER DAY x OPERATING DAYS PER WEEK x OPERATING WEEKS PER YEAR

17. Multiply kilowatts by operating hours to determine energy consumption in kWh.

Section 3

Write down the visual tasks performed in the area and how much light currently falls onto the tasks, expressed in footcandles. This will help you determine if existing light levels are serving the needs of the occupants. Then write the range of ages of the occupants in the area. Finally, indicate if there are any perceived lighting quality problems or other opportunities from the occupants or personal observation.

Sources:

Lighting Upgrade Manual, First Edition. Washington, DC: U.S. Environmental Protection Agency Green Lights Program, November 1991.

Lindsey, Jack. *Applied Illumination Engineering.* Lilburn, Georgia: The Fairmont Press, 1991.

Performing a Lighting System Audit, Revised Edition. Washington, DC: National Lighting Bureau.

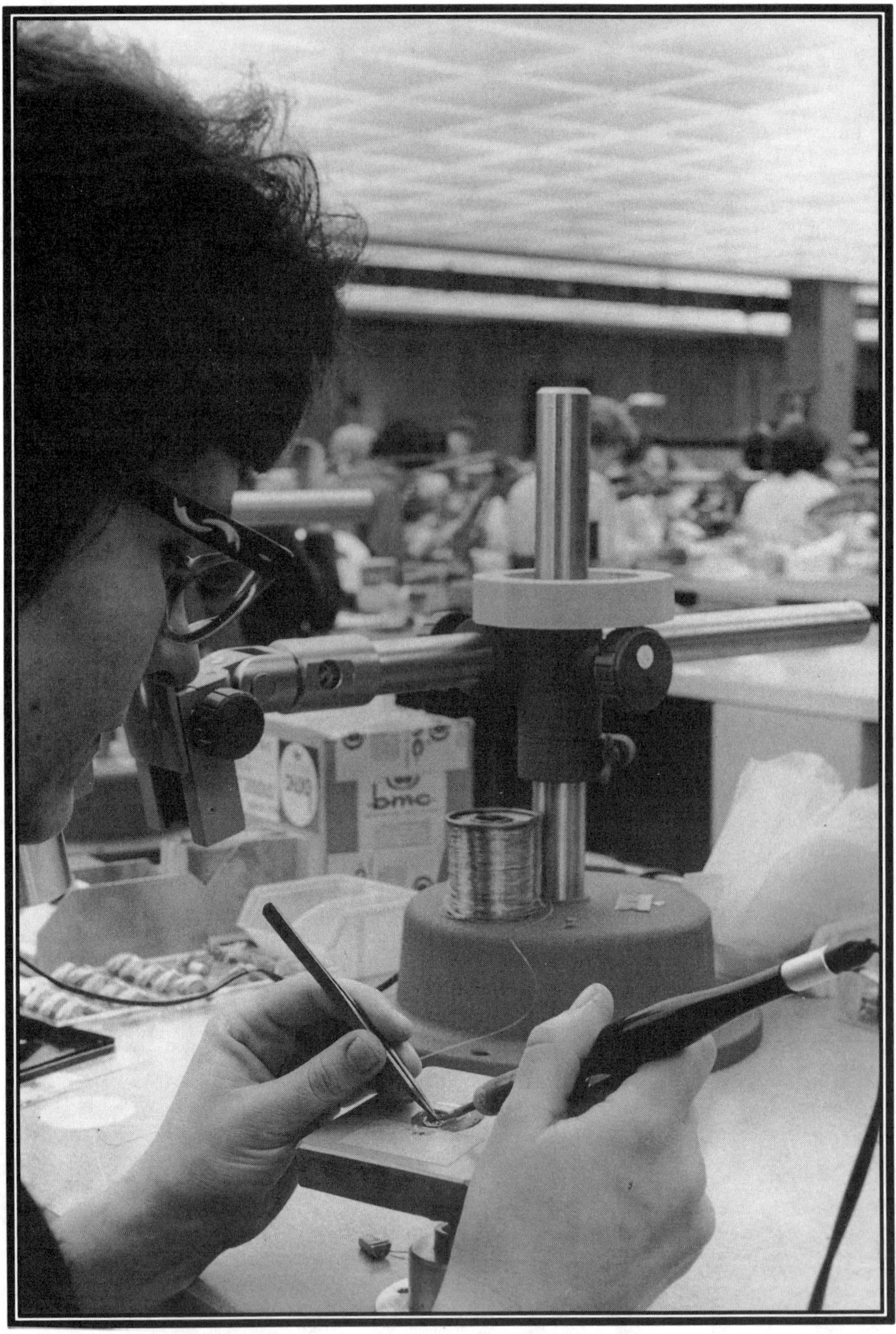

Source: National Lighting Bureau

CHAPTER 3

Effective Illumination: Meeting the "People Priority"

PEOPLE NEED LIGHT ... AND THE RIGHT LIGHTING

The process of seeing requires at least one functioning eye and visible energy (light). The human eye can adjust to a wide range of light levels, including about 10,000 footcandles on a sunny day to about 0.01 footcandles under full moonlight.

The process of performing a visual task requires at least one functioning eye and a sufficient quantity of visible light. Many activities which can be performed under sunlight cannot be performed effectively under moonlight.

The process of performing a visual task *efficiently* and *accurately* requires that ambient light levels are not too dim, which can result in eyestrain, fatigue, slower speed and error. Because some 85% of human impressions are visual, improper quantity and quality of light can result in poor human performance.

In addition, light is needed to enhance the appearance of merchandise, create a safer environment, create a secure environment, reduce errors/rejects, decrease absenteeism, improve morale, beautify a building at night, create mood, differentiate spaces and enhance corporate image. No "people priority" of the lighting system should be sacrificed for energy savings.

VALUE OF INCREASED HUMAN PERFORMANCE

The value of better human performance through effective lighting should not be overlooked. According to the Building Owners and Managers Association, the total annual cost to operate one square foot of typical office space is about $180. Of this, $150 is allocated to people (wages, salaries and benefits) and $1.00 to lighting (electricity, labor, product). Therefore, a lighting equipment upgrade which results in a 50% reduction in energy use can generate savings of about $0.50 per square foot. But an upgrade which results in only a 1% increase in worker productivity can generate $1.50 per square foot in profit. (See Chapter 9: "Success Stories: Quality Lighting Upgrades" for examples of how upgrading a lighting system can yield a profit in productivity even greater than the actual energy savings.)

The first goal of lighting management is to ensure that the lighting system consistently provides proper quantity and quality of light.

THE TOTAL ANNUAL COST FOR A TYPICAL OFFICE ($/SQ. FT./YR.): ABOUT $180

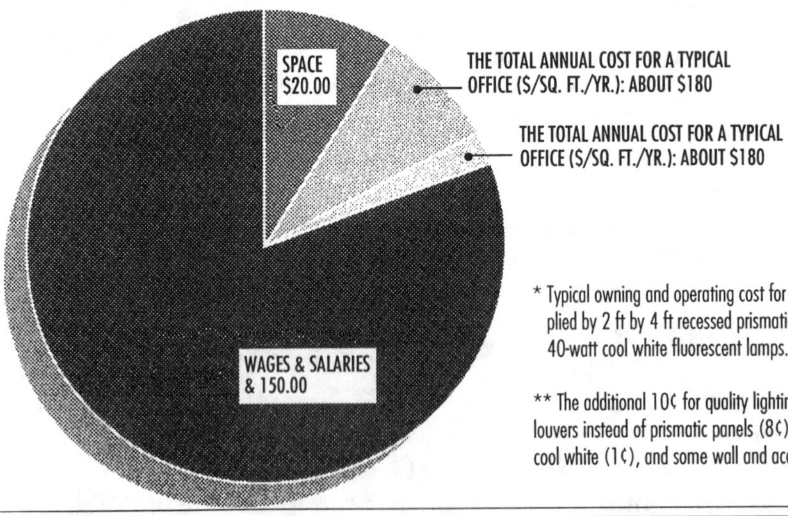

* Typical owning and operating cost for 70 footcandles supplied by 2 ft by 4 ft recessed prismatic troffers using four 40-watt cool white fluorescent lamps.

** The additional 10¢ for quality lighting assumes deep-cell louvers instead of prismatic panels (8¢), SP color instead of cool white (1¢), and some wall and accent lighting (1¢).

Figure 3-1. While the cost to light an office space is predominantly electricity, the cost to operate office space is predominantly "people costs" — wages, salaries and benefits. Therefore, a successful lighting management program makes people its first priority by providing quality lighting to occupants. *Source: GE Lighting.*

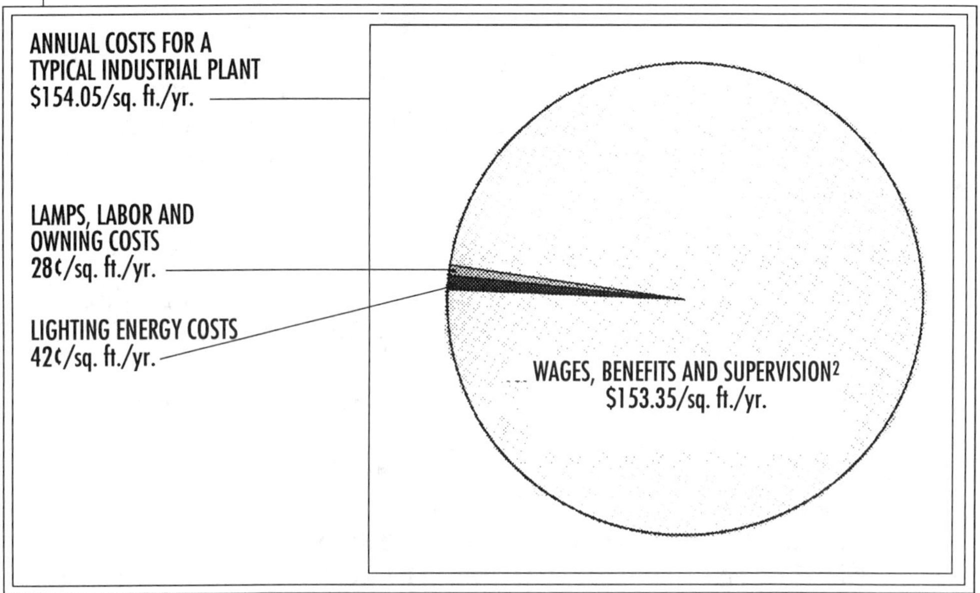

Figure 3-2. Annual costs for a typical industrial plant. Again, reducing lighting energy use can produce a satisfying return on investment; however, increasing productivity through quality lighting can be even more profitable. (Wages and benefits for typical industrial worker taken from data published by the U.S. Department of Labor Bureau of Labor Statistics, October 1989.) *Source: GE Lighting.*

SPECIFYING QUANTITY OF LIGHT

Quantity of light in an area is dependent upon the relationship between the lighting system's lumen output and how much light is actually received by the workplane. In review, lumen output is the quantity of light leaving the lamp and is measured in lumens. Light level is the amount of light impacting the workplane and is measured in footcandles with a light meter.

For the lighting system to consistently provide proper light levels, we must first determine what the minimum level is, then work our way backwards to determine the appropriate lumen output of the lighting system. The IESNA recommends ranges of light levels for people to perform some 600 visual tasks from reading a chalkboard to working on an assembly line. Each range takes into account these primary factors: time, size and contrast (in addition to brightness).

TIME

The amount of light required for a task is related to the amount of time it takes to perform it. Therefore, the amount of time available to complete a task will determine light level. If a task must be completed quickly, then a higher light level will be needed to ensure efficiency and accuracy. For example, a dimly lighted billboard next to a highway will fare poorly compared to a brightly lighted one, because the board's message is available to passing motorists only for moments.

SIZE

Generally, the larger an object appears visually, the easier it is to see it. This is why certain books and magazines published for an older audience feature larger print.

This rule recognizes that visual size can differ from physical size. For example, a car is a large object when standing next to it, but it can fit between finger and thumb when viewed at long distances. The larger the task is visually, the less light will be needed to perform it. For example, a billboard will be more effective if it carries a brightly lighted message printed in large letters, because it will be able to be seen at a longer distance and therefore for a longer period of time.

CONTRAST

Contrast is how well the task stands out from its immediate surroundings. The greater the contrast, the easier it is to see the task and subsequently the lower the light level that is needed. Imagine working in an office where all of the objects are the same color. More light is needed in this application to distinguish objects. Another example is that yellow letters printed on an orange background will require more light to be read quickly and accurately than yellow letters printed on a blue background, because the latter poses greater contrast. The benefit of contrast to vision is why many tennis players prefer a yellow or orange ball as opposed to a green ball, because the court itself is usually green.

Figure 3-3. The amount of light required for a task is related to the amount of time required to perform it with minimum error and chance of accident. The amount of time available to complete a task, therefore, is a factor in specifying light level. *Source: Philips Lighting Company.*

SPECIFYING LIGHT LEVEL

Determining minimum recommended light level takes three steps.

First, the task must be defined. Then the appropriate Illuminance Category must be determined. Within each category is a range of three light levels. The light level most appropriate for the application is then chosen. This procedure is outlined in Light Level Worksheet #1.

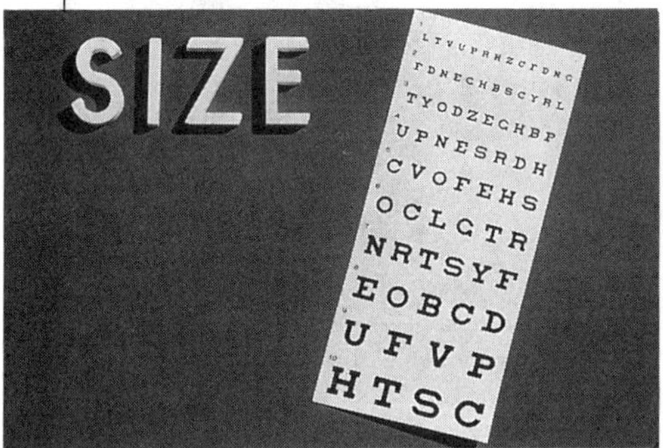

Figure 3-4. Generally, the larger a task appears visually, the easier it is to see it; less light will be needed to perform the task with minimum error and accident.
Source: Philips Lighting Company.

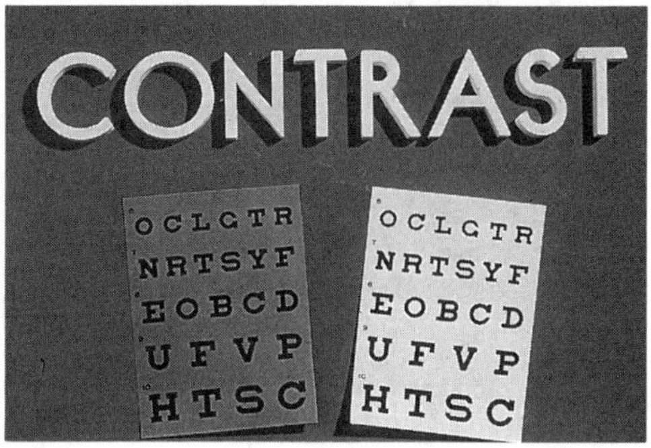

Figure 3-5. Contrast is how well a task stands out from its immediate surroundings. The greater the contrast, the easier it is to see the task and subsequently the lower the light level that is needed. *Source: Philips Lighting Company.*

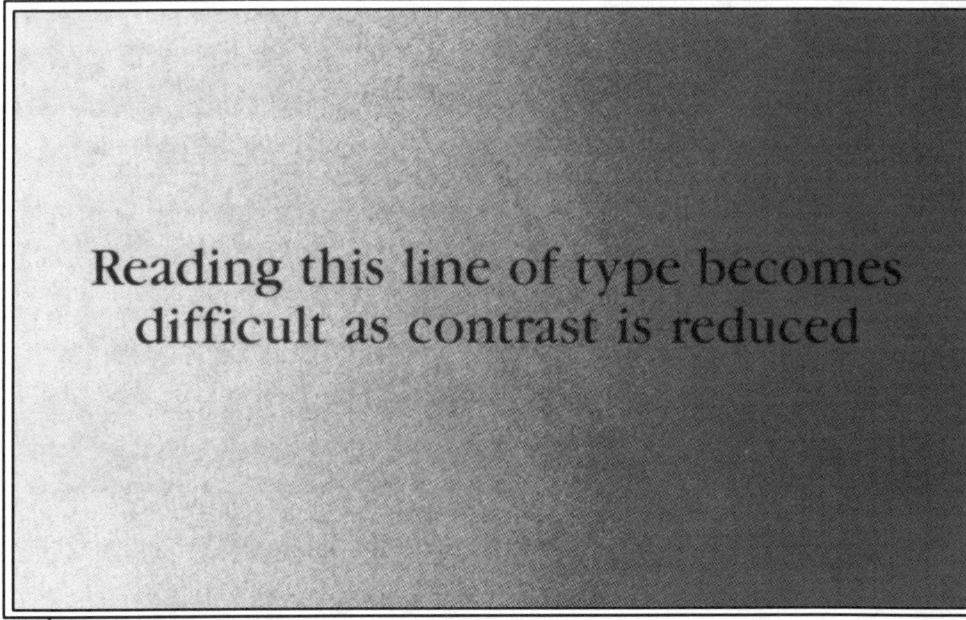

Source: National Lighting bureau.

Figure 3-6. Brightness is also considered a factor in determining light level. Brightness is the product of the task's reflectance and recorded photometric brightness, or luminance. The brighter the task is, the easier it is to see and the lower the amount of light that is required. Too little brightness decreases contrast and calls for a higher light level. Too much brightness can make the task uncomfortable to see (produce glare), which calls for a lower light level. *Source: Philips Lighting Company*.

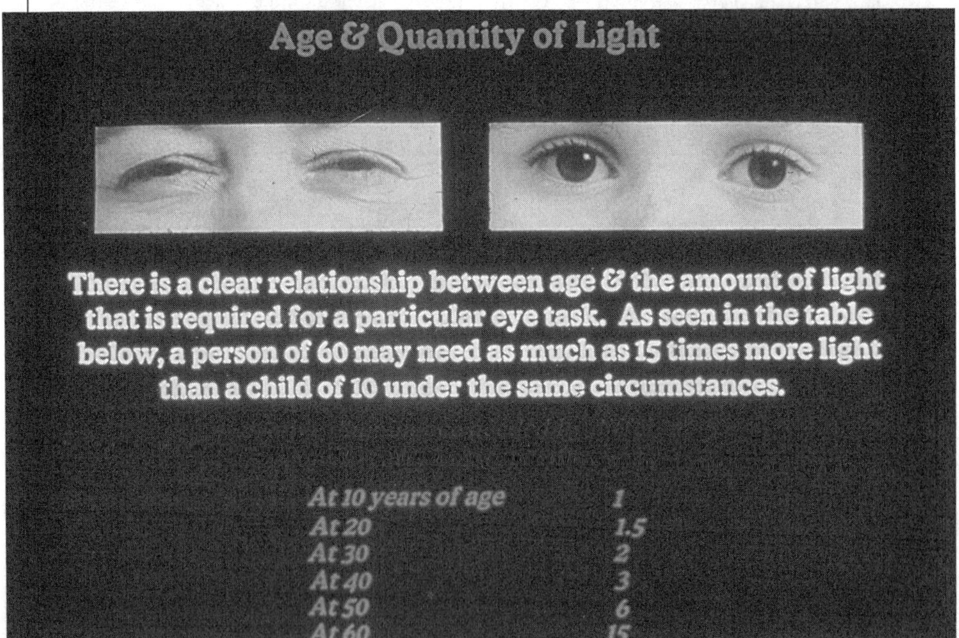

Figure 3-7. The age of the individual performing a visual task is an important consideration when specifying light level. As the eye ages, vision capabilities typically deteriorate and so higher light levels are needed for older people.
Source. Philips Lighting Company.

Light Level Worksheet #1:
Specifying Minimum Light Level

Area description: WORD PROCESSING DEPT. Area No.: 4

1. Task READING PEN-WRITTEN MATERIAL; TYPING

2. Generic task VISUAL TASK OF HIGH CONTRAST

3. Illuminance Category D

4. Illuminance range 20 - 30 - 50

Weighting Factors: If Illuminance Category is A, B or C

5. Range of occupant ages and weighting factor ______ / ______

6. Room surface reflectances + ______

7. Total weighting Factor = ______

a. For Illuminance Categories A through C			
Room and Occupant Characteristics	Weighting Factor		
	-1	0	+1
Occupants ages	Under 40	40-55	Over 55
Room surface reflectances*	Greater than 70 percent	30 to 70 percent	Less than 30 percent

Weighting Factors: If Illuminance Category is D, E, F, G, H or I

8. Range of workers' ages and weighting factor 40-60 / 1

9. Speed + 0

10. Accuracy + 1

11. Reflectance on task background + 0

12. Total weighting factor = 2

13. Recommended light level 50 Fc

b. For Illuminance Categories D through I			
Task and Worker Characteristics	Weighting Factor		
	-1	0	+1
Workers ages	Under 40	40-55	Over 55
Speed and/or accuracy**	Not important	Important	Critical
Reflectance of task background***	Greater than 70 percent	30 to 70 percent	Less than 30 percent

* Average weighted surface reflectances, including wall, floor and ceiling reflectances, if they encompass a large portion of the task area or visual surround. For instance, in an elevator lobby, where the ceiling height is 7.6 meters (25 feet), neither the task nor the visual surround encompass the ceiling, so only the floor and wall reflectances would be considered.

** In determining whether speed and/or accuracy is not important or critical, the following questions need to be answered: What are the time limitations? How important is it to perform the task rapidly? Will errors produce an unsafe condition or product? Will errors reduce productivity and be costly? For example, in reading for leisure there are no time limitations and it is not important to read rapidly. Errors will not be costly and will not be related to safety. Thus, speed and/or accuracy is not important. If however, prescription notes are to be read by a pharmacist, accuracy is critical because errors could produce an unsafe condition and time is important for customer relations.

***The task background is that portion of the task upon which the meaningful visual display is exhibited. For example, on this page the meaningful visual display includes each letter which combines with other letters to form words and phrases. The display medium, or task background, is the paper, which has a reflectance of appoximately 85 percent.

Source: IES Lighting Handbook, *1987 Application Volume*.

To complete Light Level Worksheet #1:

1. Write down the task from the Survey Worksheet completed in Chapter 2.
2. Match this task to an appropriate generic task listed in the *IES Lighting Handbook* (see Figure 3-8 for examples).
3. Determine the Illuminance Category from the *IES Lighting Handbook* listing. General tasks and appropriate categories are in Figure 3-8.
4. Write down the corresponding range of light levels from the *IES Lighting Handbook* listing. Light level ranges for the Illuminance Categories are in Figure 3-9.

5-7. Steps 5, 6 and 7 are completed only if the Illuminance Category is A, B or C. Generally, these activities are simple visual tasks such as walking through a lobby, sitting and talking in a conference room, or working at a computer display terminal which is lighted internally. These tasks are performed throughout the space, so these Illuminance Categories specify uniform light levels. If task lighting is employed for all important visual tasks in a space, any overhead lighting, although it impacts the task, can be considered general illumination and therefore in the A-C Illuminance Category range.

5. Based on the age of the oldest people who will use the space, write down the appropriate weighting factor (-1, 0 or +1).
6. Based on a weighted average of the room surface reflectances (using the ceiling only if it is low enough to be in the field of view), write down the appropriate weighting factor.
7. Calculate the sum of these two weighting factors.

8-12. Steps 8-12 are completed only if the Illuminance Category is D-I. These Categories represent tasks which are more visually demanding in order of increasing difficulty. The tasks are generally performed at fixed locations in the area, so these Illuminance Categories specify light levels impacting the task itself.

8. Obtain the range of ages of the occupants from the Survey Worksheet and write this range here. Then determine the weighting factor based on the highest end of the range.
9. Determine the importance of speed in this task. Some tasks do not require a great deal of speed, while the successful completion of others rely on it. A factory worker on an assembly line generally needs to perform his or her tasks in a set time, whereas an architect drafting a blueprint generally does not have as strict a time limit because more thought is involved.
10. Determine the importance of accuracy in this task. Like speed,

Area/Activity	Illuminance Category
Reading	
Copied tasks	
Ditto copy	E
Micro-fiche reader	B
Mimeograph	D
Photograph, moderate detail	E
Thermal copy, poor copy	F
Xerograph	D
Xerography, 3rd generation and greater	E
Electronic data processing tasks	
CRT screens	B
Impact printer	
good ribbon	D
poor ribbon	E
2nd carbon and greater	E
Ink jet printer	D
Keyboard reading	D
Machine rooms	
Active operations	D
Tape storage	D
Machine area	C
Equipment service	E
Thermal print	E
Handwritten tasks	
#2 pencil and softer leads	D
#3 pencil	E
#4 pencil and harder leads	F
Ball-point pen	D
Felt-tip pen	D
Handwritten carbon copies	E
Non photographically reproducible colors	F
Chalkboards	E
Printed tasks	
6 point type	E
8 and 10 point type	D
Glossy magazines	D
Maps	E
Newsprint	D
Typed originals	D
Typed 2nd carbon and later	E
Telephone books	E

Figure 3-8. Illuminance Categories for typical reading tasks. Some tasks are qualified. For a complete listing of activities and for footnoted qualifcations, see Page 2-20 of the 1987 Application Volume of the *IES Lighting Handbook*. *Source: Illuminating Engineering Society of North America,* IES Lighting Handbook, *1987 Application Volume*.

some tasks do not require a great deal of accuracy, whereas others rely on it. For example, a mechanic working on an automobile engine must perform his or her work with great accuracy. A doctor writing a prescription and the pharmacist reading it also need to perform their tasks accurately.

Type of Activity	Illuminance Category	Ranges of Illuminances		Reference Work-Plane
		Lux	Footcandles	
Public spaces with dark surroundings	A	20-30-50	2-3-5	General lighting throughout spaces
Simple orientation for short temporary visits	B	50-75-100	5-7.5-10	
Working spaces where visual tasks are only occasionally performed	C	100-150-200	10-15-20	
Performance of visual tasks of high contrast or large size	D	200-300-500	20-30-50	Illuminance on task
Performance of visual tasks of medium contrast or small size	E	500-750-1000	50-75-100	
Performance of visual tasks of low contrast or very small size	F	1000-1500-2000	100-150-200	
Performance of visual tasks of low contrast and very small size over a prolonged period	G	2000-3000-5000	200-300-500	Illuminance on task, obtained by a combination of general and local (supplementary lighting)
Performance of very prolonged and exacting visual task	H	5000-7500-10000	500-750-1000	
Performance of very special visual tasks of extremely low contrast and small size	I	10000-15000-20000	1000-1500-2000	

Figure 3-9. Illuminance Categories and illuminance values for generic types of activities in interiors. *Source: Illuminating Engineering Society of North America,* IES Lighting Handbook, *1987 Application of Volume.*

11. Determine the reflectance of the task background. A mechanic working on a car engine will need more light because the engine's components are likely to be metallic in color and dirty. A proofreader reading a manuscript may need less because the reflectance of the background — a white piece of paper — has a higher reflectance.
12. Calculate the sum of these four weighting factors.
13. If the sum is -2 or below, the lowest value on the range is the minimum recommended light level. If the sum is -1, 0 or +1, the middle value is recommended. And if the sum is +2 or higher, the highest value is recommended.

SPECIFYING LUMEN OUTPUT

Now that the minimum recommended light level has been determined, the required lumen output of the lighting system can be determined. This can be calculated per lamp or fixture or simply by total number of lumens.

Because the lighting system must provide minimum light level consistently, this value is considered recommended maintained illumination. Initial illumination is the light level in an area or impacting a task when lamps and/or a lighting system are clean and new. Maintained illumination is the light level in an area after light loss factors such as dirt and age take their toll on lumen output and subsequently the amount of light reaching the workplane over a given period of time.

Three formulas can be used to calculate maintained illumination (and through modification the required lumen output of the lighting system) to provide this illumination. The first is the most comprehensive, and calculates by the total number of fixtures. The second is relatively simpler, being driven by number of lamps only. The third can be the simplest formula because it calculates based on the number of lamps within one fixture and the square footage covered by one fixture (its effectiveness can be limited, however, if the upgrade entails comprehensive changes in the lighting system).

$$\text{AVERAGE MAINTAINED ILLUMINATION (FOOTCANDLES)} = \frac{\text{LAMPS/FIXTURE} \times \text{LUMENS/LAMP} \times \text{NO. OF FIXTURES} \times \text{COEFFICIENT OF UTILIZATION} \times \text{LIGHT LOSS FACTORS*}}{\text{AREA IN SQUARE FEET}}$$

$$\text{AVERAGE MAINTAINED ILLUMINATION (FOOTCANDLES)} = \frac{\text{TOTAL LAMPS} \times \text{LUMENS/LAMP} \times \text{COEFFICIENT OF UTILIZATION} \times \text{LIGHT LOSS FACTORS}}{\text{AREA IN SQUARE FEET}}$$

$$\text{AVERAGE MAINTAINED ILLUMINATION (FOOTCANDLES)} = \frac{\text{LAMPS IN ONE FIXTURE} \times \text{LUMENS/LAMP} \times \text{COEFFICIENT OF UTILIZATION} \times \text{LIGHT LOSS FACTORS}}{\text{AREA IN SQUARE FEET PER FIXTURE}}$$

* LIGHT LOSS FACTORS = BALLAST FACTOR x FIXTURE AMBIENT TEMPERATURE FACTOR x SUPPLY VOLTAGE VARIATION FACTOR x LAMP POSITION FACTOR x OPTICAL FACTOR x FIXTURE SURFACE DEPRECIATION FACTOR x LAMP BURNOUTS FACTOR x LAMP LUMEN DEPRECIATION FACTOR x FIXTURE DIRT DEPRECIATION FACTOR x ROOM SURFACE DIRT DEPRECIATION FACTOR.

These basic *maintained illumination* method formula can be reconfigured to provide the *lumen method*:

$$\text{REQUIRED LUMENS/FIXTURE} = \frac{\text{MAINTAINED ILLUMINATION} \times \text{AREA (FT}^2)}{\text{NUMBER OF FIXTURES} \times \text{COEFFICIENT OF UTILIZATION} \times \text{BALLAST FACTOR} \times \text{LIGHT LOSS FACTORS}}$$

A lighting designer will use these methods to calculate the required number of lamps or fixtures to be installed in a layout. Because we are upgrading an existing lighting system, we are only interested in determining minimum lumen output per fixture because the facility has a preset number and position of fixtures. (If new fixtures are to be installed, a lighting designer should be consulted.)

After we determine minimum lumen output per fixture, we can start identifying lamp or lamp-ballast options. First, we will determine minimum lumen output per lamp for two, three or four lamps. Now options can be compared to the recommended lumen output and feasible options chosen. If there is any doubt that these options will produce the minimum lumen output, they can be tested using the maintained illumination method.

Before we proceed, we need to review the factors that depreciate a lighting system's lumen output and the amount of light impacting the workplane. These factors are coefficient of utilization, the non-recoverable light loss factors, and the recoverable light loss factors. Understanding and measuring the impact of these factors is crucial to determine minimum lumen output and maintained light levels.

COEFFICIENT OF UTILIZATION

Coefficient of utilization (CU) is a value which expresses the percentage of the lumens produced by the lamps which reach the task after light is lost due to three factors: 1) the fixture's efficiency at transmitting light, 2) the room proportions, and 3) the ability of room surfaces to reflect light. A typical CU value might be 0.64. All other factors aside, if the lamps in one fixture together produce 10,000 lumens, then 10,000 x 0.64 = 6,400 lumens will reach the workplane. CU can be thought of as an efficiency factor or a derating factor for fixtures in the given area.

Fixture efficiency

For light to reach a task, it must leave the fixture and then be directed at the task. The physical characteristics of the fixture, such as its shape and reflectance, can affect how much light leaves the fixture as opposed to being diffused within it. Shielding material such as a lens or louver also prevents free transmission of light. In addition, the number of lamps inside the fixture also affects the amount of light leaving it.

Room proportions

A fixture close to a wall will shed much of its lumen output onto it, while almost all of the light leaving a fixture near the center of a large room will reach the workplane directly. Room surfaces absorb some light, so a large room will see a greater percentage of light reach the workplane than be absorbed by the walls.

Room surface reflectances

The ability of the ceiling, walls and floor to reflect light enhances multiple reflection of light onto the workplane. Therefore, a room with clean surfaces painted a light color will increase CU, whereas a room with dirty walls and/or dark-painted surfaces will result in a lower CU.

Procedure to determine coefficient of utilization

To determine CU in new or renovated buildings, lighting designers use the Zonal Cavity Method, which separates a room into three distinct zones, or cavities. The ceiling cavity represents the space between fixtures mounted below the ceiling and the ceiling itself. The room cavity represents the space between the fixtures and the workplane. And the floor cavity represents the space between the workplane and the floor. For retrofit design purposes, we will concern ourselves only with the room cavity. The dimensions of this space can be obtained during the area-by-area survey. In most office lighting environments, the fluorescent fixtures are not hanging fixtures, so the room cavity represents the space between the workplane and the ceiling.

Complete this procedure to determine CU:

1. *Determine Room Cavity Ratio.* This is the ratio of vertical wall area to horizontal surface area in the room, and is determined with this formula:

$$\text{ROOM CAVITY RATIO} = \frac{5 \times \text{CAVITY HEIGHT} \times (\text{LENGTH} + \text{WIDTH})}{\text{LENGTH} \times \text{WIDTH}}$$

2. *Obtain wall and ceiling reflectances* from the Survey Worksheet. Most data assumes a floor reflectance of 20%. If the actual floor reflectance in the application is not 20%, a correction factor should be applied; see the *IES Lighting Handbook* for the procedure and table of correction factors. For our purposes, we will assume an effective floor reflectance of 20%.
3. *Determine* CU. The IESNA lists generic fixture types and CU tables in the *IES Lighting Handbook.* For more specific results, fixture manufacturers' literature, which contain CU tables, should be consulted. The table for the existing fixture type is usually sufficient to determine CU, but if specular reflectors are to be installed to increase fixture efficiency, CU should be determined twice (once for

Typical Luminaire	ρcc →	80			70			50			30			10			0
	ρw →	50	30	10	50	30	10	50	30	10	50	30	10	50	30	10	0
	RCR ↓	Coefficients of Utilization for 20 Per Cent Effective Floor Cavity Reflectance (ρ_{FC} = 20)															
 Fluorescent unit with flat prismatic lens, 4-lamp 610 mm (2') wide	0	.75	.75	.75	.73	.73	.73	.70	.70	.70	.67	.67	.67	.64	.64	.64	.63
	1	.67	.64	.62	.65	.63	.61	.63	.61	.59	.60	.59	.58	.58	.57	.56	.55
	2	.59	.56	.52	.58	.55	.52	.56	.53	.51	.54	.52	.49	.52	.50	.48	.47
	3	.53	.48	.45	.52	.48	.44	.50	.46	.43	.48	.45	.43	.47	.44	.42	.41
	4	.47	.42	.38	.46	.42	.38	.45	.41	.38	.44	.40	.37	.42	.39	.37	.35
	5	.43	.37	.34	.42	.37	.33	.41	.36	.33	.39	.36	.33	.38	.35	.32	.31
	6	.39	.33	.30	.38	.33	.29	.37	.32	.29	.36	.32	.29	.35	.31	.29	.27
	7	.35	.30	.26	.35	.30	.26	.34	.29	.26	.33	.29	.26	.32	.28	.26	.24
	8	.32	.27	.24	.32	.27	.23	.31	.26	.23	.30	.26	.23	.29	.26	.23	.22
	9	.30	.25	.21	.29	.24	.21	.28	.24	.21	.28	.24	.21	.27	.24	.21	.20
	10	.27	.22	.19	.27	.22	.19	.26	.22	.19	.26	.22	.19	.25	.22	.19	.18

Figure 3-10. Coefficients of utilization for a typical fluorescent light fixture (for future reference, this is a Maintenance Category V fixture).
Source: Illuminating Engineering Society of North America, IES Lighting Handbook, *1984 Reference Volume.*

the original fixture and once for the reflectorized fixture). Reflectors are devices which increase the ability of existing fixtures to reflect and direct light.

Notes

a. If the floors are to recarpeted, or the walls to be repainted as part of the upgrade, the ability of these new surfaces to reflect light should be represented in the CU. During the survey, bring a piece of board or carpet of the new color and measure reflectance from these materials.

b. If the room is irregular-shaped, that is, not shaped like a square or rectangle, then another formula is used to determine the Room Cavity Ratio:

$$\text{ROOM CAVITY RATIO} = \frac{2.5 \times \text{CAVITY HEIGHT} \times \text{PERIMETER}}{\text{AREA}}$$

c. Room Cavity Ratios on fixture CU tables are listed as whole numbers in ascending order (0, 1, 2, 3, etc.). Actual results may not always conveniently be a whole number. In this situation, it is necessary to interpolate between two values. The best way to show the interpolation procedure is through example. Suppose we determine Room Cavity Ratio to be 1.7.

STEP 1. Determine CU for the two closest Room Cavity Ratio whole numbers, in this case 1 and 2. Suppose the CU for a Room Cavity Ratio of 1 is 0.67, and the CU for a Room Cavity Ratio of 2 is 0.6.

STEP 2. Subtract the smaller value (0.6) from the larger (0.67) and write down the difference (0.07).

STEP 3. Multiply this figure by the difference between the actual Room Cavity Ratio and the nearest lowest whole number. The actual Room Cavity Ratio is 1.7, and the nearest lowest whole number is 1; the difference is 0.7. We therefore multiply 0.07 x 0.7 and reach a result of 0.049.

STEP 4. Subtract the result (0.049) from the larger CU (0.67). The result in our example is 0.621. The interpolated Room Cavity Ratio is 0.62.

d. Like Room Cavity Ratios, ceiling and wall reflectances do not always conveniently work out to the limited values represented on the CU table. Again, we must interpolate; again, we will use an example. Suppose the ceiling reflectance is 80% and the wall reflectance is 42%; we assume the floor reflectance to be 20%. CU tables generally do not include 42% for wall reflectances, so we complete this procedure:

STEP 1. Determine CU for the two closest wall reflectances, in this case 50% (0.67) and 30% (0.65).

STEP 2. Subtract the smaller value (0.65) from the larger (0.67) and determine the difference (0.02).

STEP 3. Determine the difference between the two nearest reflectances (50% - 30% = 20%). Determine the difference between the actual reflectance (42%) and the nearest highest reflectance (50%); the result is 8%. Multiply the difference in STEP 2 (0.02) by 8 ÷ 20: 0.02 x (8 ÷ 20) = 0.008.

STEP 4. Subtract the result (0.008) from the larger CU (0.67); the result in our example is 0.0662. The CU is 0.066.

e. Sometimes, it may be necessary to interpolate between Room Cavity Ratios and reflectances. In each case, the interpolation procedure is repeated until the final CU is determined.

NON–RECOVERABLE LIGHT LOSS FACTORS

Non-recoverable light loss factors are values which express the percentage of light reaching the workplane after various characteristics and environmental factors depreciate that amount. These light loss factors are considered "non-recoverable" because preventative maintenance generally cannot reduce the extent of light loss. A non-recoverable light loss factor of 0.85 means that if the lamps in a fixture together produce 10,000 lumens, only 8,500 lumens will reach the workplane (all other factors aside). Non-recoverable light loss factors include ballast factor, ambient fixture temperature, supply voltage variation, fixture surface depreciation and lamp operating position.

Ballast factor

As discussed in Chapter 1, lamps and ballasts experience losses when operating together as a system. Different ballast models operate lamps at varying degrees of efficiency. The ability of the ballast to cause a lamp to produce a percentage of its initial rated lumens is called its ballast factor.

Most core-and-coil fluorescent ballasts meet Certified Ballast Manufacturers (CBM) performance standards, which require that the ballast cause the lamp to produce a minimum of 92.5% of initial rated lumen output (a typical ballast factor is 0.95). Ballast factors for non-CBM ballasts, such as cathode cut-out and electronic ballasts (for which ANSI standards have not yet been developed at the time of writing), may be rated with a lower ballast factor. Likewise, ballast factors for HID systems are currently not standardized (a factor of 0.9 is generally used). Ballasts factors for any brand-name ballast can be obtained from the manufacturer.

Later in this chapter, the other light loss factors will be used to determine minimum lumen output for each lamp to achieve maintained recommended light level. Ballast factor will be applied to lamp-ballast upgrade options to determine if the lumen output produced by the option complies.

Ambient fixture temperature (fluorescent systems)

Most fluorescent lamps produce their rated lumen output when operating surrounded by still air which has a temperature of 77 degrees Fahrenheit (25 degrees Celsius). Deviations above or below 77 degrees reduce lumen output. Typically, temperatures within recessed and enclosed fixtures are hotter than 77 degrees; hence, lumen output is reduced (see Figure 3-11).

Supply voltage variation

Variations in supply voltage to the fixture are the result of fluctuations in the building's power distribution system or a voltage reduction program initiated by the local electric utility. High or low supply voltage provided to the lamp (incandescent) or ballast (fluorescent and HID) affects lumen output and input watts. Generally, a lamp or lamp-ballast system receiving a high supply voltage produces higher lumen output at the expense of an increase in input wattage. Conversely, a lamp or lamp-ballast system receiving a low supply voltage produces lower lumen output with a reduction in input watts. Electronic ballasts generally are not as sensitive to small variations in supply voltage. Some newer models provide constant lumen output variations of up to ± 10% fluctuations in supply voltage.

General data pertaining to this light loss factor is listed in the *IES Lighting Handbook* for generic lamps. Information for a specific lamp can be obtained from the manufacturer's literature.

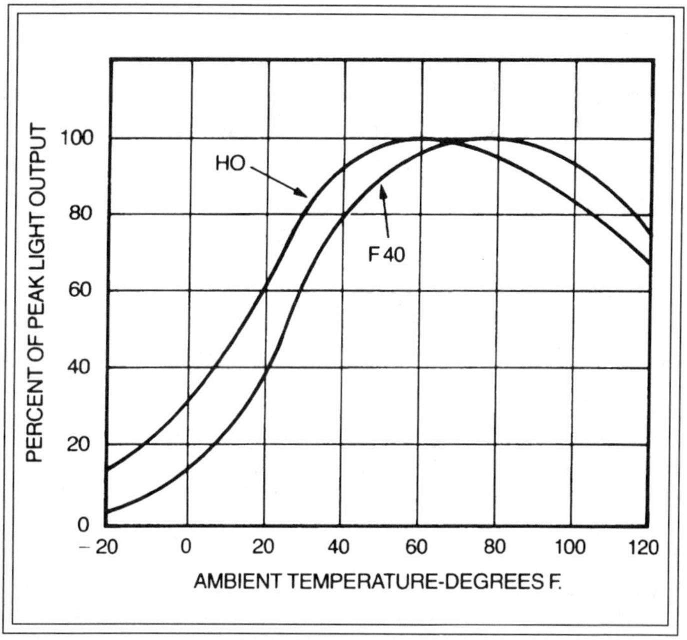

Figure 3-11. Effect of ambient temperature for bare fluorescent lamps in still air. *Source: Osram Sylvania, Inc.*

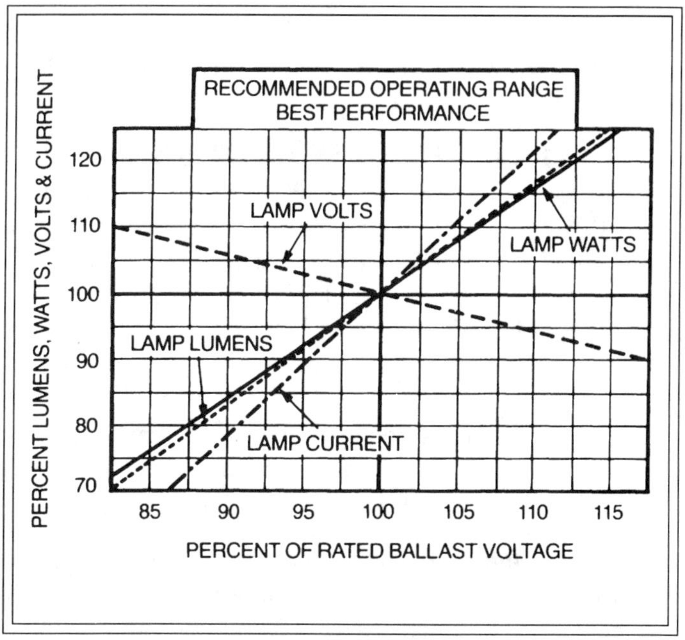

Figure 3-12. Effect of supply voltage on lamp volts, amps, watts and lumen output for a 2-lamp rapid start ballast. *Source: Osram Sylvania, Inc.*

Optical factor

Lamps contain mass, which absorbs some of their own lumen output. The amount of space they take up in the fixture also serves as an obstruction to the free transmission of light reflecting from the fixture's house-side.

For example, the "light loss" factor for T12 lamps is expressed as 1. Installing thinner-diameter T10 or T8 lamps, or removing lamps, can enhance the ability of the fixture to transmit light and increase this factor above 1.

Fixture surface depreciation

As the fixture ages, its surfaces deteriorate. Every blemish absorbs light instead of reflecting it, reducing fixture efficiency. Additionally, plastic shielding materials begin to discolor under continued exposure to heat produced by the lamps.

Fixture surface depreciation is difficult to measure and is usually not included in maintained illumination calculations for retrofit projects.

Lamp position (HID systems)

Fluorescent, incandescent, mercury and high pressure sodium lamps experience little, if any, light loss due to operating position. Many metal halide lamps, however, are affected by the "tilt factor," the angle at which the lamp is operated against a horizontal or vertical position. A decrease in lumen output occurs when the angle of the lamp shifts the cold spot on the bulb, which in turn affects the amount of metal halides entering the arc stream. The tilt factor for a particular metal halide lamp can be obtained from the lamp manufacturer.

RECOVERABLE LIGHT LOSS FACTORS

Like non-recoverable light loss factors, recoverable light loss factors are values which express the percentage of light reaching the workplane after various environmental factors depreciate that amount. These light loss factors are considered "recoverable" because curative and preventative maintenance can reduce the extent of light loss. All other factors aside, a recoverable light loss factor of 0.75 means that if the lamps in a fixture together produce 10,000 lumens, only 7,500 lumens will reach the workplane. Recoverable light loss factors include lamp burnouts, lamp lumen depreciation, fixture dirt depreciation and room surface dirt depreciation.

Lamp burnouts

A "burnout" is simply a lamp that has expired. Generally, there are two ways to treat burnouts: 1) replace them immediately and 2) replace them after a specified period of time. The lamp burnout factor is expressed as:

1 − PERCENTAGE OF LAMPS ALLOWED TO FAIL WITHOUT BEING REPLACED

If burnouts are spot-relamped promptly upon failure or group-relamped prior to failure, the burnout factor would be 1. If some 5% of the lamps are burnouts at any time, then the lamp burnout factor would be 0.95, and so on. Lighting designers often assume that burnouts will be replaced promptly, unless the fixtures are difficult (costly) to reach and/or some lamps are strategically allowed to fail before being group relamped.

Lamp lumen depreciation

Every lamp has a rated life, expressed in hours. As the lamp ages and nears end of life, it produces less and less light than it did when it was brand-new. Several factors cause this phenomenon, including deposits inside the glass bulb wall and deterioration of the phosphor coating on the inside of the bulb. Lamp lumen depreciation is characteristic of all lamps, although they vary in extent of depreciation.

Determining a lamp lumen depreciation factor would seem a simple process of cross-referencing the values on the X and Y axes in Figure 3-13. This can be done if the lamps are group relamped as part of a planned maintenance program (see Chapter 5: "Maintenance: Making the System Work for the Long Haul"). If all of the lamps in the system are to be group

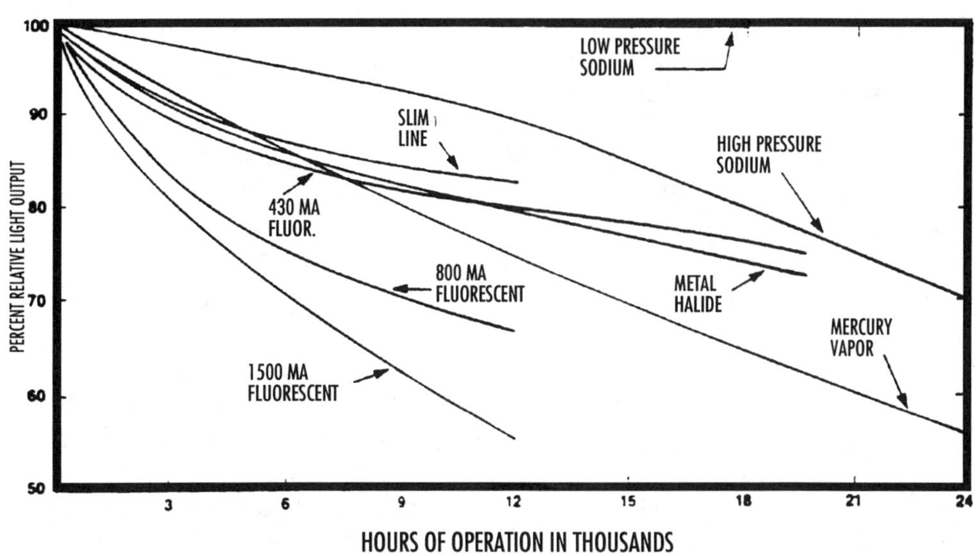

Figure 3-13. Lamp lumen depreciation of standard lamps.
Source: U.S. Department of Energy.

relamped at 70% of life, then the lamp lumen depreciation factor at 70% of life is applied.

Since lamps in most systems are replaced individually as they fail (called spot relamping), then the lighting system is not one uniform mass moving forward at the same rate of lumen depreciation. At any given time, some lamps will be nearly brand-new and producing near full lumen output, others will be less new and producing less lumen output, and so on to those that are near failure and producing a fraction of full lumen output. This means that on a lamp lumen depreciation curve, in theory, lamps will be spread across different positions from 0 on up to lumen depreciation at end of life.

To find this average number, determine rated life (taking into account hours per start). If the lumen depreciation curve were a straight line, average depreciation would be halfway between beginning and end of life. Since we are using a curve, the actual mean is at 40% of life for most fluorescent lamps (50% for some HID lamps). For example, a fluorescent lamp operating at three hours per start is rated by the manufacturer at 20,000 hours. The average lamp lumen depreciation in a spot relamping program would be that at 40% of life, or at 8,000 hours. Cross-reference the X and Y axes at 40% of life to determine the lamp lumen depreciation factor.

Fixture dirt depreciation

Dirt and dust is characteristic of any workplace, no matter how good housekeeping is. And any electrical apparatus produces heat, which attracts dust. Dirt and dust become trapped in the fixture, where it coats the lamp and fixture surfaces. Dirt and dust are darker than the clear lens/louver and the white enamel paint which typically coats the fixture's reflector, absorbing light instead of reflecting it. The extent and effect of dirt accumulation depends on the type of fixture in use, whether it is ventilated or not, the type of work performed in the area and the quality of the dirt.

To determine the fixture dirt depreciation factor, complete this three-step procedure:

STEP 1. Determine IES maintenance category. Each type of fixture falls into one of six general maintenance categories (expressed as Roman numerals I through VI). The design of some fixtures makes them more susceptible to dirt depreciation, such as "open-air" fixtures. And others are covered by a lens or cover which can be coated with dirt in addition to the lamp. The IES maintenance category of the fixtures in the application can be obtained from the manufacturer, or refer to the generic table in Figure 3-14.

STEP 2. Determine dirt conditions in the application. The quality and amount of dirt and dust impacting the lighting system in a well-kept office is different than that in a factory. Factories tend to create more dirt

Maintenance Category	Top Enclosure	Bottom Enclosure
I	1. None.	1. None
II	1. None 2. Transparent with 15 per cent or more uplight through apertures. 3. Translucent with 15 per cent or more uplight through apertures. 4. Opaque with 15 per cent or more uplight through apertures.	1. None 2. Louvers or baffles
III	1. Transparent with less than 15 per cent upward light through apertures. 2. Translucent with less than 15 per cent upward light through apertures. 3. Opaque with less than 15 per cent uplight through apertures.	1. None 2. Louvers or baffles
IV	1. Transparent unapertured. 2. Translucent unapertured. 3. Opaque unapertured.	1. None 2. Louvers
V	1. Transparent unapertured. 2. Translucent unapertured. 3. Opaque unapertured.	1. Transparent unapertured 2. Translucent unapertured
VI	1. None 2. Transparent unapertured. 3. Translucent unapertured. 4. Opaque unapertured.	1. Transparent unapertured 2. Translucent unapertured 3. Opaque unapertured

Figure 3-14. Maintenance categories based on fixture characteristics. *Source: Illuminating Engineering Society of North America,* IES Lighting Handbook, *1984 Reference Volume.*

and dust, and this matter tends to be darker. Offices attract dust but less heavy dirt. And the dirt in a factory is more likely to cling to the fixture due to oil, static and humidity.

Workplace environments are broken into five degrees of dirt conditions: Very Clean, Clean, Medium, Dirty, Very Dirty. Very Clean areas include newer commercial office buildings, healthcare facilities and industrial clean rooms. Clean areas include older commercial office buildings, a new office building with poor air conditioning system maintenance, and industrial light assembly or inspection areas. Medium areas include industrial offices, light machining or paper processing areas. Dirty areas include heavy industrial areas, foundries and high-speed printing areas. And Very

	Very Clean	Clean	Medium	Dirty	Very Dirty
Generated Dirt	None	Very little	Noticeable but not heavy	Accumulates rapidly	Constant accumulation
Ambient Dirt	None (or none enters area)	Some (almost none enters)	Some enters area	Large amount enters area	Almost none excluded
Removal or Filtration	Excellent	Better than average	Poorer than average	Only fans or blowers if any	None
Adhesion	None	Slight	Enough to be visible after some months	High—probably due to oil, humidity or static	High
Examples	High grade offices, not near production; laboratories; clean rooms	Offices in older buildings or near production; light assembly; inspection	Mill offices; paper processing; light machining	Heat treating; high speed printing; rubber processing	Similar to Dirty but luminaires within immediate area of contamination

Figure 3-15. Five degrees of dirt conditions. *Source: Illuminating Engineering Society of North America,* IES Lighting Handbook, *1984 Reference Volume.*

Dirty areas are similar to Dirty areas except outside dirt is allowed to enter the building and there is no system for removing dirty air.

Figure 3-15 shows the five dirt conditions and the characteristics of the types of areas that present these conditions. Use this table in conjunction with best judgment in completing Figure 3-16 to determine dirt conditions for the application.

STEP 3. Determine the factor. Now that we have established the fixture maintenance category and the dirt conditions in the area, we can refer to depreciation curves to determine the fixture dirt depreciation factor. Six graphs match the six fixture maintenance categories. Each graph contains five curves, each for a different dirt condition.

Each graph is limited to 36 months, or three years. To gain the fixture surface depreciation factor for a lighting system that is not cleaned in a scheduled maintenance program within a three-year interval, use the formula:

$$\text{FIXTURE DIRT DEPRECIATION} = \text{FOOTCANDLES}^{-A(t^B)}$$

$$\text{FIXTURE DIRT DEPRECIATION} = \frac{1}{\text{FOOTCANDLES}^{A(t^B)}}$$

A = CONSTANT; FOUND IN TABLE 3-18.
B = CONSTANT; FOUND IN TABLE 3-18.
t = TIME IN DECIMAL YEARS

Factors for Use in Table Below

1 = Cleanest conditions imaginable
2 = Clean, but not the cleanest
3 = Average
4 = Dirty, but not the dirtiest
5 = Dirtiest conditions imaginable

Type of Dirt	Area Adjacent to Task Area						Filter Factor (per cent of dirt passed)		Area Surrounding Task						Sub Total
	Intermittent Dirt		Constant Dirt		Total				From Adjacent		Intermittent Dirt		Constant Dirt		
Adhesive Dirt		+		=		×		=		+		+		=	
Attracted Dirt		+		=		×		=		+		+		=	
Inert Dirt		+		=		×		=		+		+		=	
														Total of Dirt Factors	

0–12 = Very Clean	13–24 = Clean	25–36 = Medium	37–48 = Dirty	49–60 = Very Dirty

Figure 3-16. Evalutation of dirt conditions. *Source: Illuminating Engineering Society of North America,* IES Lighting Handbook, *1984 Reference Volume.*

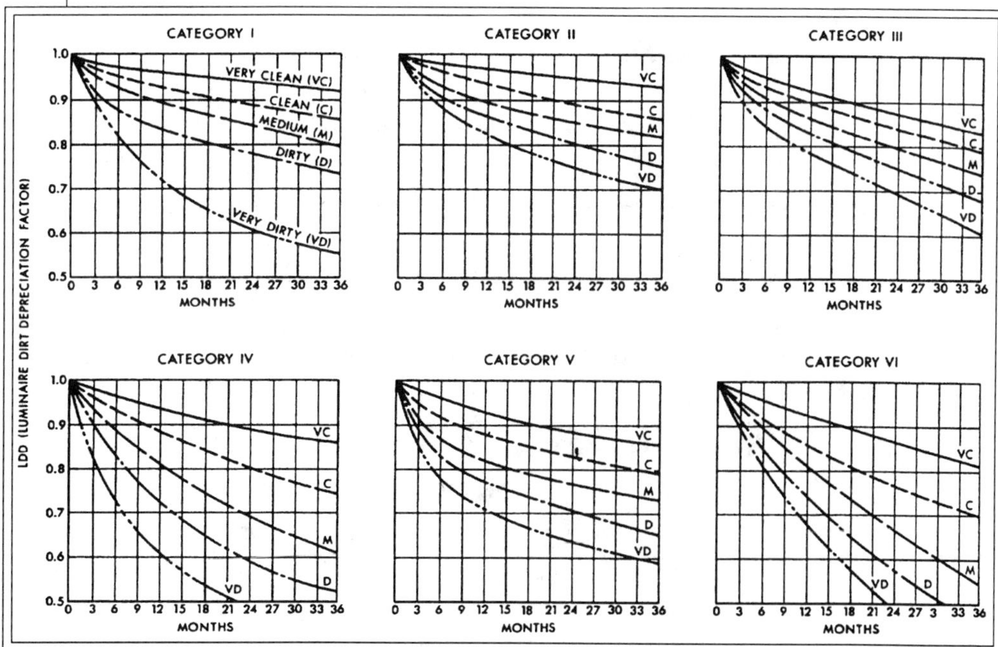

Figure 3-17. The fixture dirt depreciation factor can be determined by finding the point for the maintenance category and dirt condition when the fixture will be cleaned.
Source: Illuminating Engineering Society of North America, IES Lighting Handbook, *1984 Reference Volume.*

Room surface dirt depreciation

Room surface dirt depreciation is the amount of light loss due to absorption by dirt and dust accumulated on room surfaces. It also affects the coefficient of utilization as the ability of room surfaces to reflect light at the start of the upgrade.

Luminaire Maintenance Category	B	A				
		Very Clean	Clean	Medium	Dirty	Very Dirty
I	.69	.038	.071_	.111	.162	.301
II	.62	.033	.068	.102	.147	.188
III	.70	.079	.106	.143	.184	.236
IV	.72	.070	.131	.216	.314	.452
V	.53	.078	.128	.190	.249	.321
VI	.88	.076	.145	.218	.284	.396

Figure 3-18. Fixture dirt depreciation constants. *Source: Illuminating Engineering Society of North America,* IES Lighting Handbook, *1984 Reference Volume.*

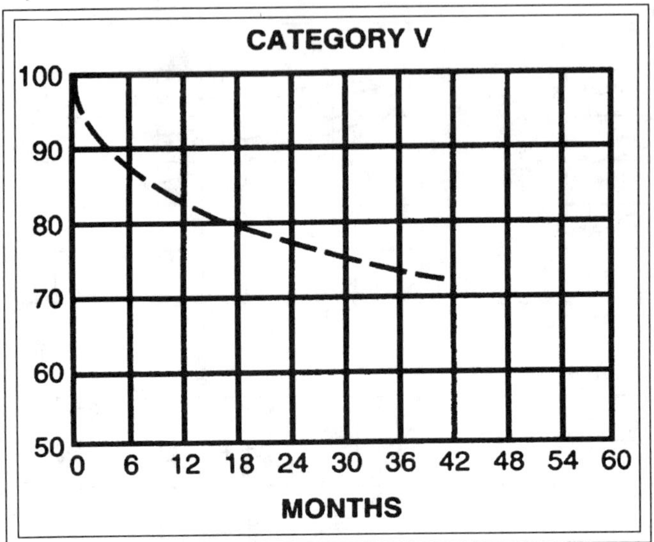

Figure 3-19. Typical fixture dirt depreciation for a four-lamp lensed fluorescent fixture. *Source: Philips Lighting Company.*

To determine the factor, complete this five-step procedure:

STEP 1. Determine the Room Cavity Ratio as an integer.

STEP 2. Determine the fixture maintenance category.

STEP 3. Determine the percentage of expected fixture dirt depreciation.

STEP 4. Determine the fixture distribution type. Specific information can be obtained from the manufacturer, but fixture types producing typical light distribution patterns are shown in Figure 3-20.

STEP 5. Use the results of steps 1-4 in Figure 3-21 to determine the room surface dirt depreciation factor.

COMPLETE LIGHT LEVEL WORKSHEET #2

Now that we know what factors will affect the amount of lumens reaching the workplane, we can complete Light Level Worksheet #2. This worksheet will reveal the minimum lumen output per fixture (and per lamp) to produce the desired maintained light level.

Note that lighting design computer software, and other helpful software, is available from manufacturers and other sources to make the process faster and easier.

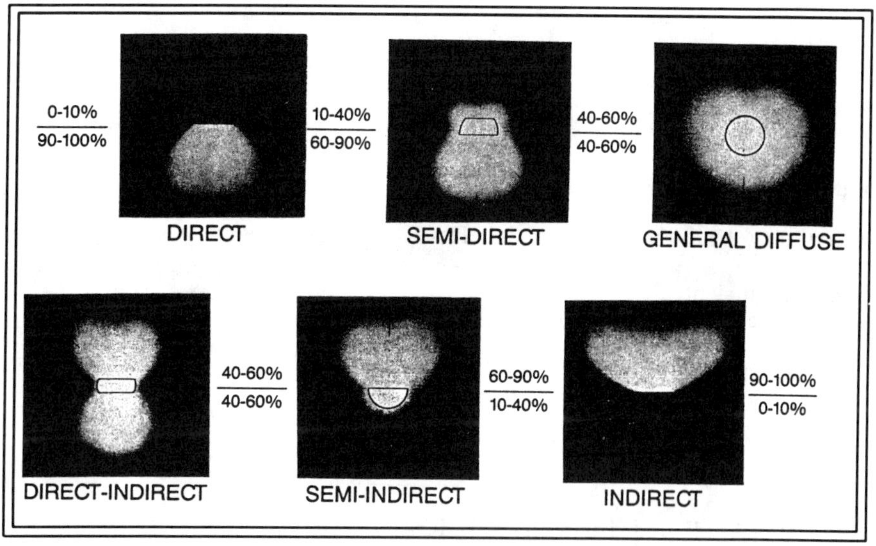

Figure 3-20. Typical light distribution patterns for fixture types used for general lighting. *Source: Illuminating Engineering Society of North America,* IES Lighting Handbook, *1987 Application Volume.*

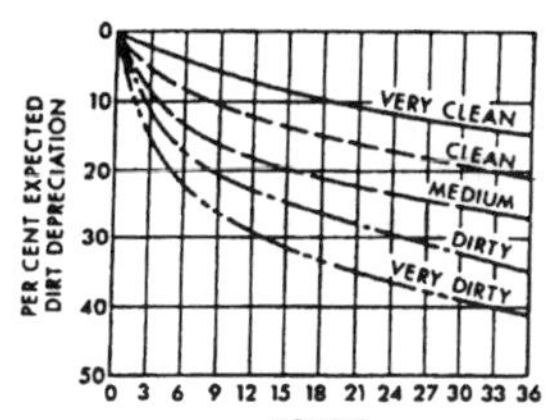

Figure 3-21. Room surface dirt depreciation factors. *Source: Illuminating Engineering Society of North America,* IES Lighting Handbook, *1984 Application Volume.*

	Luminaire Distribution Type																			
	Direct				Semi-Direct				Direct-Indirect				Semi-Indirect				Indirect			
Per Cent Expected Dirt Depreciation	10	20	30	40	10	20	30	40	10	20	30	40	10	20	30	40	10	20	30	40
Room Cavity Ratio																				
1	.98	.96	.94	.92	.97	.92	.89	.84	.94	.87	.80	.76	.94	.87	.80	.73	.90	.80	.70	.60
2	.98	.96	.94	.92	.96	.92	.88	.83	.94	.87	.80	.75	.94	.87	.79	.72	.90	.80	.69	.59
3	.98	.95	.93	.90	.96	.91	.87	.82	.94	.86	.79	.74	.94	.86	.78	.71	.90	.79	.68	.58
4	.97	.95	.92	.90	.95	.90	.85	.80	.94	.86	.79	.73	.94	.86	.78	.70	.89	.78	.67	.56
5	.97	.94	.91	.89	.94	.90	.84	.79	.93	.86	.78	.72	.93	.86	.77	.69	.89	.78	.66	.55
6	.97	.94	.91	.88	.94	.89	.83	.78	.93	.85	.78	.71	.93	.85	.76	.68	.89	.77	.66	.54
7	.97	.94	.90	.87	.93	.88	.82	.77	.93	.84	.77	.70	.93	.84	.76	.68	.89	.76	.65	.53
8	.96	.93	.89	.86	.93	.87	.81	.75	.93	.84	.76	.69	.93	.84	.76	.68	.88	.76	.64	.52
9	.96	.92	.88	.85	.93	.87	.80	.74	.93	.84	.76	.68	.93	.84	.75	.67	.88	.75	.63	.51
10	.96	.92	.87	.83	.93	.86	.79	.72	.93	.84	.75	.67	.92	.83	.75	.67	.88	.75	.62	.50

Light Level Worksheet #2, Page 1:
Calculating Required Lumen Output Per Fixture

Area description: WORD PROCESSING DEPT. Area No.: 4

				Reflector retrofit
1. Average light level to be maintained		50	Fc	
2. Area	x	7,200	ft^2	
3. **Variable A**	=	360,000		
4. Number of fixtures		90		
5. Coefficient of utilization	x	0.62	x	
NON-RECOVERABLE LIGHT LOSS FACTORS				
6. Fixture ambient temperature factor	x	—		
7. Supply voltage variation factor	x	—		
8. Lamp position factor	x	—		
RECOVERABLE LIGHT LOSS FACTORS				
9. Lamp burnouts factor	x	1		
10. Lamp lumen depreciation	x	0.81		
11. Fixture dirt depreciation	x	0.76	x	
12. Room surface dirt depreciation	x	0.84		
13. **Variable B**	=	28.85	=	
14. Required lumens per fixture **Variable A ÷ Variable B**	=	12,478	=	

Light Level Worksheet #2, Page 2:
Calculating Required Lumen Output Per Lamp

Area description: WORD PROCESSING DEPT. Area No.: 4

14. Required lumens per fixture	12,478	
	÷ 2 lamps	÷ 2 lamps
15. Required lumens per lamp	= 6,239	=
16. Required lumens per fixture	12,478	
	÷ 3 lamps	÷ 3 lamps
17. Required lumens per lamp	= 4,159	=
18. Required lumens per fixture	12,478	
	÷ 4 lamps	
19. Required lumens per lamp	= 3,120	

	Option #1	Option #2	Option #3	Option #4
	4 F32T8 LAMPS/ 1 ELECTRONIC BALLAST	4 F32T8 LAMPS/ 2 T8 MAG. BALLASTS	4 F32T8 LAMPS/ 2 T8 HYBRID BALLASTS	4 F40T12/ES TRICHROMATIC LAMPS/ 2 F.L.O. HYBRID BALLASTS
20. Lumens/Lamp	2,900	2,900	2,900	3,300
21. Ballast factor	x 0.93	x 0.93	x 0.87	x 0.88
22. Optical factor 1	x 1.05	x 1.05	x 1.05	x 1
23. Optical factor 2	x 1	x 1	x 1	x 1
24. Lumen maintenance adjustment factor	x 1.09	x 1.09	x 1.09	x 1.03
25. End lumens/lamp for upgrade option	= 3,087	= 3,087	= 2,888	= 2,991
26. Does it comply?	YES	YES	NO	YES
	(1% DROP IN LIGHT LEVEL →	)	(CONSIDER ANYWAY; 9% DROP)	(4% DROP)

Completing Light Level Worksheet #2

The first step to completing Light Level Worksheet #2 is to supply the necessary backup information about the room for identification purposes. Then follow the procedure, using the operating characteristics of F40/CW lamps as the base:

1. Indicate the average light level, in footcandles, to be maintained.
2. Multiply by the space's area in square feet.
3. The result is Variable A, which will be used in later calculations.
4. Indicate the number of fixtures in the space.
5. Determine the CU. If reflectors are being considered, determine CU for the reflectorized fixture. Obtain from the reflector manufacturer. Multiply the number of fixtures by the CU.
6. In this section, we will determine non-recoverable light loss factors. Write the fixture ambient temperature factor. Multiply the result of Step 5 by this light loss factor, if appropriate.
7. Multiply by the light loss factor resulting from variation in supply voltage, if appropriate.
8. If this is an HID system, multiply by the light loss factor resulting from lamp position, if appropriate.
9. Now multiply the result of Steps 4 through 8 by a series of recoverable light loss factors. Multiply by the lamp burnouts factor first.
10. Multiply by lamp lumen depreciation factor. If planned maintenance will be employed after the upgrade, use the factor depicting lamp lumen depreciation at the point of group relamping (move ahead to Chapter 5, which covers lighting system maintenance). As this book progresses, by the end of Chapter 5 we will have determined the lamp lumen depreciation factor after choosing an appropriate maintenance option; the text then refers the reader back to complete this worksheet.

If no planned maintenance option will be chosen, assume the lamps will be replaced individually upon failure (spot relamping). The lamp lumen depreciation factor for use in this worksheet will be that at 40% of lamp life (the mean).

11. Multiply by the fixture dirt depreciation factor. If planned maintenance will be employed after the upgrade, use the factor depicting fixture dirt depreciation at the point of fixture cleaning as determined in Chapter 5. If no planned maintenance will be employed after the upgrade, use the fixture dirt depreciation factor at the point depicting this light loss factor at 10 years.

12. Multiply by the room surface dirt depreciation factor. Room surfaces are not cleaned in most applications until repainting; even in planned maintenance programs, they are not cleaned as often as fixtures. Use the room surface dirt depreciation factor at the point depicting this light loss factor at 10 years.
13. The result is Variable B. Two blanks are available, one for standard fixture and one for a reflectorized fixture if reflectors are being considered.
14. Divide Variable A by Variable B to determine the lumen output required by the lamps in each fixture to provide the necessary light level, in footcandles, for the area.
15. Divide by 2 to determine how many lumens each lamp will have to produce if the fixture is to be retrofitted to contain two lamps.
16. Repeat required lumens per fixture.
17. Divide by 3 to determine how many lumens each lamp will have to produce if the fixture is to be retrofitted to contain three lamps.
18. Repeat required lumens per fixture.
19. Divide by 4 to determine how many lumens each lamp will have to produce if the fixture is to be retrofitted to contain four lamps. There is no blank for a reflector retrofit because the purpose of the reflector retrofit is generally to delamp.
20. Now set up several lamp-ballast options. Indicate the lumen output of the lamp in each option.
21. Multiply lumen output by the ballast factor for the ballast in this option. See Chapter 1 for sample lamp-ballast systems with lumen output and typical ballast factors listed.
22. Multiply by optical factor 1, which accounts for the effect on lumen output due to installation of thinner-diameter lamps such as T8 or T10 lamps. For T12 lamps, use a factor of 1. The factors for T8 and T10 lamps can be obtained from the lamp manufacturer.
23. Multiply by optical factor 2, which accounts for the effect on lumen output due to the installation of fewer lamps in the fixture. Obtain this factor from a fixture manufacturer.
24. Multiply by an adjustment factor to account for the effect on lumen output due to the improved lumen maintenance characteristics of T8 and T10 lamps. This can be gained with this method:

Divide the lamp lumen depreciation factor of the T8 or T10 lamp at the group relamping interval (determined in Chapter 5) — or at 40% of life (the mean) if spot relamping is chosen as the maintenance option — by the lamp lumen depreciation factor of the F40/CW in Step 10. Lamp lumen depreciation factors can be obtained from the

LUMEN MAINTENANCE ADJUSTMENT FACTOR	=	LAMP LUMEN DEPRECIATION FACTOR OF T8 OR T10 LAMP @ TIME OF GROUP GROUP RELAMPING OR @ 40% OF LIFE (THE MEAN) IF SPOT RELAMPED	÷	LAMP LUMEN DEPRECIATION FACTOR OF F40/CW (T12) FROM STEP 10

lamp manufacturer selected.

25. The result is the lumen output per lamp interacting with the ballast (and the fixture) as a system.

26. Determine whether the lumen output produced by the proposed lamp-ballast option complies with the average light level to be maintained. A quantity within 20% above or below the figure in Step 25 is considered by some to be acceptable. It is recommended, however, to consider the figure in Step 25 as a rough minimum. If a strong deviation is to occur, it is often desirable to consult a lighting professional.

COMPLETING LIGHT LEVEL WORKSHEET #3

Light Level Worksheet #3 is similar to Worksheet #2, using the maintained illumination method instead of the lumen method. It can be completed as a double-check to make sure lamp-ballast options comply. It can also be completed to determine actual footcandle levels produced by individual lamp-ballast options at the end of a given period of operation.

Light Level Worksheet #3: Determining Maintained Light Levels

Area description: WORD PROCESSING DEPT. Area No.: 4

	Option #1	Option #2	Option #3	Option #4
	4 F32T8 LAMPS/ 1 ELECTRONIC BALLAST	4 F32T8 LAMPS/ 2 T8 MAGNETIC BALLASTS	4 F32T8 LAMPS/ 2 T8 HYBRID BALLASTS	4 F40T12/ES TRICHROMATIC LAMPS/ 2 F.L.O. HYBRID BALLASTS
1. Number of lamps per fixture	4	4	4	4
2. Lumen output per lamp	x 2,900	x 2,900	x 2,900	x 3,300
3. Lamp lumen output per fixture	= 11,600	= 11,600	= 11,600	= 13,200
4. Total number of fixtures	x 90	x 90	x 90	x 90
5. Total lamp lumen output	= 1,044,000	= 1,044,000	= 1,044,000	= 1,188,000
6. Coefficient of utilization (CU)	x 0.62	x 0.62	x 0.62	x 0.62
7. Ballast factor	x 0.93	x 0.93	x 0.87	x 0.88
8. Fixture ambient temperature factor	x .	x .	x .	x .
9. Supply voltage variation factor	x .	x .	x .	x .
10. Lamp operating position factor	x .	x .	x .	x .
11. Optical factor1	x 1.05	x 1.05	x 1.05	x 1
12. Optical factor2	x .	x .	x .	x .
13. Burnout factor	x 1	x 1	x 1	x 1
14. Lamp lumen depreciation	x 0.88	x 0.88	x 0.88	x 0.83
15. Fixture dirt depreciation	x 0.76	x 0.76	x 0.76	x 0.76
16. Room surface dirt depreciation	x 0.84	x 0.84	x 0.84	x 0.84
17. End lumens	= 355,091	= 355,091	= 316,364	= 343,449
18. Area in square feet	÷ 7,200	÷ 7,200	÷ 7,200	÷ 7,200
MAINTAINED LIGHT LEVEL (FOOTCANDLES)	= 49	= 49	= 44	= 48

(@ 10 YEARS ⟶)

QUALITY OF LIGHT

Quality of light describes how light leaving the lamp interacts with the work environment — room surfaces such as desk tops, computer screens and the like — and directly with occupants' eyes. Just as people need the right amount of light to perform tasks productively, so do they need good lighting quality to perform tasks comfortably. Ensuring good lighting quality primarily involves controlling luminance (photometric brightness) and color rendition. Shadows should also be corrected unless they are effected for aesthetic ambience.

LUMINANCE

Luminance, or photometric brightness, is the illumination leaving room surfaces after those surfaces have been struck with light. Brightness is the light we can actually see and that allows us to perceive objects. Brightness within the field of view which causes irritation or impaired vision is called glare. As indicated by its definition, there are two effects of glare: discomfort and disability. There are two ways to perceive glare: directly and reflected.

Discomfort glare

Discomfort glare is glare that does not prevent vision but nonetheless makes seeing uncomfortable. It can be experienced when driving toward a car which has a bright spot of sunlight reflecting off its windshield. In an office application, discomfort glare can be caused by sunlight striking the windows or by a bright spot of light reflecting from a computer screen. While discomfort glare does not prevent vision, it can reduce speed, accuracy and satisfaction with the environment.

Disability glare

Disability glare is glare that disables visual performance. This does not necessarily mean that the person is blinded; it simply means that performance of a certain visual task(s) is impaired. One example of disability glare is light reflecting off the glossy pages of a magazine, which prevents a person from reading the print.

Direct glare

Direct glare comes straight from the light source itself. If possible, remove the fixture from the person's field of vision. This can be accomplished by either physically moving the fixture or moving the occupant. Another way to combat direct glare is to install shielding material that diffuses light leaving the fixture.

Reflected glare

Reflected glare is perceived on the performed task itself, such as a piece of paper or a computer screen. One type of reflected glare is called a veiling reflection, whereby the task is veiled by reflected light. An example of a veiling reflection is glare on a glossy magazine page which obscures the print. The problem can be combated by moving the fixture, task or occupant. Moving the task can mean either physically moving the entire task or positioning it at a different angle so that it does not produce glare. Another solution to reflected glare is to install shielding material that more effectively redistributes light leaving the fixture, such as a parabolic louver.

Visual Comfort Probability (VCP)

Visual comfort probability (VCP) is a rating given to indoor fixtures to indicate how well accepted they will be by the area's occupants. The VCP scale runs from 0 to 100. Any given rating indicates what percentage of the occupants in the poorest location in the area would not be bothered by direct glare caused by a uniform lighting system of identical fixtures. A VCP of 75, therefore, means 75% of the occupants in the poorest location would not be bothered by glare. Generally, office environments require that fixtures have a VCP rating of 70 or more, although this figure has been revised by some in recent years to 80 or more for environments where new vertical-task computers are used. According to information researched by

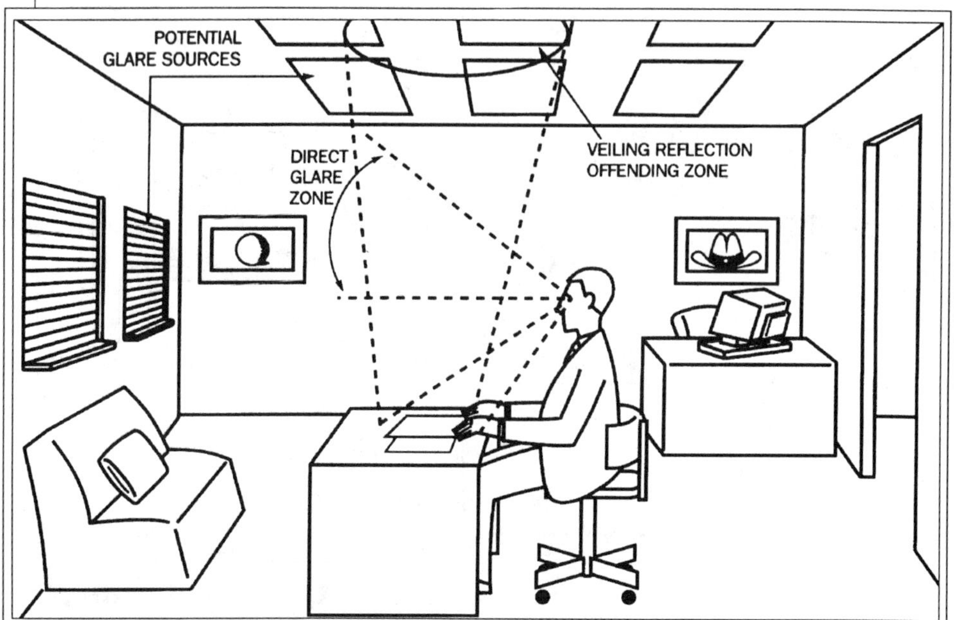

Figure 3-22. In the typical office, veiling reflections can be caused when the fixture is in front of and above the person performing the task. *Source: National Lighting Bureau.*

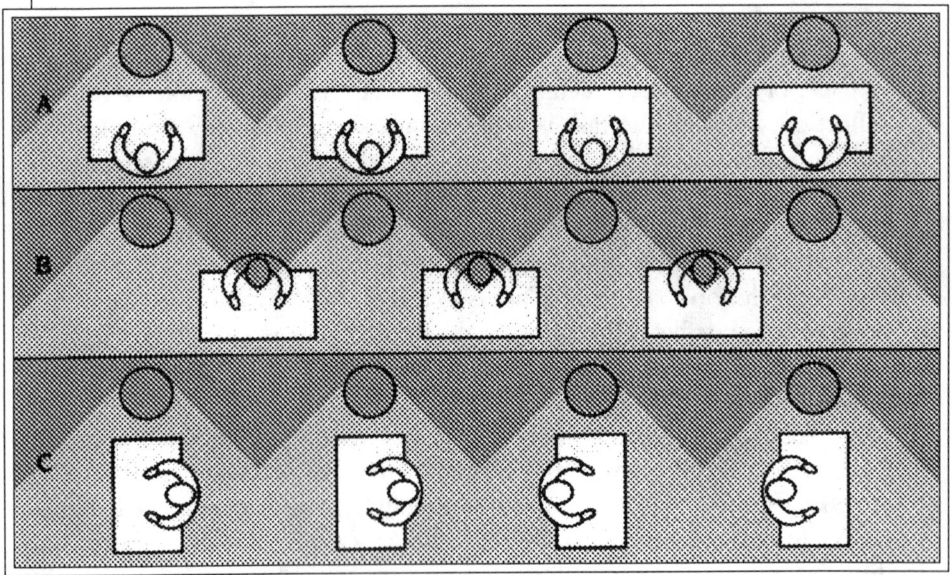

Figure 3-23. Veiling reflections are experienced by workers in Layout A because the light source is directly in front of and above them. This situation can be remedied by rearranging the workplace into Layout B or Layout C. *Source: National Lighting Bureau.*

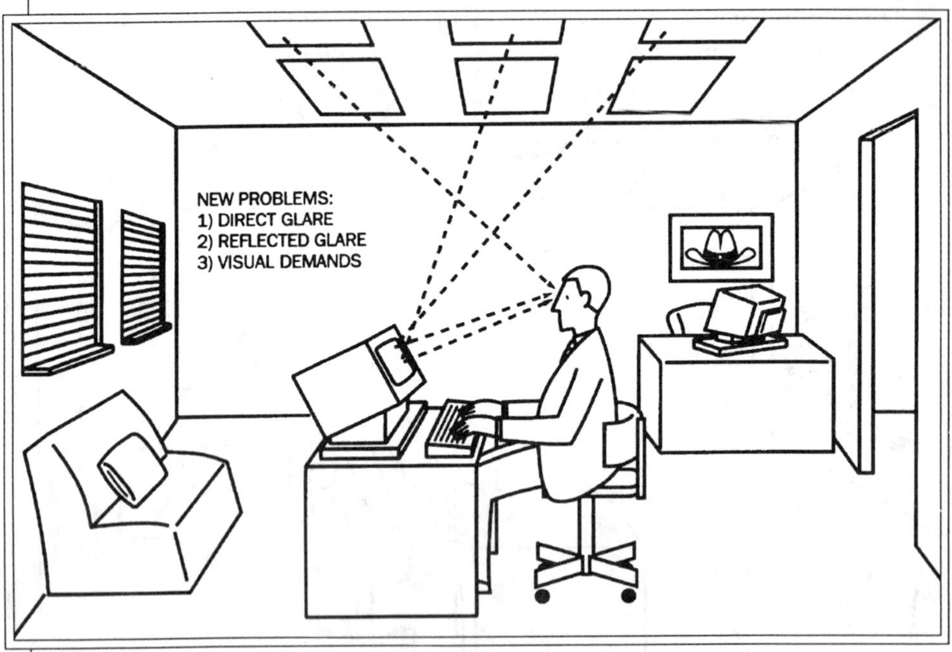

Figure 3-24. The mass introduction of video display terminals (VDTs) into the office workplace has led to a new array of glare problems. *Source: National Lighting Bureau.*

the National Lighting Bureau, installation of a parabolic louver can increase the VCP rating of a typical 2'x4' fluorescent fixture to as high as 99, while a clear lens is in the 50-85 range, a plastic louver is in the 50-70 range and a diffuser is in the 40-50 range. Many other types of shielding material are available.

Most often, fixture efficiency suffers as VCP is increased; this is not always true, however.

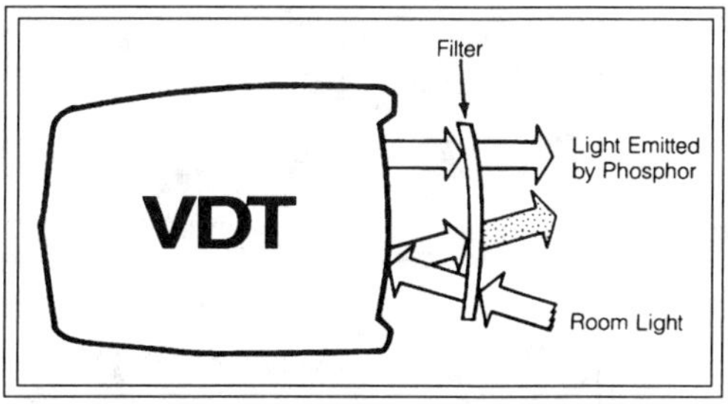

Figure 3-25. Light can reflect off the VDT screen's surface, causing reflected glare which can discourage or even impair vision unless the person performing the task moves himself or the VDT. Another problem: the typical VDT screen filter absorbs a portion of the light impacting the screen and reflecting from the screen. This can result in lowered contrast on the screen. *Source: National Lighting Bureau.*

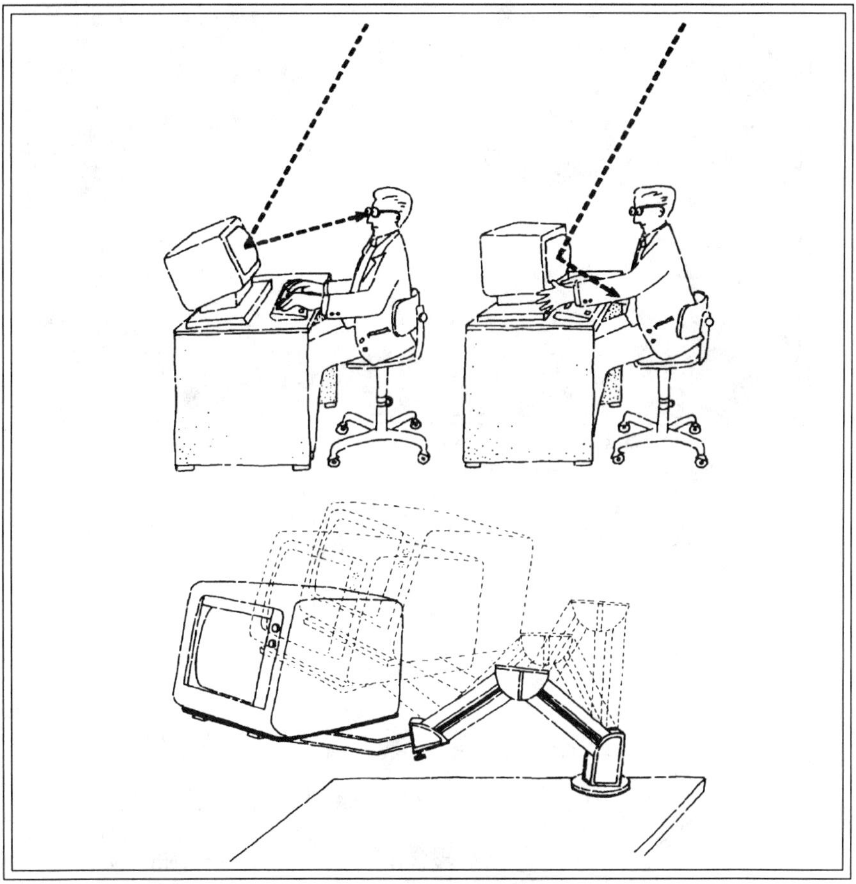

Figure 3-26. In Figure 3-23, veiling reflections were remedied by moving the people performing the task. Another remedy is to move the task. With VDTs, this can be accomplished by adjusting the angle of the VDT if it has an adjustable stand. The second illustration shows the most effective type of VDT stand, which permits nearly unlimited adjustment. *Source: National Lighting Bureau.*

Figure 3-27. Task lighting is becoming more popular. In a task lighting system, small adjustable light fixtures are positioned at the workstation to provide light for the task. This leaves the overhead fixtures with the responsibility for general lighting for orientation. This in turn means that less light will be required from the overhead system, so these fixtures can be delamped, or relamped with energy-saving lamps.
Source: GE Lighting.

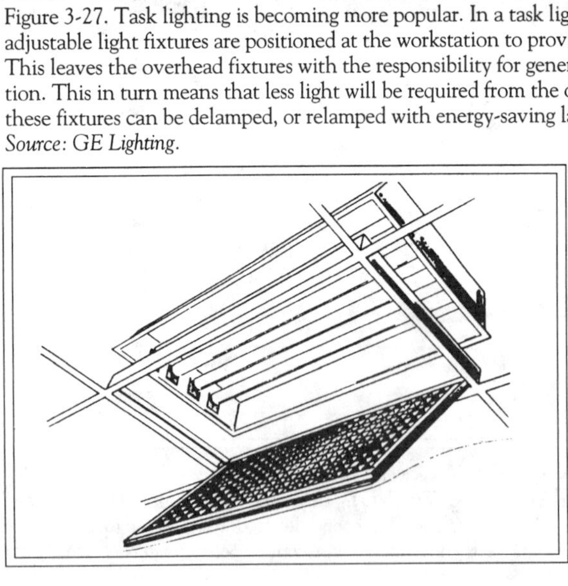

Figure 3-28. Parabolic louvers have become a popular means of redistributing light leaving a fixture. They soften the light, reducing glare on VDT screens and other tasks. Typically, however, the coefficients of utilization for fixtures with such shielding material are lower than for standard fixture types.
Source: A.L.P. Lighting & Ceiling Products.

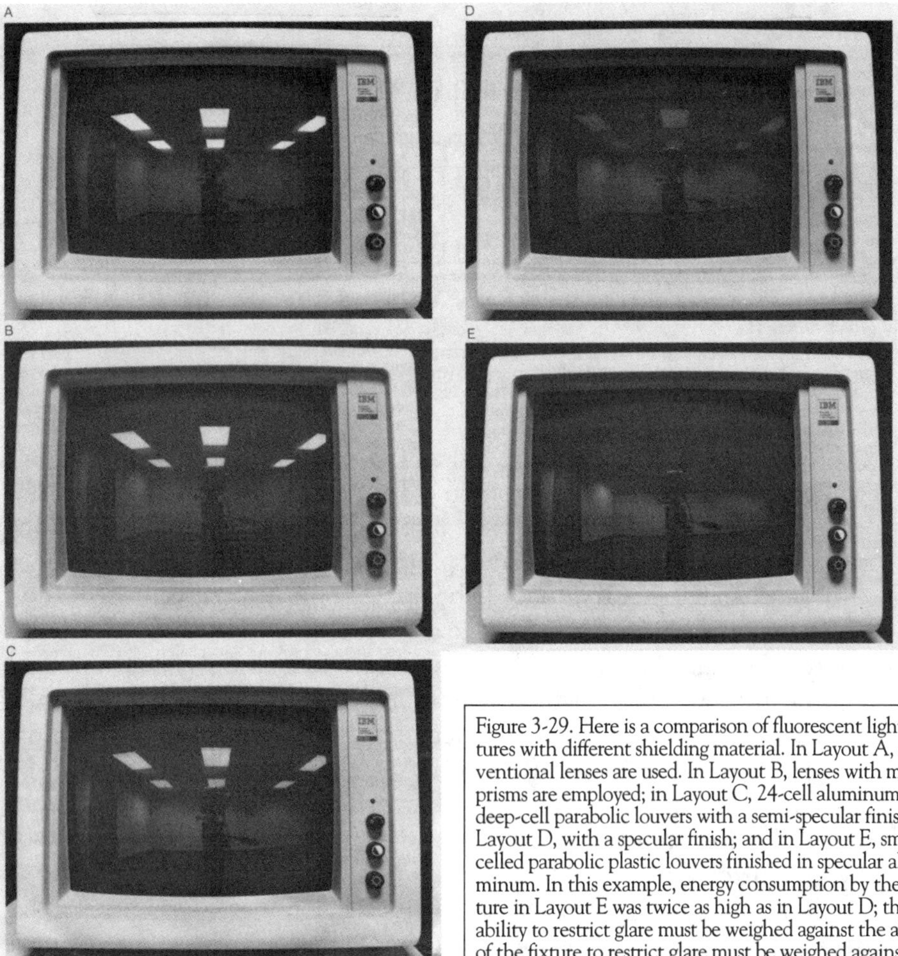

Figure 3-29. Here is a comparison of fluorescent light fixtures with different shielding material. In Layout A, conventional lenses are used. In Layout B, lenses with male prisms are employed; in Layout C, 24-cell aluminum deep-cell parabolic louvers with a semi-specular finish; in Layout D, with a specular finish; and in Layout E, small-celled parabolic plastic louvers finished in specular aluminum. In this example, energy consumption by the fixture in Layout E was twice as high as in Layout D; the ability to restrict glare must be weighed against the ability of the fixture to restrict glare must be weighed against the ability of the fixture to deliver glare must be weighed against the ability of the fixture to deliver light to the workplace. *Source: National Lighting Bureau.*

COLOR AND QUALITY OF LIGHT

Earlier in this book, we have used the term "visible light." It is true that light can excite the eye's retina, producing a visual sensation. Visible light, however, cannot be seen while traveling through space; it must reflect or be emitted from an object's surface. Even the beam of light reaching into the night sky from a powerful flashlight is not visible to us as light waves — we are actually seeing light reflected from a multitude of tiny dust particles floating in air.

Light is represented on the electromagnetic spectrum, which represents all forms of radiated energy. White visible light comprises a small portion of the spectrum, and can be further reduced to another spectrum. The visible light spectrum is composed of different colors/wavelengths, from violet (380 nanometers) to red (about 620 to 760 nanometers). An

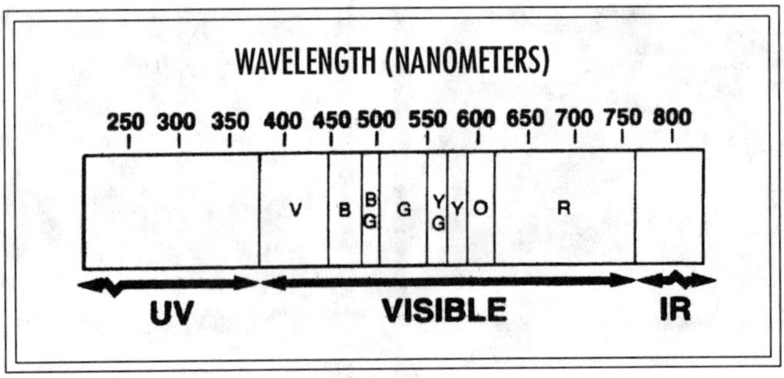

Figure 3-30. The visible light spectrum.
Source: Osram Sylvania, Inc.

even balance of these light waves, as the spectrum indicates, composes white visible light. The composition of light can be physically demonstrated by shining a beam of white light through a glass prism; the light will be refracted into a rainbow of colors appearing from the other side.

We already know that once light strikes an object, we see its reflection and hence the object itself. We see an object as a certain color, however, because its surface is chemically oriented to absorb all wavelengths of light except its "own color," which it reflects. A red object struck by white light, for example, absorbs all wavelengths of light except a segment of the 620-760 nanometer range, which is reflected. We therefore perceive the color red. A pure white object struck by white light will reflect all colors, since all colors are represented in the white object. A black object struck by white light will absorb all colors, so we see black, the absence of color.

Therefore, for an object to be a certain color, that color must be present in the object and produced by the light striking it. This point is important to understand when considering artificial light sources. Only a few lamp types produce light that is approximate to pure white light. Some lamps produce light that is saturated with red and orange wavelengths, others with blue and green. Low pressure sodium lamps produce light saturated with yellow wavelengths, meaning all objects lighted by this HID source appear yellow, black or a shade of gray. The wavelength composition of the light a particular lamp produces can be viewed on a spectral power distribution curve. This curve shows concentrations of energy at each color band on the visible light spectrum.

The spectral power distribution curve is limited in its ability to show how a lamp will render colors, however. That is why two methods of measuring a lamp's color quality have been established. Correlated color temperature describes the color of the light produced and the color of the lamp itself. Color rendering describes how well the light accurately depicts an object's colors.

Because colors — like proper quantity of light and elimination of glare

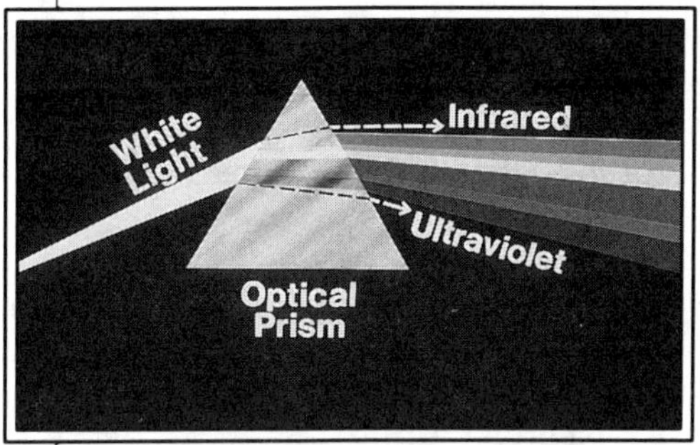

Figure 3-31. The composition of light can be physically demonstrated by shining a beam of white light through a glass prism; the light will be refracted into a rainbow of colors appearing from the other side. The converse is also true.
Source: Philips Lighting Company.

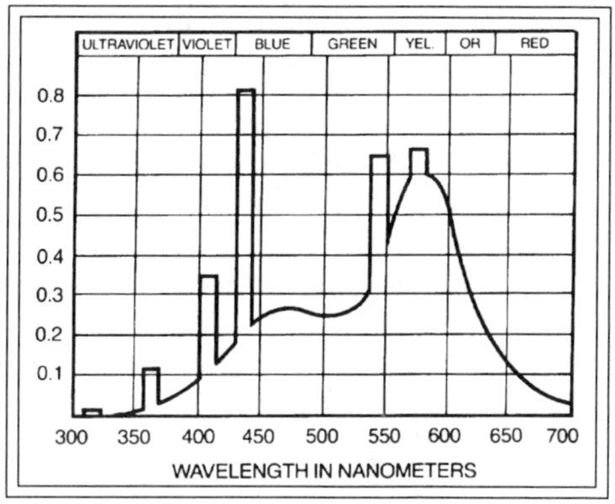

Figure 3-32. Spectral power distribution curve for a F40/CW lamp.
Source: Osram Sylvania, Inc.

— can affect human feelings, a lamp's color temperature and color rendering are key components of lighting quality. While providing recommended quantity of light has been shown to increase speed and accuracy, so can proper selection of color temperature and color rendering ability of lamps influence human behavior.

Correlated color temperature

Color temperature is the color appearance of the light produced by a lamp and the color appearance of the lamp itself. Color temperature is expressed on the Kelvin scale. To determine a lamp's color temperature, a theoretical reference source called a blackbody radiator is used. Imagine a block of steel in a laboratory which is completely black at room temperature. Electricity is passed through the block, which causes it to heat until it glows reddish orange. At higher temperatures, its color appearance becomes orange, then yellow, and so on until the metal glows blue or bluish white.

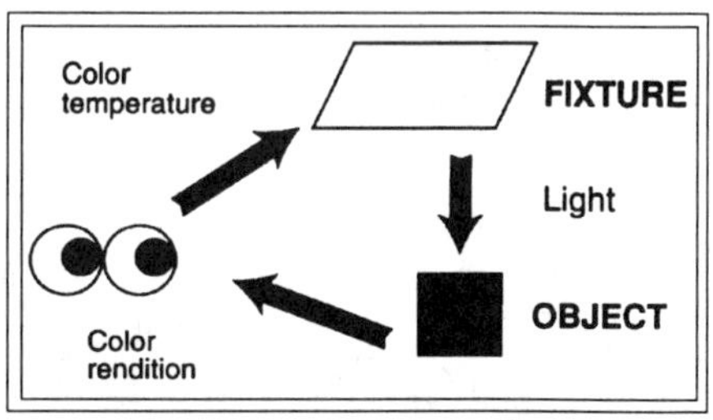

Figure 3-33. Illustration demonstrating relationship between correlated color temperture and color rendering.
Source: InterNational Association of Lighting Management Companies.

At any time during the heating, the block's temperature could be measured and its color at that time recorded. The color produced by various lamps are assessed against this reference source as color temperature; if the lamp is an incandescent lamp, the result is called the lamp's "color temperature." If the lamp is a fluorescent or HID lamp, it is approximate and therefore called "correlated color temperature."

Example: A given cool white fluorescent lamp is rated at 4100 Kelvin correlated color temperature, which means it gives off a similar color appearance when the blackbody radiator is heated to 4100 degrees Kelvin (Celsius plus 273).

Table 3-1. General categories of color temperature.

3500 Kelvin or less	Warm (Orangish white)
3500 Kelvin to 4000 Kelvin	Neutral (White)
4000 Kelvin to 6000 Kelvin	Cool (Bluish white)

As seen in Table 3-1, the color temperature of lamps lead them to be generally categorized as "warm," "neutral" or "cool" light sources. Warm light is more heavily laden with red and orange wavelengths, bringing out some fleshtones and reddish objects more than a neutral or cool light source; the lamp itself will appear orangish white during operation. A neutral, or white, light source is more balanced in wavelengths; the lamp itself will appear white. Cool light is more heavily laden with blue and green wavelengths, bringing out blue and green objects more than warmer colors. The lamp itself will appear bluish white during operation.

Choice of color temperature is driven by psychological concerns. Warm light sources are generally preferred for residential use and some retail applications, while neutral or cool light sources are usually preferred for office work and similar applications.

Color rendering

Color rendition describes how a light source makes the color of objects appear individually. It also describes how well the light produced by a lamp shows subtle variations in color shades in an object. Color rendering can be distinguished from color temperature by considering an example:

Two objects are illuminated by a light source. One is red, and the other is blue. The light source itself has a high correlated color temperature, making it more saturated with blue and green and therefore a cool light source. The red object in our experiment, therefore, is dulled while the blue object appears a rich blue. Now our cool light source is exchanged for another that has the same cool color temperature but also a high color rendering ability. The blue object is still a rich blue, but now the red object appears a rich red as well.

The color rendering ability of a light source can be assessed based on its Color Rendering Index (CRI) rating on a scale of 0 to 100. A lamp's CRI rating is determined through a laboratory test (performed through

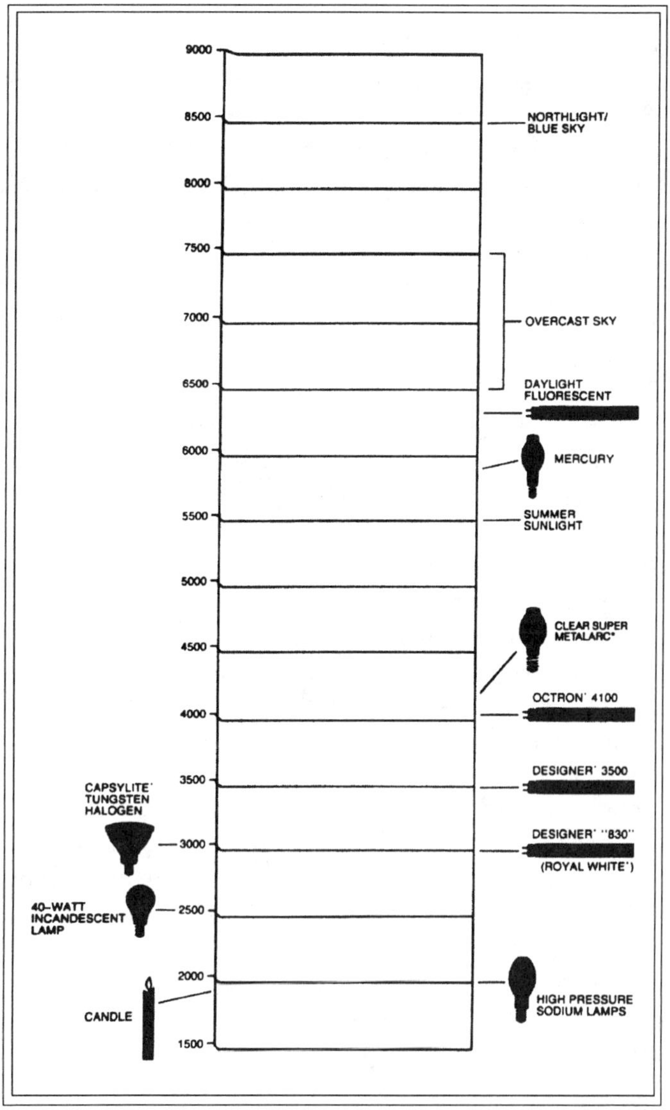

Figure 3-34. Range of color temperatures and correlated color temperatures for various lamp types. *Source:Osram Sylvania, Inc.*

computer analysis). First, the appearances of eight standardized colors are measured under the light of a standard-source incandescent lamp, which is given a rating of 100 CRI by definition. The tested lamp lights the same colors and the shifts in color appearance are again recorded and then averaged; the result is the lamp's CRI rating. The higher a lamp's CRI rating is, the higher its color rendering ability is.

Note that during this test, both the incandescent standard and the tested lamp must have approximately the same color temperature for the results to be meaningful. When comparing the CRIs of two lamps for an application, make sure they have the same color temperature. Comparing two lamps with different CRIs but also different color temperatures is meaningless.

For years, office environments have been lighted by fluorescent lamps which produced a light many people find unfavorable to reddish colors such as fleshtones. A traditional F40/CW fluorescent lamp, for example, holds a CRI rating of 62.

Today, lamps are available which enhance our ability to perceive the "true" color of objects and subtle variations in shade. These lamps, first discussed in Chapter 1, are labeled "triphosphor" or "trichromatic" lamps. An F32T8 trichromatic lamp, for example, holds a CRI rating of 70 to 85. And while the color rendering abilities of fluorescent lamps continues to advance, so do the abilities of various HID light sources such as metal halide and high pressure sodium lamps.

Let the application determine color quality

In most cases, the efficiency of lamps increases as their color rendering ability is boosted with a second coat of triphosphors. High color rendering light sources are typically more expensive, however; therefore, the application should determine the need for high color rendition. The application should also dictate the color temperature of the light — whether a warm, neutral or cool light source is best-suited.

In retail applications, for example, color is a prime concern because customers need to be able to choose clothes and other products of the correct color; additionally, bright colorful merchandise and produce are more likely to be purchased than products which are dulled or given unflattering color by poor lighting.

Conversely, in applications where color rendering is not important, such as a parking lot, low-color-rendering but highly efficient sources such as high pressure sodium lamps can be chosen.

COMPLETE THE LIGHTING QUALITY WORKSHEET

On the next page is the Lighting Quality Worksheet. To complete, fill in the identification section, then detail the problems (noted in Chapter 2) and solutions as shown in the example. For some problems, it may be desirable to seek the consultation of a professional.

Lighting Quality Worksheet

Area description: WORD PROCESSING AREA Area No.: 4

Problem:	Corrective action:
1. OCCUPANTS COMPLAIN AREA LOOKS "DRAB"	1. INSTALL TRICHROMATIC LAMPS TO INCREASE COLOR RENDITION
2. ROWS B AND C COMPLAIN OF OVERHEAD GLARE	2. RECOMMEND MOVING WORKSTATIONS 90° TO W.P. DEPT. MANAGER

Sources:

Engineering Bulletins. Danvers, Massachusetts: Sylvania Lighting Division.

IES Lighting Handbook, Reference Volume. New York: Illuminating Engineering Society of North America, 1984.

Lindsey, Jack, *Applied Illumination Engineering*. Lilburn, Georgia: The Fairmont Press, 1991.

Office Lighting and Productivity. Washington, DC: National Lighting Bureau, 1988.

Profiting from Lighting Modernization. Washington, DC: National Lighting Bureau, 1987.

Thumann, Albert, PE, "How to Design a Lighting System," from *Lighting Efficiency Applications*, Second Edition. Lilburn, Georgia: The Fairmont Press, 1992.

Further reading:

Getting the Most from Your Lighting Dollar, Second Edition. Washington, DC: National Lighting Bureau.

Industrial Lighting Handbook. Washington, DC: National Lighting Bureau.

Jones, Richard, "Parabolic Louvers Beat the VDT Glare Problem," *MainLighter: The Lighting Management & Maintenance Magazine*, February 1992.

Knott, Roger L., "Selection Criteria for Lighting Energy Management," from *Lighting Efficiency Applications*, Second Edition, Albert Thumann, PE, editor. Lilburn, Georgia: The Fairmont Press, 1992.

Lighting for Safety and Security. Washington, DC: National Lighting Bureau.

Snider, R.E., "Practical Selection of Fluorescent Lamps with Emphasis on Efficiency and Color," from *Lighting Efficiency Applications*, Second Edition, Albert Thumann, PE, editor. Lilburn, Georgia: The Fairmont Press, 1992.

Solving the Puzzle of VDT Viewing Problems. Washington, DC: National Lighting Bureau, 1987.

Source: National Lighting Bureau

CHAPTER 4

Retrofit Economics: Finding the Bottom Line

MAKE THE SYSTEM EFFICIENT

Between 1982-1992, the average cost of electricity rose 130-150%. Successful strategic lighting management ensures that the lighting system consistently provides proper quantity and quality of light at the lowest operating and maintenance cost. In this chapter, we will discuss how to minimize operating cost. Numerous lighting technologies and techniques are now available which can match, increase or strategically reduce the existing lighting system's performance with the goal of saving energy. The greatest energy savings are usually realized in the east and west coastal regions, where energy rates are high and continuing to increase.

Regardless of geography, however, business trends indicate increasing global competition, increasing expectations from consumers, and an increasing need to stay on the cutting edge of new technology. The U.S. currently consumes twice as much energy to produce one unit of gross national product (GNP) than the less energy-intensive economies of Japan and former West Germany. For example, the U.S. spent 10% of its GNP on energy in 1989; Japan and West Germany spent 5%. This allows Japanese and German companies to produce goods more efficiently, making them more competitive.

By committing to lighting energy efficiency, a company is not only cutting costs, but generating revenue which can be invested in marketing, research and development, automation and other projects. All of this helps the company become stronger in a fiercely competitive domestic and global economy.

In this chapter, we will review how utilities charge for their product, generic upgrade approaches, measuring payback and return on investment, conducting a simple payback analysis, and how to calculate heating, ventilation and air conditioning (HVAC) savings. Now that we have reviewed upgrade options and tested them to see if they produce minimum lumen output, we can review their ability to pay for themselves in energy savings.

HOW UTILITIES CHARGE FOR SERVICE

The cost of electricity is not purely a simple use charge. The electric bill is the result of intensive calculations which account for use, demand for power and other factors.

ENERGY USE CHARGE

Utilities charge by the kilowatt-hour (kWh) consumed by the facility's electrical systems. For this charge, the utility may use a sliding block rate structure. This means that the first group of kWh consumed would cost a certain dollar amount per kWh, the second block a different figure, and so on. Generally, the more energy is consumed, the cheaper energy becomes.

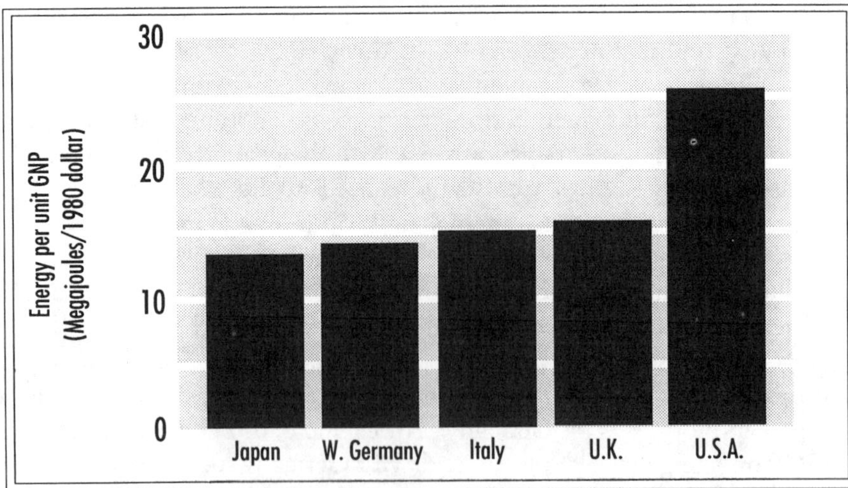

Figure 4-1. Energy intensity of the economics of Japan, (old) West Germany, Italy, the United Kingdom and the United States. The U.S. consumes twice the amount of energy to produce one unit of gross national product (GNP) as relatively efficient Japan. *Source:* R.M. *White, "The Great Climate Debate,"* Scientific America, *July 1990, pp.* 36-43.

DEMAND CHARGE

The utility may charge for demand, based on the cost to the utility to build and maintain the production, transmission and distribution equipment necessary to service a building's entire load (in kilowatts). If a building contains 1,000 fixtures and only 900 are in use at any one time, the utility nonetheless must invest in the equipment necessary to potentially service all 1,000 fixtures.

In an effort of fairness, the utility uses a special demand-meter to measure actual demand and charges the facility accordingly. Each day, the meter measures a building's average demand (in kilowatts) in increments of time called the demand interval, typically 15 or 30 minutes. The utility charges a dollar amount per kilowatt for the interval when demand was at its highest during a given month (called the demand "peak").

Note that demand charges may also be based on kilovolt-amperes.

Demand charge and the ratchet clause

The utility revenue structure may feature a "ratchet clause." In short, this means that the demand charge will reflect maximum demand measured in the recent past. The highest demand charges are usually experienced during the summer season, when air cooling systems are in operation. The ratchet clause might state, therefore, that any demand charge levied during the winter months cannot be lower than a percentage of the maximum demand charge levied during the previous summer.

Example: A utility which uses 80% as its percentage levies $100 as the maximum demand charge during the summer season for a building. During a winter month in the early part of the next calendar year, the demand charge is $50. The utility charges the building $80 instead of $50 for that month.

POWER FACTOR CHARGE

The utility may charge for use of low power factor equipment. Fluorescent ballasts may be rated as either high power factor or low ("normal") power factor. They require nearly the same wattage, but low power factor ballasts draw roughly twice as much electrical current. To supply this current, the utility must invest in production, transmission and distribution equipment. To users of low power factor equipment, this charge may be formulated:

A = ENERGY CONSUMPTION (IN kWh) ÷ ACTUAL POWER FACTOR FOR BUILDING

$$B = \text{ENERGY CONSUMPTION (IN kWh)} \div \text{UTILITY STANDARD (USUALLY 0.9)}$$

$$\text{POWER FACTOR CHARGE (\$)} = A - B \times \text{kWh RATE (\$)}$$

For this reason, it is generally advantageous to use high power factor ballasts. There are times, however, when the cost of this charge is outweighed by the benefit of a type of equipment; for example, many types of compact fluorescent units are driven by a low power factor ballast. The energy savings gained by replacing incandescent lamps with compact fluorescents exceeds the power factor charge for using these ballasts.

FUEL ADJUSTMENT CHARGE

The cost of fossil fuels (coal, oil and natural gas), which are the primary source of energy produced by most utilities, sometimes rise faster than public service commissions can complete the long and difficult process of approving an increase in rates. Many public service commissions, therefore, allow utilities to bypass the process and bill the additional cost of fuel as a variable charge.

OTHER CHARGES

The reader can check with the local electric utility to obtain information about other charges and to gain information specific to his or her facility. Appendix I may prove helpful to conduct a rough economic analysis; it is a listing of states with corresponding average commercial and industrial utility revenue per kWh published by the Energy Information Administration (U.S. Department of Energy). Note that the figures in Appendix I do not represent cost per kWh for energy use only; it is the utility's average revenue per kWh, representing a blend of all of the above charges.

REDUCE THE ELECTRIC BILL

Because lighting management means obtaining the best lighting for the lowest operating cost, the lighting upgrade plan should contain one or more objectives for reducing the electric bill. Up to three basic goals can be established. One goal is to reduce demand for power, measured in kilowatts (kW). A related goal is to reduce unit power density (UPD), expressed in watts per square foot. A third goal is to reduce energy consumption, measured in kWh.

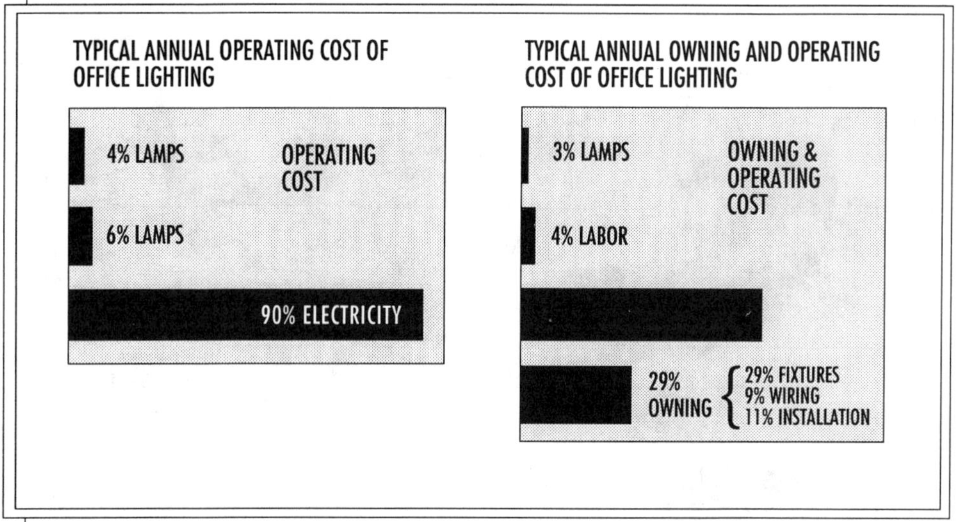

Figure 4.2. Some 90% of the cost of operating a lighting system in a typical office is electricity. converting to energy— efficient lighting may present a significant initial cost, but can permanently lower operating costs. When equipment is properly selected, the energy savings outweigh the initial cost, creating a payback period. *Source: GE Lighting.*

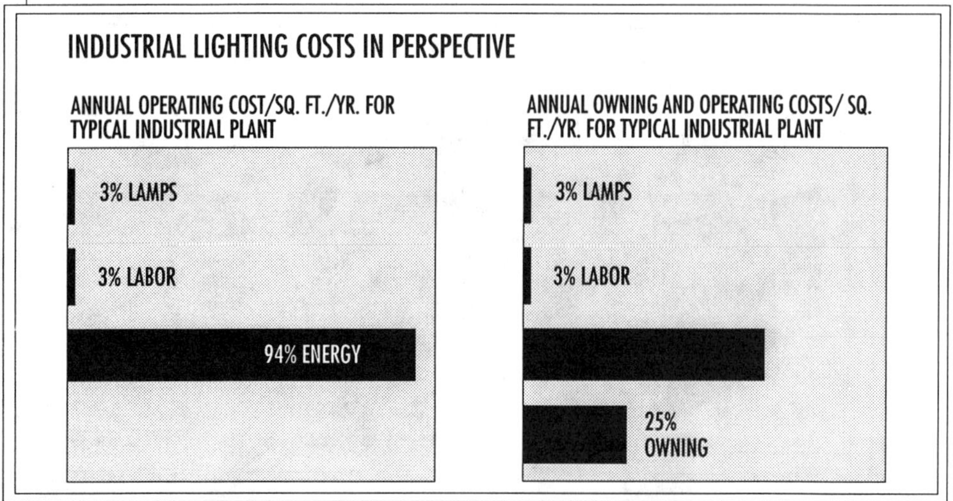

Figure 4.3. Typical annual cost of owning and operating a lighting system in an industrial facility. Again, the bulk of the cost is energy. (Figures represent Quality Lighting Costs, Annual Operating, and Annual Owning and Operating Costs based on typical industrial lighting installations and data from General Electric Company General Lighting Cost Analysis, November 1988). *Source: GE Lighting.*

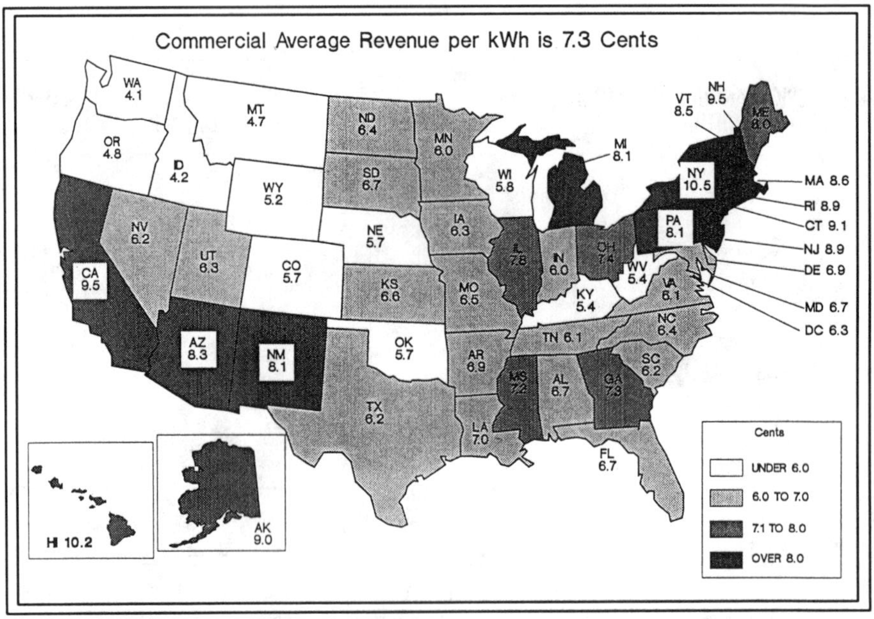

Figure 4-4. Average revenue per kWh for the commercial sector by state, 1990. Note that the average revenue per kWh of electricity is calculated by dividing revenue by kWh purchased. *Source: U.S. Department of Energy, Energy Information Administration, Form EIA-861, "Annual Electric Utility Report."*

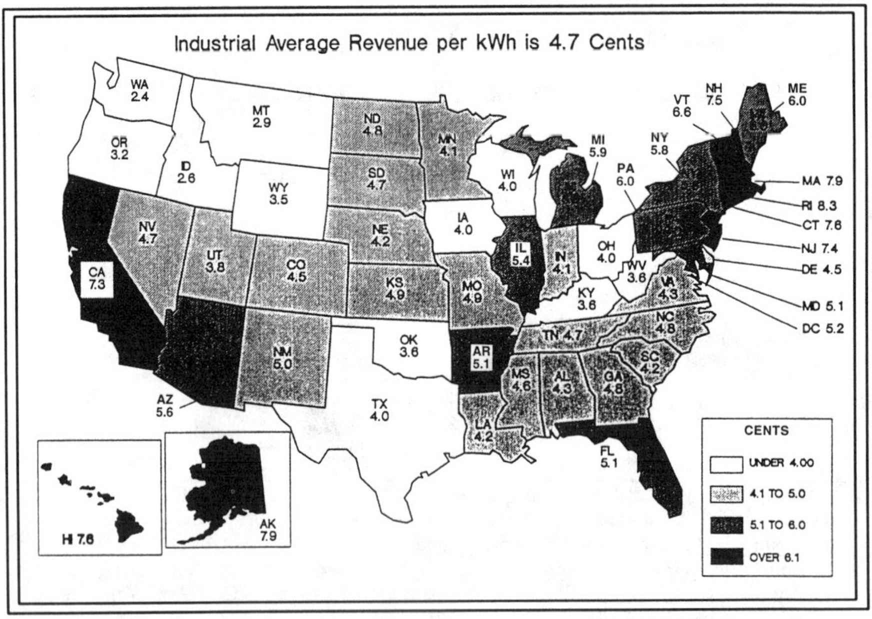

Figure 4-5. Average revenue per kWh for the industrial sector by state, 1990. Note that the average revenue per kWh of electricity is calculated by dividing revenue by kWh purchased. *Source: U.S. Department of Energy, Energy Information Administration, Form EIA-861, "Annual Electric Utility Report."*

DEMAND FOR POWER

Demand is the average amount of power required by the facility's lighting system. The total kW need for a facility is simple to calculate:

TOTAL LOAD (kW) = TOTAL WATTS CONSUMED BY SYSTEM ÷ 1,000

Many utilities charge for demand as well as actual energy use. The demand charge is based on the peak amount of power required by the customer during a given month. Reducing demand as part of the energy objective can be important to reduce these charges, especially when a ratchet clause is in effect.

UNIT POWER DENSITY (UPD)

Many government energy offices and agencies use UPD as a means to limit energy consumption; it is the basis for their energy codes for new buildings. One reason why they do so is because a watts-per-square-foot figure can be determined simply:

UNIT POWER DENSITY = TOTAL SYSTEM WATTAGE ÷ TOTAL AREA (SQUARE FEET)

It also provides a common denominator for systems in all buildings, which is often essential in developing a lighting energy code. The Lighting Research Center publishes a booklet listing a state-by-state review of energy codes with contact information (see Chapter 8: "Getting Help: Associations, Trade Groups, Professionals, Government, Publications").

Generally, these codes apply to new construction and not for retrofit of existing buildings. However, it may be desirable to make compliance with state energy standards an energy objective. This entails establishing a power budget for each area according to how much light is required by the tasks performed there.

ENERGY CONSUMPTION

Energy consumption (use) is the amount of power a system consumes over a given period of time. One kWh is the amount of energy consumed by a system operating for one hour.

SYSTEM kWh CONSUMPTION	=	WATTS CONSUMED BY SYSTEM	x	OPERATING HOURS	÷	1,000

All utilities charge a dollar amount per kWh, making reduced energy consumption an important energy objective.

UPGRADE THE LIGHTING SYSTEM FOR ENERGY SAVINGS

Upgrading a lighting system can save energy in two ways. It can reduce electrical input, or it can reduce hours of use. Electrical input is measured in watts and kW, and electrical consumption reflects both demand and hours per use as the kWh. As we have seen, the kW and the kWh are the basic products for which the utility charges.

Numerous technologies are commercially available which can reduce kW demand and kWh consumption. Planned maintenance can also play a central or supporting role in reducing both demand and consumption. Upgrades to reduce load and save energy can take several basic forms. Those directly reducing energy consumption are positioned here based on approaches to light level:

REDUCE LIGHT LEVELS

Some new lighting equipment options can be chosen to reduce lumen output for the resultant energy savings. An example of this type of equipment is a system containing F40T12/ES fluorescent lamps powered by premium magnetic or cathode cut-out ballasts, which reduces energy consumption while producing fewer lumens than a traditional system. Applications for this type of option include areas that are presently overlighted or will be overlighted because of a planned lighting maintenance program. An example of such an application is a corridor which is overlighted for the simple task of orientation.

INCREASE LIGHT LEVELS

Some upgrade options increase room reflectances and/or the performance of the fixture, lamp-ballast system or overall lighting system. By increasing light level, lamp(s) can be removed or a current limiter can be installed to reduce electrical input. One example of increasing light levels is to repaint room surfaces a lighter color and to properly maintain the system. An example of a technology-based upgrade is to replace four lamps in a four-lamp fixture with two F40T10 lamps and a reflector. By increasing fixture efficiency and lumen output of the lamps, light levels can be increased to a point where lamps and ballasts can be removed.

MAINTAIN LIGHT LEVELS

It may be important for an upgrade option to provide a light level comparable to the old system. Numerous upgrade options can produce similar lumen output to the old system for significantly less energy. One example is an F32T8-electronic ballast system, which can result in similar light levels for up to 40% less energy. Another example is replacing a 60-watt incandescent light bulb with a 13-watt compact fluorescent unit.

DECREASE HOURS OF USE

While most upgrade options involve reducing electrical input, others reduce hours per use and hence kWh consumption. Controls such as energy management systems can be programmed to turn lights on and off at preset intervals (reducing the amount of time lights are on without an area being occupied). Another control is the occupancy sensor, which monitors a space to determine whether an area is occupied and therefore whether the lights should be on or off. A behavior-based option can be chosen in which occupants are educated, encouraged and required to turn lights off when they are not in use.

PROPER MAINTENANCE

Planned maintenance, to be covered in the next chapter, is the practice of group lamp replacement and cleaning of the fixtures at scheduled periodic intervals to raise average light level and to maximize labor efficiency.

Planned maintenance can play a significant supporting role in an upgrade. By reducing depreciation of light level via planned maintenance, fewer lumens are required from the system to provide desired maintained illumination. A planned maintenance program therefore basically acts as an option which increases performance for the same amount of energy. This provides more flexibility in choosing options which decrease, maintain, or increase performance.

CHOOSING THE OPTION

Factors which influence ultimate choice of the option include: 1) its relative ability to meets the needs of the occupants and the application, 2) the amount of energy savings produced through its employment, 3) the length of time before the option pays for itself in energy savings, 4) its initial cost, 5) its ability to be cost-effectively maintained, 6) occupant acceptance, and 7) the reported reliability and warranted performance of the equipment.

Figure 4-6. Computer program packages, such as L.E.A.P. (Light Economic Analysis Package) from GE Lighting, can simplify the upgrade process. *Source: GE Lighting.*

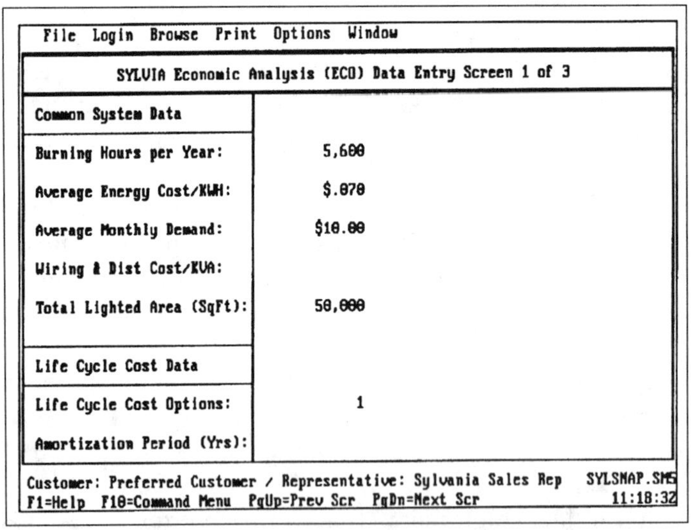

Figure 4-7. The Sylvia computer lighting analysis package from Osram Sylvania, Inc. Sylvia allows the user to choose from five program models — Lighting Life Cycle Cost, Group Relamping, Relight Energy Analysis, Return on Investment and Economic Analysis — to enhance speed and accuracy in the design, selection and maintenance of upgraded lighting systems. *Source: Osram Sylvania, Inc.*

Each of these factors should be weighed against each other and corporate priorities determined in Chapter 2.

MEASURING ENERGY SAVINGS

Energy savings produced by a system can be estimated and/or measured and monitored using a variety of monitoring equipment. Before conducting a full economic analysis of any of the options which were able to produce appropriate light levels in Chapter 3, a simple payback analysis should be performed. The simple payback analysis is a relatively simple procedure compared to a full life-cycle costing analysis, which takes into account the full costs of installing, operating and maintaining a lighting system over its lifetime. It is relatively short and uncomplicated, producing both the projected payback period in decimal years and the internal rate of return on investment. These values can be compared to the corporate investment policy and an option can be chosen for full analysis and a trial installation.

Payback and ROI formula

Calculating simple payback. The simple payback formula is used to determine the amount of time in decimal years which will pass before an upgrade option's energy savings cover the initial cost. If the option produces any other benefits, such as a measurable increase in occupant productivity, these can be added to the energy savings. Most companies, however, assess energy projects purely on payback on energy savings.

PAYBACK (YEARS) = NET INSTALLATION COST ($) ÷ ANNUAL ENERGY SAVINGS ($)

Five-year cash flow can be determined:

5-YEAR CASH FLOW=5 YEARS–PAYBACK (YEARS) x ANNUAL ENERGY SAVINGS

Calculating simple return on investment. The simple return on investment (ROI) formula is used in conjunction with the payback formula to screen energy projects. It produces an internal rate of return on investment based on the relationship between annual energy savings and the net installation cost.

RETURN ON INVESTMENT (%) = ANNUAL ENERGY SAVINGS ($) ÷ NET INSTALLATION COST ($)

Determining annual energy savings. Energy savings are determined by calculating, based on reduced electrical input and/or hours of use, kWh savings resulting from the option. This can be conservatively determined purely on lighting energy savings, or a combination of lighting energy savings and the related net HVAC savings.

Determining net installation cost. Installation cost reflects the cost of purchasing all equipment and services required to implement an upgrade. The net installation cost is the installation cost minus funding received from the local electric utility if that utility offers rebates as part of a demand side management (DSM) program (see Chapter 6: "Retrofit Financing: Putting Money to Work").

If planned maintenance is to support an upgrade option, its cost is typically not figured into the simple payback formula and is instead assessed separately against the cost of unplanned maintenance.

COMPLETE THE RETROFIT ECONOMICS WORKSHEET

To determine energy savings, simple payback, simple return on investment and other economic factors for up to four upgrade options, complete the Retrofit Economics Worksheet on the following pages for each area. To speed up the process, the worksheet can be computerized or obtained as computer software from different sources in various versions. In addition, entire pages and/or sections can be bypassed. The Retrofit Economics Worksheet was designed to be as comprehensive as possible to handle as many different upgrade options and situations as possible. For example, if no utility financial incentives are available, then Page 4 can be bypassed.

Note that the product prices used in this example are likely to be higher than those encountered at one's local distributor. This is due to several factors, one of which is the discount advantage of purchasing in volume.

Retrofit Economics Worksheet:
Determining Simple Payback Per Fixture/Per Facility, Page 1

Area description: WORD PROCESSING DEPT. Area No.: 4

Fixture type: FLUORESCENT RAPID START

LIGHTING ENERGY SAVINGS	Existing	Option #1	Option #2	Option #3	Option #4
1. Number of lamps per fixture and type	4 F40/CW LAMPS	4F32T8 LAMPS	4F32T8 LAMPS	4F32T8 LAMPS	4F40T12/ES TRICHROMATIC LAMPS
2. Number of ballasts per fixture and type	2 MAGNETIC LAMPS	1 ELECTRONIC BALLAST	2T8 MAGNETIC BALLASTS	2T8 HYBRID BALLASTS	2 F.L.O. HYBRID BALLASTS
3. Other equipment (occupancy sensors, reflectors, etc.)	STANDARD LIGHTSWITCH				
4. System input watts per fixture	192	112	142	124	132
5. Operating hours per year	x 3,900 ÷ 1,000	x 3,900 ÷ 1,000	x 3,900 ÷ 1,000	x 3,900 ÷ 1,000	x 3,900 ÷ 1,000
6. Lighting energy consumption (kWh)	= 748.8	= 436.8	= 553.8	= 483.6	= 514.8
7. Lighting energy consumption (kWh) for existing system		− 748.8	− 748.8	− 748.8	− 748.8
8. Lighting energy savings (kWh)per fixture(make positive)		= 312	(+)− 195	(+)− 265.2	(+)− 234
9. Number of fixtures		x 90	x 90	x 90	x 90
10. Total lighting energy savings (kWh)		= 28,080	= 17,550	= 23,868	= 21,060
11. Operating hours from **Step 5**		÷ 3,900	÷ 3,900	÷ 3,900	÷ 3,900
12. Total demand savings (kW)		= 7.2	= 4.5	= 6.12	= 5.4
13. Total lighting energy savings (kWh) from **Step 10**		28,080	17,550	23,868	21,060
14. Cost per kWh		x $ 0.10	x $ 0.10	x $ 0.10	x $ 0.10
15. TOTAL LIGHTING ENERGY SAVINGS PER YEAR ($)		= $ 2,808	= $ 1,755	= $ 2,386.80	= $ 2,106

Retrofit Economics Worksheet:
Determining Simple Payback, Page 2

Area description: ______________________ Area No.: ________

	Existing	Option #1	Option #2	Option #3	Option #4
A/C ENERGY SAVINGS					
16. Total lighting energy savings (kWh) from **Step 10**		______	______	______	______
17. Cooling season in weeks ÷ 52		x______	x______	x______	x______
		x 0.9	x 0.9	x 0.9	x 0.9
		÷ 2.7	÷ 2.7	÷ 2.7	÷ 2.7
18. Total A/C energy savings (kWh)		=______	=______	=______	=______
19. Cost per kWh from **Step 14**		x______	x______	x______	x______
20. Total A/C energy savings ($)		= $_____	= $_____	= $_____	= $_____
21. Heating season (1–Step 17)		______	______	______	______
		x 0.2	x 0.2	x 0.2	x 0.2
22. **Variable A**		=______	=______	=______	=______
23. Total lighting energy savings (kWh) from **Step 10**		______	______	______	______
		x 3,414	x 3,414	x 3,414	x 3,414
		x **Var. A**	x **Var. A**	x **Var. A**	x **Var. A**
		÷ 0.8	÷ 0.8	÷ 0.8	÷ 0.8
24. Increase in needed heat load in BTU		=______	=______	=______	=______
		÷ 100,000	÷ 100,000	÷ 100,000	÷ 100,000
25. Increase in needed heat load in therms		=______	=______	=______	=______
26. Utility cost per therm		x $_____	x $_____	x $_____	x $_____
27. Additional heating cost		= $_____	= $_____	= $_____	= $_____
28. A/C energy savings from **Step 20**		$ ______	$ ______	$ ______	$ ______
29. Additional heating cost from **Step 27**		– $_____	– $_____	– $_____	– $_____
30. TOTAL NET H.V.A.C. SAVINGS		= $_____	= $_____	= $_____	= $_____

Retrofit Economics Worksheet:
Determining Simple Payback Per Fixture/Per Facility, Page 3

Area description: WORD PROCESSING DEPT. Area No.: 4

	Existing	Option #1	Option #2	Option #3	Option #4
INSTALLATION COST					
31. Cost per lamp		$3.53	$3.53	$3.53	$3.36
32. Number of lamps per fixture		x 4	x 4	x 4	x 4
33. Lamp cost per fixture		= $14.12	= $14.12	= $14.12	= $14.12
34. Cost per ballast		$60.00	$20.00	$30.00	$27.00
35. Number of ballasts per fixture		x 1	x 2	x 2	x 2
36. Ballast cost per fixture		= $60.00	= $40.00	= $60.00	= $54.00
37. Reflector cost		$.	$.	$.	$.
38. Equipment cost per fixture (33 + 36 + 37)		= $74.12	= $54.12	= $74.12	= $67.44
39. Electrical labor rate per hour		$28.00	$28.00	$28.00	$28.00
40. Hours to install components in one fixture		x 0.4	x 0.6	x 0.6	x 0.6
41. Labor cost per fixture		= $11.20	= $16.80	= $16.80	= $16.80
42. Total cost per fixture (38 + 41)		= $85.32	= $70.92	= $90.92	= $84.24
43. Number of fixtures from **Step 9**		x 90	x 90	x 90	x 90
44. Total installation cost for fixtures		= $7,678.80	= $6,382.80	= $8,182.80	= $7,572.60
45. Cost of other:________		$.	$.	$.	$.
46. Number of other		x .	x .	x .	x .
47. Total equipment cost of other		= $.	= $.	= $.	= $.
48. Electrical labor rate per hour from **Step 39**		$.	$.	$.	$.
49. Time to install all other		x .	x .	x .	x .
50. Labor cost of other		= $.	= $.	= $.	= $.
51. Total cost of other (47 + 50)		= $.	= $.	= $.	= $.
52. TOTAL INSTALLATION COST (44 + 51)		= $7,678.80	= $6,382.80	= $8,182.80	= $7,572.60

Retrofit Economics Worksheet:
Determining Simple Payback Per Fixture/Per Facility, Page 4

Area description: WORD PROCESSING DEPT. Area No.: 4

	Existing	Option #1	Option #2	Option #3	Option #4
UTILITY REBATES					
53. Total demand savings (kW) from **Step 12**		=	=	=	=
54. Rebate per kW		x $	x $	x $	x $
55. kW rebate		= $	= $	= $	= $
56. Total lighting energy savings (kWh) from **Step 10**					
57. Rebate per kWh		x $	x $	x $	x $
58. Total kWh rebate (assumes no rebate for A/C savings)		= $	= $	= $	= $
59. Rebate per lamp		$	$	$	$
60. Total number of lamps (lamps per fixture x fixtures from **Step 9**)		x	x	x	x
61. Total lamp rebate		= $	= $	= $	= $
62. Rebate per ballast		$	$	$	$
63. Total number of ballasts (ballasts per fixture x no. of fixtures from **Step 9**)		x	x	x	x
64. Total ballast rebate		= $	= $	= $	= $
65. Rebate per reflector		$	$	$	$
66. Number of reflectors		x	x	x	x
67. Total reflector rebate		= $	= $	= $	= $
68. Rebate per other:_______		$	$	$	$
69. Number of other					
70. Total other rebate		= $	= $	= $	= $
71. TOTAL INSTALLATION COST from **Step 52**		$7,678.80	$6,382.80	$8,182.80	$7,572.60
72. TOTAL REBATES (55 + 58 + 61 + 64 + 67 + 70)		– $ 0	– $ 0	– $ 0	– $ 0
73. NET TOTAL INSTALLATION COST		= $7,678.80	= $6,382.80	$8,182.80	= $7,572.60

Retrofit Economics Worksheet:
Determining Simple Payback Per Fixture/Per Facility, Page 5

Area description: WORD PROCESSING DEPT. Area No.: 4

	Existing	Option #1	Option #2	Option #3	Option #4
SIMPLE PAYBACK					
74. TOTAL LIGHTING ENERGY SAVINGS from **Step 15**		$2,808	$1,755	$2,386.60	$2,106
75. NET HVAC SAVINGS from **Step 30**		+ $302.56	+ $189.11	+ $257.18	+ $226.93
76. TOTAL ENERGY SAVINGS		= $3,110.56	= $1,944.11	= $2,643.98	= $2,332.93
77. NET TOTAL INSTALLATION COST from **Step 73**		$7,678.80	$6,382.80	$8,182.80	$7,572.60
78. TOTAL ENERGY SAVINGS from **Step 76**		÷ $3,110.56	÷ $1,944.11	÷ $2,643.98	÷ $2,332.93
79. PAYBACK (YEARS)		= 2.46	= 3.28	= 3.09	= 3.25
SIMPLE RETURN ON INVESTMENT					
80. TOTAL ANNUAL ENERGY SAVINGS from **Step 76**		$3,110.56	$1,944.11	$2,643.98	$2,332.93
81. NET TOTAL INSTALLATION COST from **Step 73**		÷ $7,678.80	÷ $6,382.80	÷ $8,182.80	÷ $7,572.60
		x 100	x 100	x 100	x 100
82. RETURN ON INVESTMENT (%)		= 40.5 %	= 30.5 %	= 32.3 %	= 30.8 %
83. Does the option comply with company financial policy?		YES	NO	YES	NO

Retrofit Economics Worksheet, Page 1

To complete the Worksheet, first write the room's description and number. Then write the fixture type. We can determine the total lighting energy savings per fixture and for the area by completing the rest of Page 1. By the end of Page 1, the total lighting energy savings for the area are given a dollar value.

1. Write the number and type of lamps per fixture for this fixture type. If delamping is being considered in an upgrade option, indicate this accordingly.

2. Write the number and type of ballasts per fixture for this fixture type. If delamping is being considered in an upgrade option, ballasts can be removed from the lighting system. Indicate this accordingly.

3. Write any other equipment that is in use, such as occupancy sensors or reflectors. In the case of controls, it can be useful to write "light switch" if no other controls are present.

4. Determine the system input watts per fixture. Obtain from manufacturers' catalogs.

5. Multiply by the number of hours per year the lighting system is in use. This information can be estimated (see the Survey Worksheet) or a more accurate pre-upgrade measurement can be taken using occupancy monitoring equipment. If occupancy sensors are being considered in one or more of the options, adjust the operating hours accordingly through an engineering estimate, information gained through a trial installation in a similar room, the advice of a professional, available published literature, or the advice of manufacturer representatives.

The number "5" is in bold print because the figures calculated in this step will be referenced and used in later calculations.

Divide the result of this step by 1,000.

6. The result is the lighting energy consumption per year, expressed as a relationship between demand for power and amount of time power is used: the kWh.

7. Subtract the lighting energy consumption for the existing system from the energy consumption of each upgrade option.

8. The result is the lighting energy savings (kWh) per fixture. *On the calculator, it will appear as a negative number. Write it as a positive number.*

9. Multiply by the total number of fixtures in the room to be upgraded.

10. The result is the total lighting energy savings (kWh) for the room. To reduce legwork, the Worksheet can be modified so that savings for all fixtures in the building or facility being considered for the same upgrade options can be calculated. This can be highly desirable, but remember that each option must provide light levels appropriate to tasks in various rooms.

11. Divide by operating hours from Step 5.

12. The result is the total lighting demand savings for the room or building in kilowatts. This step is taken to provide information to complete other steps, such as calculating the sum of utility rebates if the utility offers a dollar amount for each kW removed.

13. Repeat the total lighting energy savings (kWh) from Step 10.

14. Multiply by the cost per kWh ($). Obtain from the local utility. For a general engineering estimate, use the average commercial or industrial kWh costs provided in Appendix I.

15. The result is the dollar value of the total annual lighting energy savings for the room or building.

Retrofit Economics Worksheet, Page 2

On Page 2, we will calculate the impact on the net HVAC savings that removing heat through saving lighting energy presents.

16. Repeat the total lighting energy savings (kWh) for the room or building from Step 10 again.

17. Multiply by the fraction of the year during which the cooling system is in full use. The interiors of most large buildings are cooled all year, but we will calculate only the energy savings for the cooling season. The fraction of the year is determined by dividing the cooling season in weeks by 52. The cooling season in weeks can be obtained from Appendix I, where a list of more than 150 cities is available.

Multiply this figure by 0.9, which is the estimated fraction of the heat load produced by the lighting system which must be met by mechanical cooling. Note: These factors and how to calculate the air cooling benefit of conversion to an energy-efficient lighting system are discussed in detail later in this chapter.

Multiply this figure by 2.7, which is the estimated efficiency of a typical cooling system.

18. The result is the total air cooling energy savings (kWh) for the room or building.

19. Multiply by the utility cost per kWh from Step 14.

20. The result is the total air cooling energy savings expressed as a dollar amount.

21. Because heat is being removed from the room, the heating system must provide additional heat to make up for the loss (although it can be argued that the addition of computers, fax machines, copiers and other electronic office equipment generates enough heat to make up for the loss).

Determine the heating season as a fraction of the year by subtracting the cooling season in Step 17 from 1.

Multiply by 0.2, the estimated typical fraction of lighting reduction that has be made up by additional heating.

22. The result is Variable A, which will be used in the next series of calculations.

23. Repeat the total lighting energy savings (kWh) for the room or building from Step 10.

Multiply by 3,414 to determine the annual British Thermal Unit (BTU) equivalent of the lighting energy savings.

Multiply by Variable A.

Divide by 0.8, the estimated efficiency of a typical heating system.

24. The result is the total increase in heat needed in the space, expressed in BTU.

Divide by 100,000.

25. The result is the equivalent needed heat increase expressed in therms.

26. Multiply by the utility cost per therm ($). Obtain from the local utility.

27. The result is the additional heating cost posed by removing heat produced by the lighting system, expressed as a dollar amount.

28. Repeat the air cooling savings ($) from Step 20.

29. Subtract the additional heating cost ($) from the air cooling savings.

30. The result is the net HVAC savings, expressed as a dollar amount. Note that in some cases, this figure may be negative.

Retrofit Economics Worksheet, Page 3

On this and the next page, we will determine the total installation cost. Note that disposal costs for lamps and ballasts are not represented; they are often not considered real costs of an upgrade to be incorporated into the payback assessment since the existing lamps and ballasts will have to be replaced and disposed of upon failure anyway. The cost can be assessed separately for informational purposes by multiplying the number of lamps and ballasts to be replaced by the unit disposal cost presented by the chosen disposal option — recycling, landfill, etc. If the reader desires to incorporate disposal costs into the net installation cost, add the resulting figure to the total installation cost in Step 52 and proceed normally.

31. Write the cost of the lamp, including appropriate sales tax. Obtain from the supplier or contractor.

32. Multiply by the number of lamps per fixture.

33. The result is the lamp cost per fixture.

34. Write the cost per ballast, including appropriate sales tax. Obtain from the supplier or contractor.

35. Multiply by the number of ballasts per fixture.

36. The result is the ballast cost per fixture.

37. If installation of reflectors is being considered in one or more of the options, write the reflector cost per fixture. Obtain from the supplier or contractor.

38. Add the lamp cost per fixture from Step 33, the ballast cost per fixture from Step 36, and the reflector cost per fixture from Step 37 to determine total equipment cost per fixture.

39. Write the electrical labor rate per hour. Obtain from the contractor.

40. Multiply the estimated number of hours (expressed as a decimal)

to install the equipment in one fixture.

41. The result is the labor cost per fixture.

42. Add the total equipment cost per fixture in Step 38 with the total labor cost per fixture in Step 41 to determine the total installation (and disposal) cost per fixture.

43. Multiply by the total number of this fixture type in the room or building, from Step 9.

44. The result is the total installation cost for fixtures.

45. Since other equipment such as occupancy sensors is not installed on a per-fixture basis, the cost of the equipment and installing it is calculated separately. Write the type of equipment in the available blank and the cost per unit.

46. Multiply by the number of this equipment.

47. The result is the total equipment cost for this equipment type.

48. Repeat the electrical labor rate per hour from Step 39 or indicate an appropriate labor rate.

49. Multiply by the number of hours required to install all of these units.

50. The result is the total labor cost to install this equipment.

51. Add the total cost of equipment from Step 47 and the total labor cost to install it from Step 50 to determine the total cost of this equipment's installation.

52. Add the installation cost for the fixtures in Step 44 to the installation cost of all other equipment in Step 51 to determine the total installation cost.

Retrofit Economics Worksheet, Page 4

On this page, we will determine the impact of available utility financial incentives to upgrade. Numerous electric (and several gas) utilities, concentrated most heavily on the east and west costs, offer rebates based upon a percentage of the installed cost of qualifying technologies, upon kW removed or kWh saved, or upon a combination of these. The presence of a rebate can dramatically reduce the payback period. For more information about utility rebates, see Chapter 6: "Retrofit Financing: Putting Money to Work."

The Worksheet is not designed to account for the financial incentive of discount utility rates. If the local utility offers this less-often-employed incentive to upgrade, adjust Page 1 accordingly to compare the cost of operating lighting system options before comparing savings.

53. Repeat the total reduction in load savings in kW from Step 12.

54. Multiply by the rebate per kW, if offered. Obtain from the local utility.

55. The result is the kW rebate.

56. Repeat the total lighting energy savings (kWh) from Step 10.

57. Multiply by the rebate per kWh, if offered. Obtain from the local utility.

58. The result is the kWh rebate. Air cooling energy savings are excluded from this step. Check with the local utility to determine the requirements and procedure for gaining a rebate on air cooling savings.

59. Write the rebate per lamp, if offered. Obtain from the local utility.

60. Multiply by the number of lamps qualifying for the rebate. This means all qualifying lamps in the space, a total gained by multiplying the number of lamps per fixture by the number of fixtures from Step 9.

61. The result is the total lamp rebate for the room or building.

62. Write the rebate per ballast, if offered. Obtain from the local utility.

63. Multiply by the number of rebate-qualifying ballasts. This means all rebate-qualifying ballasts in the area, a total gained by multiplying the number of qualifying ballasts per fixture by the number of fixtures from Step 9.

64. The result is the total ballast rebate for the area or building.

65. Write the rebate per reflector, if offered. Obtain from the local utility.

66. Multiply by the number of reflectors to be installed.

67. The result is the total reflector rebate.

68. Write the rebate for other equipment such as occupancy sensors, if offered. Obtain from the local utility.

69. Multiply by the number of this equipment to be installed.

70. The result is the total rebate for other equipment.

71. Repeat the total installation cost from Step 52. We will now calculate the rebates' impact on the installation cost.

72. Add all appropriate rebates, if any are offered, including the rebate for kW removed and kWh saved, and lamp, ballast, reflector and other equipment rebates. These were determined in Steps 55, 58, 61, 64, 67 and 71. Subtract the resulting figure from the total installation cost indicated in Step 52.

73. The result is the net installation cost for each option after the impact of rebates have been assessed.

Retrofit Economics Worksheet, Page 5

This primary purpose of this page of the Worksheet is to use the information gained so far to determine simple payback and return on investment.

74. Repeat the total lighting energy savings, expressed as a dollar amount, from Step 15.

75. Add to the total HVAC energy savings, expressed as a dollar amount, from Step 30.

76. The result is the total energy savings for the area or building, expressed as a dollar amount.

77. Repeat the total net installation cost from Step 73.

78. Divide by total annual energy savings from Step 76 — as indicated by the simple payback formula (payback [years] = installation cost [$] ÷ annual energy savings [$]).

79. The result is the time period which will pass before the option pays for itself (payback period), expressed in decimal years. If payback is desired in months, multiply by 12 and write "months" after the figure.

80. Repeat the total annual energy savings from Step 76.

81. Divide by the total net installation cost from Step 73, then multiply the result by 100 — as per the simple return on investment formula (return on investment [%] = annual energy savings [$] ÷ installation cost [$] x 100).

82. The result is the simple return on investment, expressed as a percentage.

83. Compare the payback period and return on investment against the company's requirements for investment in an energy project. Indicate which, if any, of the options comply.

CALCULATING NET H.V.A.C. SAVINGS

As electrical systems, lighting systems produce infrared energy, or heat. The incandescent lamp converts only 10-12% of its electrical input into light, while the rest is released directly as heat. Traditional fluorescent lamp-ballast systems convert about 20% — again, the rest is released directly as heat. Even the visible light, once it impacts the workplane, is dissipated as heat.

While some of this heat escapes the building through windows, most of it is trapped inside. That is why any time energy use is reduced, less air cooling is needed to remove the associated heat generation. Lighting retrofit is no exception.

DETERMINING ACTUAL SAVINGS

Actual HVAC energy savings depends on many factors, including geometric and thermal characteristics of the building; type and efficiency of the HVAC system; heating and cooling loads; building location; and lighting system operating hours. The U.S. Department of Energy (DOE) and the Electric Power Research Institute (EPRI) are currently researching the interaction of lighting and HVAC systems.

For years, one watt of air cooling energy savings saved for every saved three watts of lighting energy was considered the general rule of thumb measuring air cooling savings. While a good starting point, this simple figure does not account for dramatic differences in climate and additional heating requirements.

Robert Rundquist, PE, offers a formula with supporting data and estimates to assess a "more accurate"

Figure 4-8. Lighting systems produce light and heat. Even the visible light, once it impacts objects on the workplane, is dissipated as heat.
Source: National Lighting Bureau

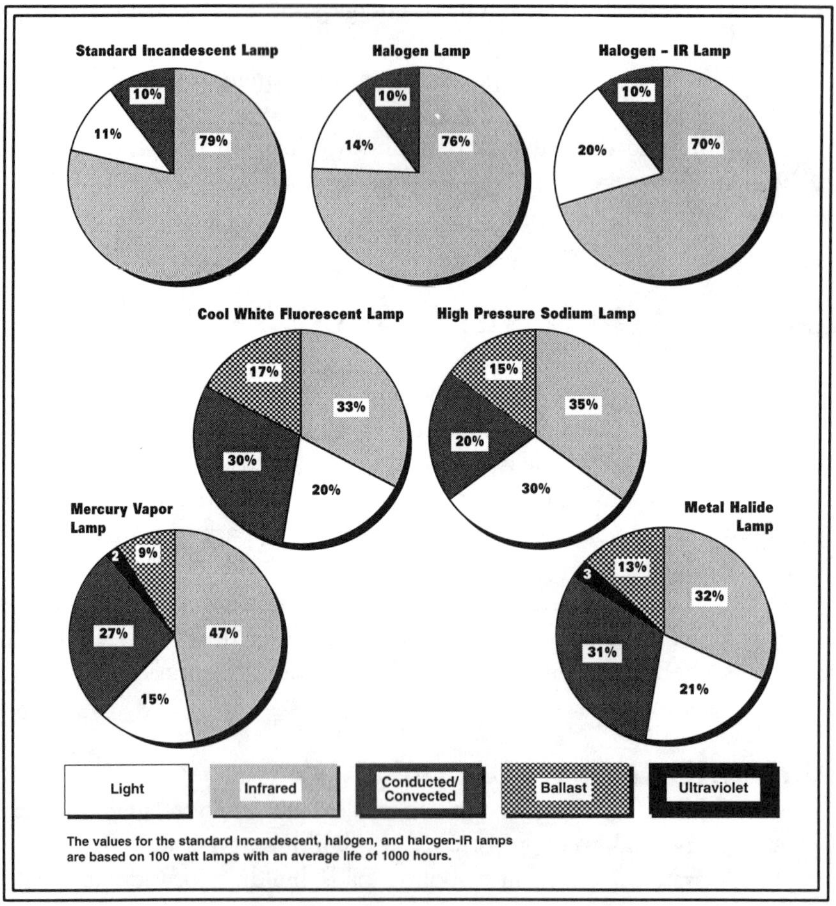

Figure 4-9. Heat and energy conversions for various lamps. Lighting systems convert only a fraction of their electrical input into visible light. *Source: General Electric, Lamp Technology Division.*

figure for air cooling savings. He offers another formula (again with supplementary information) to account for additional heat required, which yields net HVAC savings. Rundquist, president of R.A. Rundquist Associates of Northampton, Massachusetts, is a professional engineer with some two decades' experience in HVAC design, energy analysis and energy calculation research. His formula was derived from independent research and research conducted by the American Society of Heating, Refrigeration and Air Conditioning Engineers (ASHRAE). It has been validated by DOE-2 computer runs and other methods.

MEASURING AIR COOLING ENERGY SAVINGS

To measure air cooling energy savings, lighting energy use must be reduced by a specific amount that stays constant throughout the year. A simple example would be to replace an incandescent light bulb with a compact fluorescent unit.

FRACTION OF LIGHTING SAVINGS AS AIR COOLING SAVINGS	=	FRACTION OF THE YEAR OF THE COOLING SEASON	x	LIGHTING LOAD MET BY MECHANICAL COOLING	÷	SYSTEM COEFFICIENT OF PERFORMANCE

Fraction of the year of the cooling season. Appendix I shows a list of 153 cities in the United States with their cooling seasons expressed in weeks. Express this figure as a fraction of the year (weeks ÷ 52).

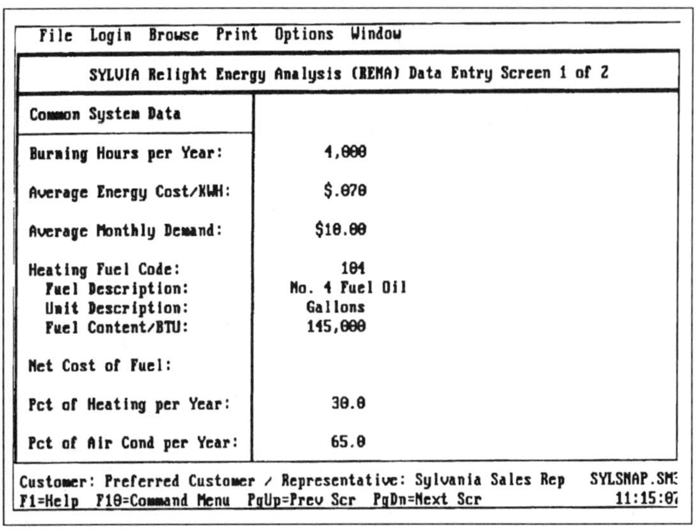

Figure 4-10. Some computer upgrade analysis programs, such as Sylvia from Osram Sylvania, Inc., incorporate the ability to calculate air cooling savings. *Source: Osram Sylvania Inc.*

Fraction of the daily lighting load met by mechanical cooling. Usually about 90%, this is the amount of heat generated by the lighting system which is met by air cooling. The other 10% of the heat is dissipated at night or into a plenum above the workspace.

Coefficient of performance. Tests on cooling systems — air conditioners working together with fans and pumps — show that for every kW put into the system, 2.7 kW of cool air is produced. The system gains this efficiency with the help of many system and environmental factors which assist cooling. At peak efficiency, the system can produce about a 3.2 kW equivalent of cool air. Coefficient of performance can vary by type and age of cooling system; for optimum results, use the actual coefficient of performance for the system being used.

Example

Suppose we choose Option #1 from the Retrofit Economics Worksheet, which entails installation of four T8 lamps powered by one electronic ballast, a reduction in electrical input of 80 watts per fixture (192 – 112) or 312 kWh of energy consumption. The cooling season for the local region is 23.5 weeks per year, or about 45% of the year.

$$0.45 \times 0.90 \div 2.7 = 0.15$$

For every kWh saved from this retrofit, an average 0.15 kWh of air cooling energy is saved. (Note that in this case, the cooling savings are half the three-for-one ratio which is the rule of thumb.) Air cooling energy savings are then calculated:

312 kWh x 0.15 = 46.8 kWh OF AIR COOLING SAVINGS PER FIXTURE

The local electric utility charges $0.10 per kWh (about 2-3 cents above the national commercial average), resulting in an additional $4.68 saved per fixture as a result of this retrofit.

MEASURING ADDITIONAL HEAT REQUIRED AFTER RETROFIT

In most large buildings, the additional heat would only be needed in the perimeter zone, because the interior spaces must be cooled all year. This cooling is produced by an economizer in most buildings. Before even calculating the additional heat required, take a moment to consider what new electrical equipment is being added to the facility, such as computers, fax machines and printers; remember that this equipment generates heat, possibly enough to make up the difference and more.

To calculate additional heat required, Rundquist supplies another formula with supporting data and estimates:

EXTRA HEAT REQUIRED IN BTU	=	(1- FRACTION REPRESENTING COOLING SEASON)	x	FRACTION OF LIGHTING REDUCTION THAT HAS TO BE MADE UP BY HEATING	x	ANNUAL BTU EQUIVALENT OF LIGHTING SAVED	÷	SEASONAL HEATING EFFICIENCY

Many of the figures used in this formula are estimates or averages; for optimum results, use figures specific to the system in use.

Fraction of the year as heating season. This simply represents the portion of the year not designated as the cooling season. Take the fraction used for the cooling season and subtract it from one (1) to gain the fraction of the year representing the heating season. This is a rough estimate of the heating season; there are times during the year when the building is neither actively heated nor cooled. Therefore, the cost of additional heating resulting from this formula may be somewhat artificially high. This in turn will make the total HVAC savings (after the cost of additional heating is subtracted from the commonly larger air cooling savings) a somewhat more conservative estimate.

Fraction of the lighting reduction that has to be made up by heating. The portion of the lighting reduction that must be made up by heating comes from observing what fraction of lighting heat is released at night when the building "really" pays for heating ("really" meaning that at this time in most buildings, nobody is in the space to use the heating, hence money is spent without truly realizing value). This figure has been estimated at about 20%.

Annual BTU equivalent of lighting saved. This figure represents the lighting reduction in kWh multiplied by 3,414 British Thermal Units (BTU) per kWh.

Seasonal heating efficiency. As in the cooling formula, this represents an estimate of the basic efficiency of the heating system: 80%. Heating system efficiency can vary from about 65-95%, depending on use, type and technology.

Example

Our chosen retrofit option yields 46.8 kWh of air cooling savings per fixture after saving 312 kWh of lighting energy per fixture, but it is determined that additional heat is now required.

0.55 x 0.2 x (312 kWh x 3,414 BTU/kWh) ÷ 0.8 = BTU PER FIXTURE
0.55 x 0.2 x 1,065,168 BTU ÷ 0.8 = 146,460.6 BTU PER FIXTURE

The BTU result is broken into therms:

146,460.6 BTU ÷ 100,000 = 1.47 THERMS PER FIXTURE

Suppose the local utility charges $0.90 per therm (a relatively high figure):

1.47 THERMS PER FIXTURE x $0.90 PER THERM =	$1.32	ADDITIONAL HEATING COST

Air cooling is generally more expensive, so the air cooling savings in dollars generally exceeds the cost of the additional heat requirement. In this case, the additional heating cost is $1.32 per fixture, while the air cooling savings total $4.68 per fixture, netting annual HVAC savings of $3.36 per fixture.

FREEZING NEW COOLING REQUIREMENTS

According to the National Lighting Bureau, about two out of three nonresidential buildings in the United States are at least 25 years old. At that time, mechanical engineers could not realistically predict demand on the cooling system presented by more than two decades of growth in office technology. Copiers, printers, computers, fax machines and other office equipment generate heat which the existing cooling system may not be able to handle effectively on hot summer days.

About 85% of every kW of lighting shows up as peak cooling load. Each resulting kW requires 3,414 BTU of cooling per hour. In turn, one ton of cooling requires 12,000 BTU per hour. Determine air cooling "savings" in tons:

A/C SAVINGS IN TONS = LIGHTING kW REDUCTION x 0.85 x 3,414 ÷ 12,000

Example

Our chosen retrofit option reduces energy consumption per fixture by 312 kWh; for all of the 90 fixtures in this area, total demand for power is reduced about seven kW. Suppose this retrofit is part of a larger retrofit which results in a reduction in load totaling 100 kW.

100 kW x 0.85 x 3,414 ÷ 12,000 = 24.2 TONS OF A/C SAVINGS

Suppose we had planned to add 20 tons of air cooling to the existing air cooling system to remove heat generated by new office equipment. By reducing the heat produced by the lighting system through lighting retrofit, this capacity can be deferred. Assuming a marginal cost of $1,500 per ton, this would defer a one-time installation cost of $30,000.

Sources:

Andreas, John C. *Energy-Efficient Electric Motors: Selection and Application*, Second Edition. New York: Marcel Dekker, Inc., 1992.

Hathaway III, Alden, "Global Warming in the Office Environment," *MainLighter: The Lighting Management & Maintenance Magazine*, June 1992.

Lighting Upgrade Manual, First Edition. Washington, DC: U.S. Environmental Protection Agency Green Lights Program, November 1991.

Lindsey, Jack L. *Applied Illumination Engineering*. Lilburn, Georgia: The Fairmont Press, 1991.

Mendelsohn, Cary, "The Domino Effect — Lighting/Air Conditioning/Energy/Environment," *Strategic Planning for Energy and the Environment*, Summer 1991.

Rundquist, Robert, PE, President, R.A. Rundquist Associates of Northampton, Massachusetts.

Further reading to learn about life-cycle costing:

Getting the Most Out of Your Lighting Dollar, Second Edition. Washington, DC: National Lighting Bureau.

Lighting Economics: An Intermediate Approach to Economics as Applied to Lighting Practice. New York: The Illuminating Engineering Society of North America, 1991.

Profiting from Lighting Modernization. Washington, DC: National Lighting Bureau, 1987.

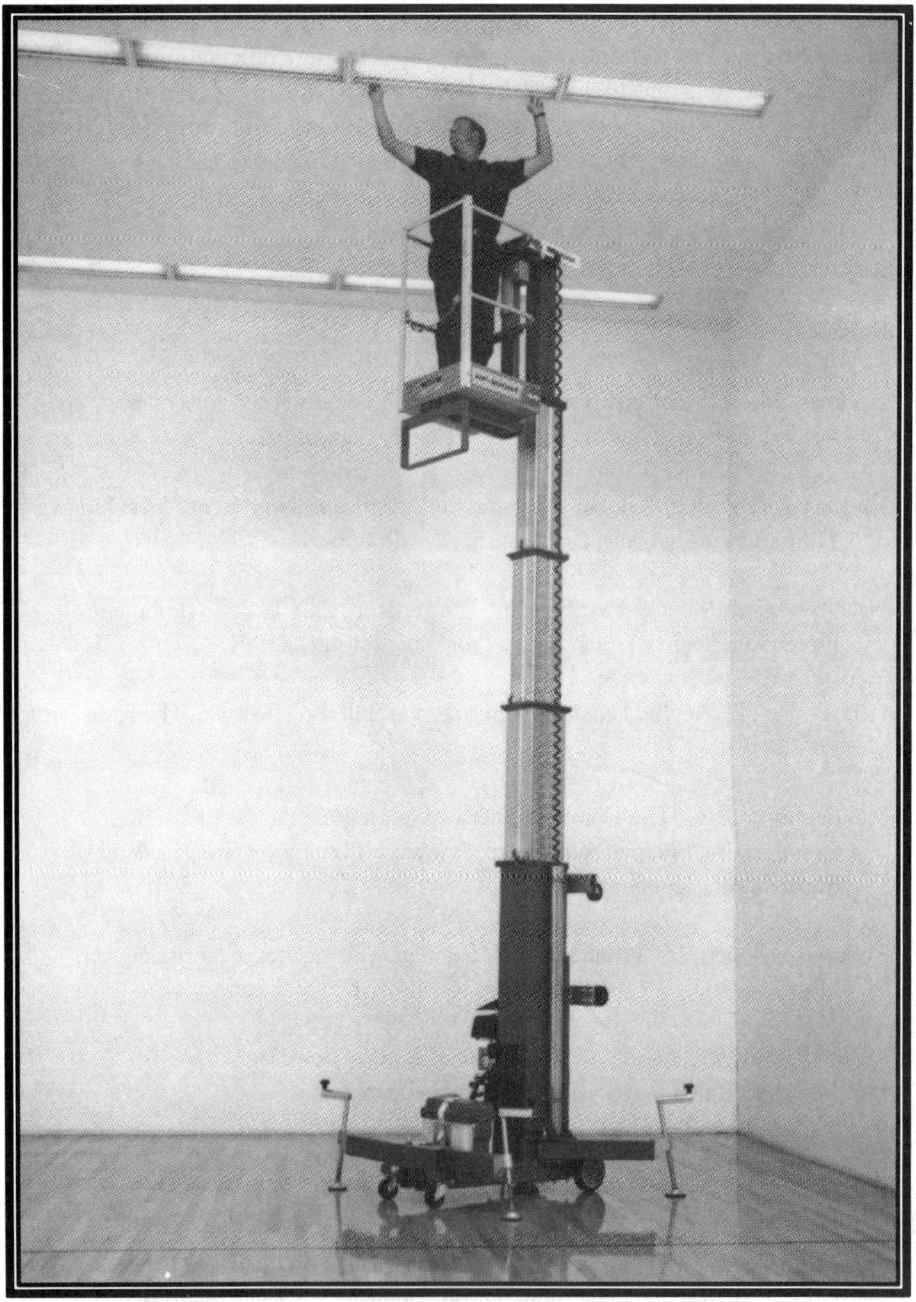

Source: Upright, Inc.

CHAPTER 5

Maintenance: Making the System Work for the Long Haul

THE MISSION OF PLANNED LIGHTING MAINTENANCE

A lighting upgrade does not end with installation. No matter how energy-efficient or quality-sensitive a newly installed lighting system may be, it nonetheless succumbs to depreciation and must be maintained. Acquiring new lighting equipment can pose a sizable investment; it pays to take care of it properly so that it consistently provides the best possible performance.

ANATOMY OF POOR MAINTENANCE

Most lighting systems are poorly maintained. Lamps are replaced individually as they fail. Defective components such as broken sockets (lampholders) and loose wires, in addition to failed ballasts, are replaced after they cause the fixture to malfunction. Fixtures are usually cleaned only after dirt buildup or light loss is noticed — or not at all. And relamping and cleaning are usually performed with little attention to proper methods and equipment.

Overall, maintenance without a defined plan, trained personnel and proper equipment is relatively uneconomical and wasteful in terms of labor and energy use.

ANATOMY OF PLANNED MAINTENANCE

To properly grasp the concept of planned lighting maintenance, it must be understood both in principle and application.

Maintenance in principle

Planned lighting maintenance is the discipline of organizing labor and resources to achieve two primary goals. First, it ensures consistent recommended light levels for those who need the light to perform at their best. Second, it consistently ensures the most economical use of the lighting system, providing the most light for the lowest cost.

Maintenance in application

Planned lighting maintenance means organizing labor and resources to combat recoverable light loss factors such as lamp lumen depreciation, fixture dirt depreciation, burnouts and room surface depreciation. Combined, these factors can reduce light level by 20-60% over time, while the money spent on electricity continues to reflect the full wattage of the system.

Effectively maintaining light level typically involves four strategies:

Group relamping: As opposed to spot relamping (replacing burned-out lamps individually), group relamping entails replacing all lamps in a system at scheduled periodic intervals when lumen depreciation and the lamp failure rate make it economical to do so.

Cleaning fixtures: Scheduled periodic cleaning of the fixtures is performed to remove accumulated dirt and dust and retard future accumulation. Fixtures are typically cleaned at time of relamping, although more frequent cleanings may be economical in dirtier environments or with fixtures more susceptible to the effects of dirt depreciation.

Inspection: Scheduled periodic inspection of fixtures reveals defective components such as broken sockets, inoperative ballasts and problems such as "flickering," "cycling" and "fluttering." Any inoperative components can be spot-replaced during the inspection. As problems occur between inspections, trained and qualified personnel should be available to troubleshoot them immediately.

Cleaning room surfaces: Scheduled periodic cleaning and/or repainting of room surfaces (walls and ceilings) maintains maximum reflectance.

BENEFITS OF PLANNED LIGHTING MAINTENANCE

Effective lighting maintenance can produce numerous benefits, including ensuring consistent proper light level as well as the most effective ongoing use of labor, resources and energy.

Consistent proper light level

In Chapter 3, we discussed the effect of proper light level on human morale and productivity. It is also essential to enhance image, mood, safety, security and the fulfillment of other needs. Because so much care was taken to *specify* correct light level, equal care must be taken to *maintain* it on a continual basis through planned lighting maintenance.

Effective use of labor

As we will discuss later in this chapter, standardized maintenance techniques such as group relamping can cut the nominal labor cost of lamp replacement by as much as 90%. We will go step by step through conducting an economic analysis of this benefit later in this chapter.

Effective use of resources

Group relamping, cleaning and inspecting the lighting system all help ensure that the company gains the most value from its resources.

Figure 5-1. Proper light levels can affect human morale and productivity. Because so much care is taken in a lighting management program to specify correct light level, equal care must be taken to maintain it on a continual basis through planned maintenance.
Source: National Lighting Bureau.

Group relamping ensures that lamp costs are less expensive per unit, because they are bought and installed in volume instead of in smaller qualities to satisfy spot relamping needs. Additionally, less inventory room is needed because the lamps are installed and disposed *en masse*, with fewer lamps kept in the closet for occasional spot relamping. And fewer lamps will be stored in inventory, reducing "inventory shrinkage" due to lamps breaking or "disappearing." Inventory costs are therefore reduced.

Inspecting the lighting system periodically enables the maintenance crew to spot and replace failed ballasts and broken or damaged components. They will also troubleshoot problem fixtures before the problems become bigger ones, such as premature ballast failure.

Cleaning and group relamping increase average maintained light level, which means the lighting system needs to produce fewer lumens to provide the same light level than an unmaintained system (see next section). Properly performed maintenance can also extend the life of the fixture by slowing fixture surface depreciation.

EFFECTIVE USE OF ENERGY

An often overlooked benefit of planned lighting maintenance is the potential to reduce energy use by installing energy-saving lamps (if lamps have not already been replaced with energy-efficient counterparts) and strategically remove lamps. Current limiters can also be installed. These are devices which restrict electrical input and therefore energy consumption and lumen output by the desired amount.

Depreciation cushion

When a lighting designer creates a lighting system, he or she must ensure that despite all light loss factors, recommended light levels will be provided to the area's occupants. The difference in the number of footcandles produced by a brand-new lighting system and the number of footcandles produced by the system after depreciation by recoverable light loss factors is called the depreciation cushion. Spaces are "overlighted" so that after recoverable light loss, recommended light levels will still be provided.

Older lighting systems have larger depreciation cushions. This is because the design community believed that to efficiently perform various tasks, a higher light level was needed than is recommended for those same tasks today.

Reduce depreciation cushion with maintenance

It follows then that if the depreciation cushion can be reduced while minimum proper footcandle levels are maintained, then fewer lumens are required to produce that additional light. If planned lighting maintenance increases average light levels by 10% (decreasing the depreciation cushion by 10%), then it follows that approximately 10% fewer lumens are needed to provide the additional light. In turn, this means 1) lamps can be installed which produce less lumen output and save energy (if energy-efficient lamps have not already been installed), 2) a number of lamps (and ballasts and fixtures) producing those footcandles can be removed from the design (and the associated cost of their operation and replacement), and/or 3) other retrofit equipment, such as current limiters, can be installed to save energy.

Plan ahead

Of course, this opportunity is best exploited by keeping planned lighting maintenance in mind during upgrade planning, not as an afterthought. The choice of any new lighting equipment should reflect all aspects of the upgrade. In addition, fewer lamps and reduced energy use will impact the payback analysis and position the lighting investment better to decision-makers during the proposal phase. In fact, in cases where a good upgrade option fails to meet the company's internal rate of return, the impact of maintenance may be enough to make the option qualify.

GROUP RELAMPING

Group relamping is the backbone of effective lighting maintenance. It is defined as the replacement of all lamps in a system at the same time when it is more economical to replace than to continue operating and spot-replacing the existing lamps. Group relamping is generally considered a viable economic option if the labor cost of spot-replacing one lamp, minus the cost of group-replacing it, is greater than the new cost of the lamp.

The "economic group relamping interval" depends on the application, although in fluorescent systems it is typically when 70% of the lamps' rated life is spent. At this time, the remaining lamps that have not already failed and been spot-relamped are treated as if they have burned out — because from an economic point of view, they are considered as having burned out.

THE LIGHTING SYSTEM AT END OF LIFE

In Chapter 3, we reviewed the four recoverable light loss factors: lamp lumen depreciation, fixture dirt depreciation, lamp burnouts and room surface dirt depreciation. Group relamping combats the effects of lamp lumen depreciation and burnouts.

Cost of lamp lumen depreciation

A lamp's lumen output gradually decreases as operating hours increase. Because wattage remains constant while this happens, the dollars spent per footcandle increase gradually as lumen output decreases gradually. Suppose a fluorescent lamp-ballast group initially provides 75 footcandles for a space at an operating cost of $16 per year. When the lamps reach rated life, they produce only about 80% of their original lumen output, causing light level to drop to 60 footcandles at a cost of $16 per year. The cost of light rose from $0.21 to $0.26 per footcandle. This cost is subtle because instead of seeing a real rise in cost for light, the same bill is paid for less light.

Cost of burnouts

For fluorescent systems, lamp failure rates rise sharply past 70% of rated life. By this time, nearly 10% of a large group of lamps (installed at the same time) are expected to fail. An additional 40% will fail between then and end of rated life.

If a fluorescent lamp is rated at 20,000 hours, 40% of a system's lamps can be expected to fail during the 6,000 hours of operation preceding rated life. The rest will fail soon after. The amount of time spent on lamp replacement rises sharply with the failure rate. So does the amount of productivity lost by employees who are under fixtures being spot-relamped or who pause to watch a nearby fixture being spot-relamped. And so does the possibility of damage to machinery, carpeting, furniture or merchandise. Of these, an accurate cost can be assigned to lamp replacement labor.

The cost of replacement labor is determined by assigning a dollar value (based on wages) to the time spent replacing each lamp. The cost-difference between group relamping and spot relamping a fixture is usually a nominal rather than a hard cost, as these personnel are typically salaried in-house employees. This cost actually represents lost opportunity — the cost of human resources being committed to one inefficient activity for a period of time instead of one or more useful activities during that same period. This cost of labor can be much higher when fixtures are hard to reach, such as high-bay HID lamps in factories and parking lot pole-fixtures.

The cost of damaged components comes from lamps left in fixtures after burnout. Ballasts attempting to start failed lamps may fail prematurely. On some circuits, a lamp failure can cause rectification, which is likely to cause premature ballast failure. In high pressure sodium and low-wattage metal halide systems, the starter component may fail if a failed lamp is left in the fixture.

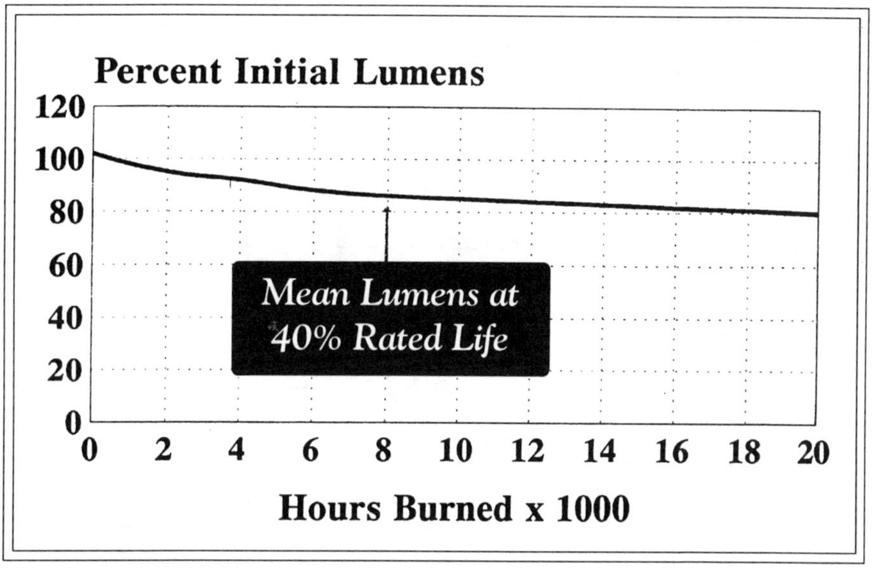

Figure 5-2. Lumen maintenance curve showing lumen depreciation for a traditional F40/CW fluorescent lamp. *Source: Osram Sylvania, Inc.*

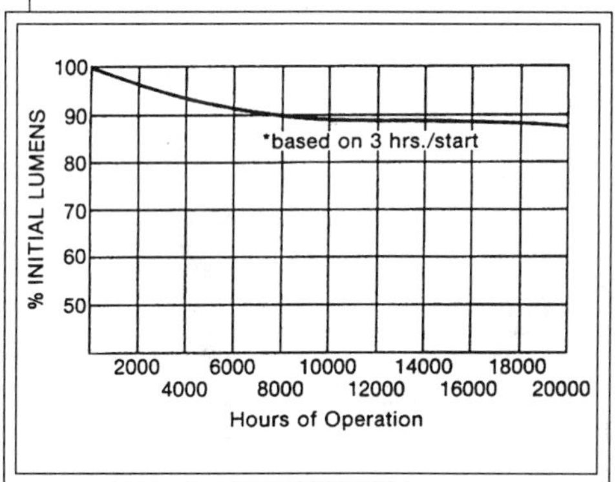

Figure 5-3. Lumen maintenance, showing lumen depreciation for a typical F17, F25, F32 or F40 T8 lamp. Notice that the superior construction of the T8 gives it better lumen maintenance than traditional F40/CWs.
Source: Osram Sylvania, Inc.

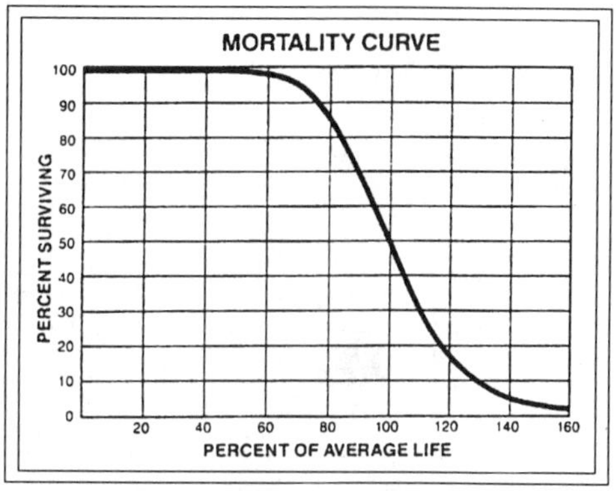

Figure 5-4. A typical mortality curve for fluorescent lamps. Large groups of lamps tend to follow this curve.
Source: Osram Sylvania, Inc.

The largest cost posed by both lumen depreciation and burnouts, of course, is attributable to lost human efficiency, accuracy, morale, safety and security, as well as tarnished corporate image and appeal of merchandise. This cost is subtle and difficult to quantify, but is nonetheless present.

Figure 5-5. In the next six panels, we see the effect of burnouts on a lighting system containing 100 fresh fluorescent lamps (each rectangle represents one lamp). As the lamps near end of rated life, they begin to fail at a higher rate. Soon after rated life, the rest will fail. If these lamps are rated at 20,000 hours and are spot relamped, this will mean some 40 trips to the fixtures in the 6,000 hours of operation preceeding rated life. Each trip presents a nominal labor cost.
Source: GE Lighting.

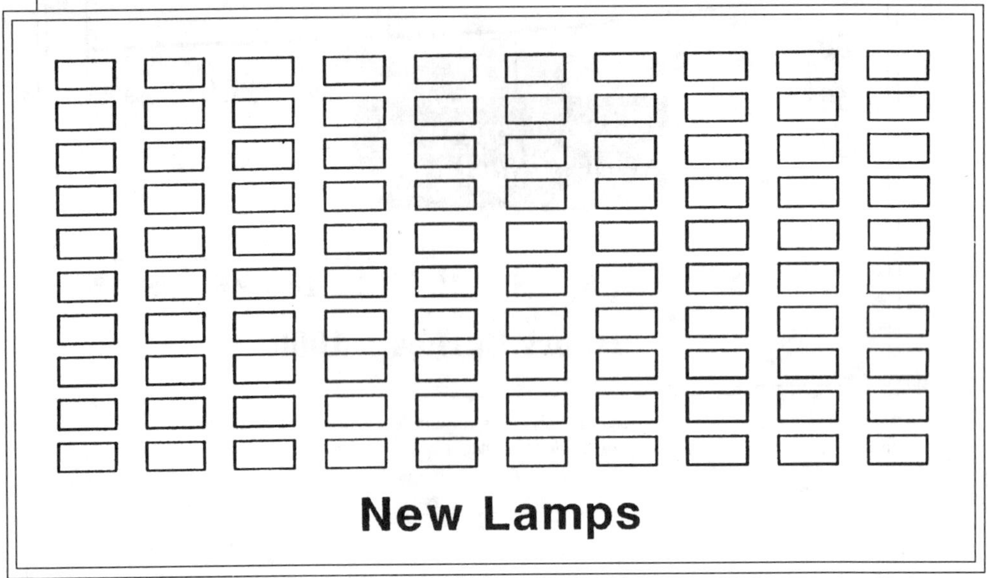

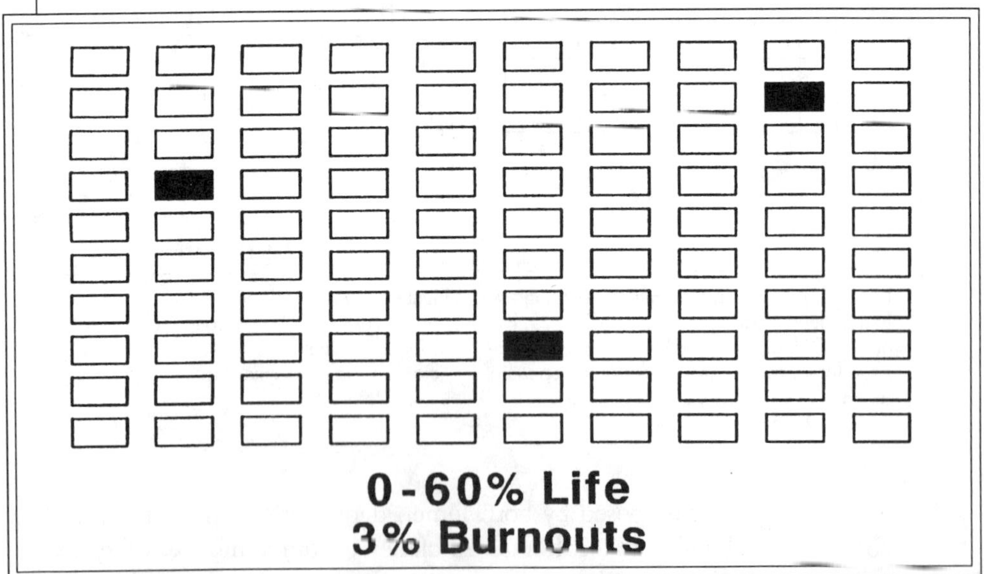

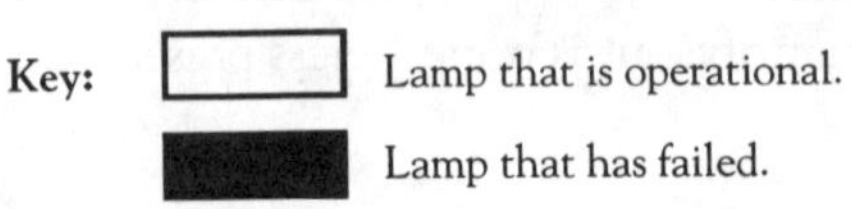

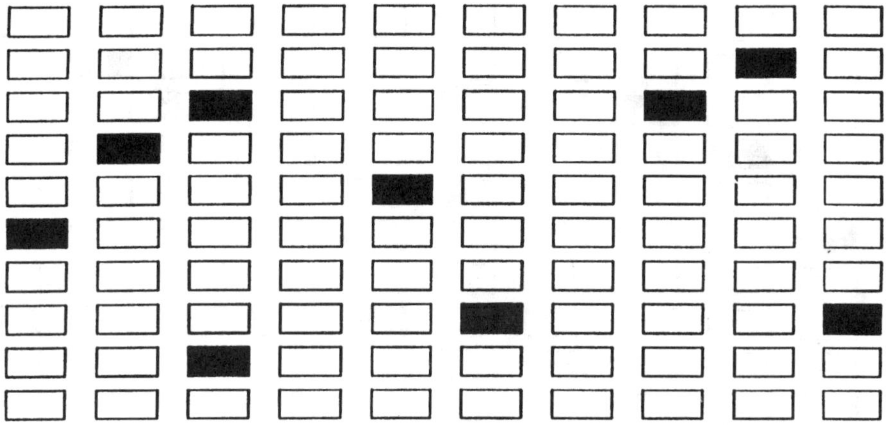

60-70% Life
6% Burnouts
(9% Total)

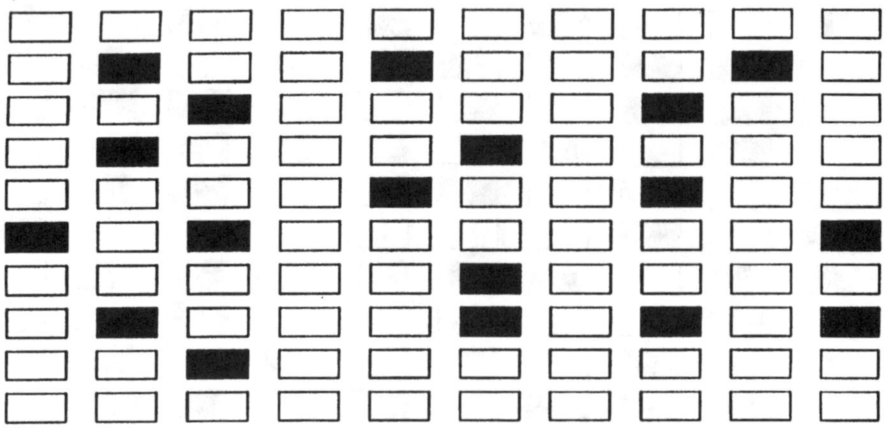

70-80% Life
9% Burnouts
(18% Total)

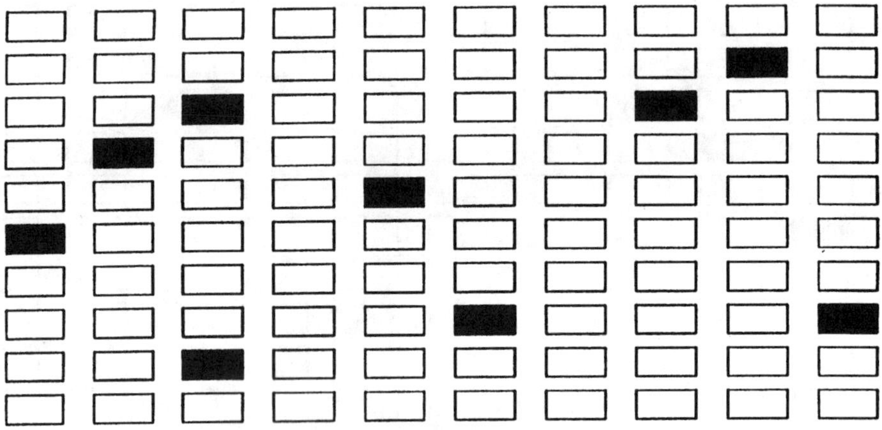

60-70% Life
6% Burnouts
(9% Total)

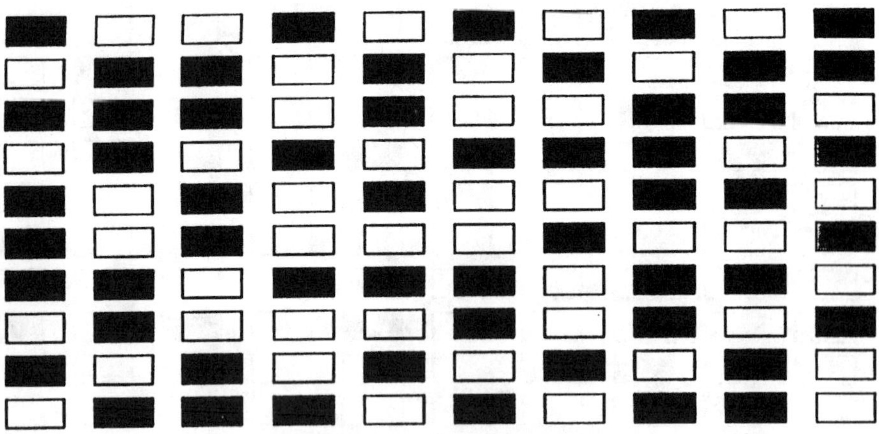

90-100% Life
18% Burnouts
(50% Total)

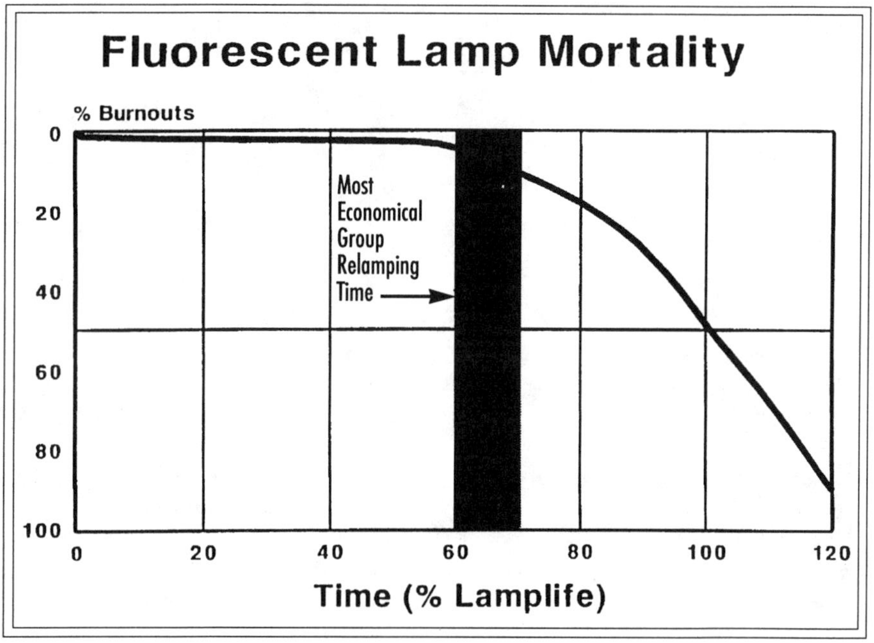

Figure 5-6. The most economical group relamping time is typically at 60-70% of rated life. *Source*: *GE Lighting*.

GROUP RELAMPING VERSUS SPOT RELAMPING

Most building maintenance personnel spot relamp, meaning they replace the lamps individually as they burn out. Ideally, these lamps will be replaced immediately and therefore present only brief and minor light loss. But even so, consider the sequence of events after a fluorescent lamp fails: 1) the complaint is reported and logged, 2) a ladder is brought out of the maintenance closet and a lamp is pulled from inventory, 3) the equipment is hauled to the fixture, 4) furniture may be moved, 5) the lamp is replaced, 6) the furniture is moved back, 7) the equipment is returned to the closet, and 8) the failed lamp is disposed.

According to the National Lighting Bureau, the typical amount of time committed to this process for a fluorescent lamp is about 30 minutes. It can take even longer if scaffolding or other heavy equipment must be set up. If we assume a labor rate of $15 per hour (including wages and benefits), the cost to replace one lamp is $7.50.

Under these circumstances, it is as easy to change 10 lamps as it is to change one. That is the inspiration behind group relamping, to change the lamps in an assembly-line fashion when a crew and all of the equipment is in place. And to change them before 60-70% of rated life, when the failure rate accelerates and before lumen depreciation reaches 20% at 20,000 hours (rated life at three hours per start).

Figure 5-7. Spot relamping is generally a wasteful practice. In most applications, it is much more cost-effective to group relamp all of the fixtures at scheduled periodic intervals when all personnel, lamps and equipment are in place. *Source: Philips Lighting Company.*

According to the interNational Association of Lighting Management Companies (NALMCO), two or three minutes are required to replace one lamp during group relamping. Again, at a labor rate of $15 per hour, group relamping one lamp would cost about about 75 cents — a savings of up to 90% on labor replacement costs.

CLEANING

Cleaning is an integral part of any effective lighting maintenance program. Cleaning most commonly entails washing fixtures. Lamps will be cleaned with the fixtures if they are not scheduled to be relamped. At longer intervals, room surfaces may be dusted, washed or repainted. The goals of cleaning are 1) to maximize reflectance of fixture and room surfaces and 2) to minimize obstruction of transmitted light due to dirt on the lamp and/or shielding material. Of course, the overall goal is to increase light level on the task and/or workplane.

(Note that cleaning and group relamping do not have to take place separately. It is true that depending on dirt conditions in the environment, cleaning may be performed between relamping intervals. But cleaning can often be performed with group relamping, making both services more economical.)

Figure 5-8. Cleaning commonly entails washing fixtures and in some maintenance programs, the lamps as well. An integral part of any effective maintenance program, cleaning ensures a higher range of light levels and produces other benefits. Cleaning is often economically combined with group relamping. *Source: Philadelphia Lighting Maintenance.*

THE LIGHTING SYSTEM AT END OF LIFE

In Chapter 3, we reviewed the four recoverable light loss factors. We discussed how group relamping combats the effects of lamp lumen depreciation and burnouts. Cleaning combats the effects of fixture dirt depreciation and room surface dirt depreciation.

Effects of dirt depreciation

Most often, fixture dirt depreciation is much more threatening to light level than room surface dirt depreciation. Dirt on room surfaces interferes with light that is to be reflected off of walls and other surfaces, whereas dirt

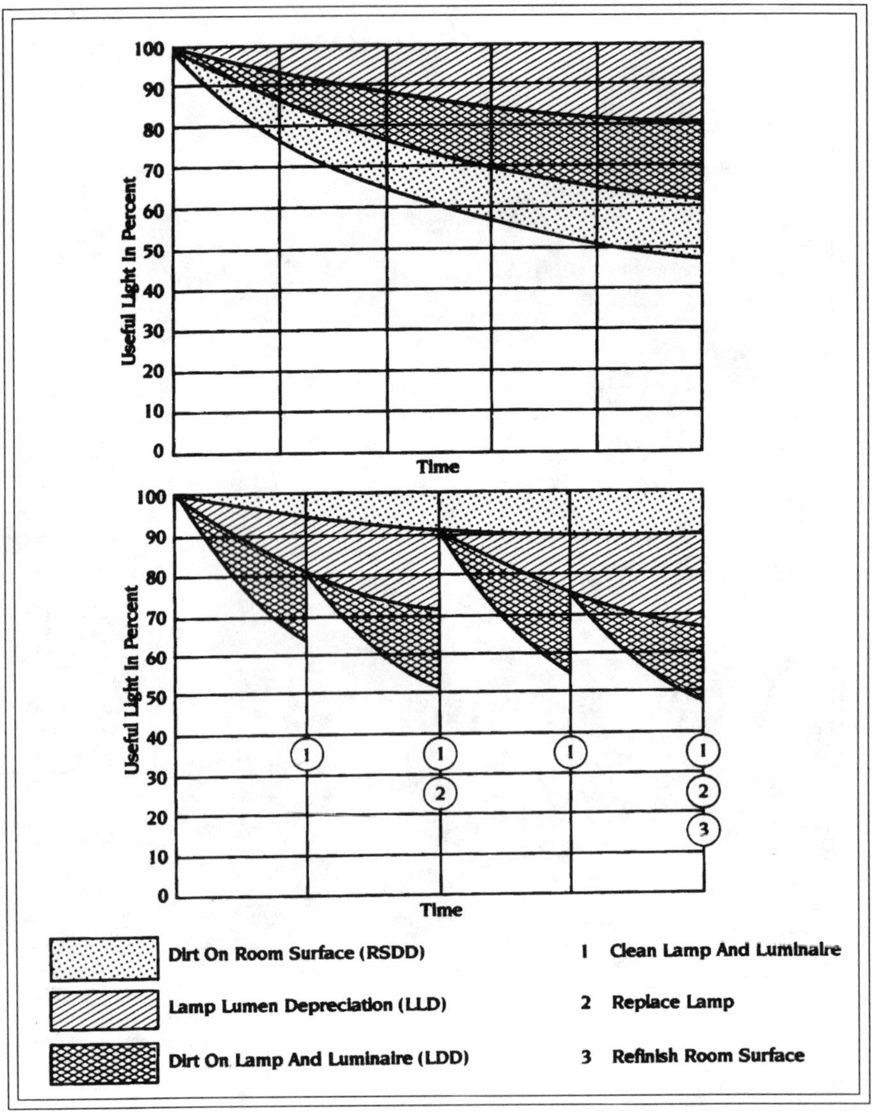

Figure 5-9. The top graph shows standard depreciation of useful light in a poorly maintained system. The bottom graph illustrates the effects of scheduled lighting maintenance in preserving the amount of useful light. *Source: National Lighting Bureau.*

on fixture surfaces interferes with light output directly at the source. Room surface depreciation should not be discounted, however, because room surface reflectances are primary factors in determining coefficient of utilization (CU) during an upgrade. Poor room reflectances' effect on CU means that an application may require more fixtures or higher lumen output than would otherwise be necessary to provide sufficient illumination. In some environments, the room surfaces should be cleaned as aggressively as if they were fixture surfaces. An example is a lighting system that incorporates indirect lighting (light is diffused and reflected off the ceiling from the light source), in which light level relies more heavily on the reflectance of the room surfaces.

Together, fixture dirt depreciation and room surface dirt depreciation can cause light level to wane by about 10% over time in extremely clean environments to as much as 50% or more in others. As in the case of lamp lumen depreciation, this again makes the cost of light go up, as light level decreases while input wattage remains the same. And again, the greatest cost of dirt depreciation is in lost human efficiency, accuracy, morale, safety, security and corporate image. The cost of these consequences of light loss is subtle and difficult to quantity, but is nonetheless present.

MAINTENANCE OPERATIONS AND METHODS

Performing a successful maintenance program requires attention to all appropriate safety rules and regulations; choosing the right equipment and using it correctly; and having knowledgeable, qualified and responsible people do the work. Only qualified personnel who are properly trained and who observe safety rules and regulations should be permitted to perform work on an electrical system.

Methods

Group relamping and cleaning strategies should be tailored to the environmental aspects of the application. Relamping and cleaning office fixtures requires a different approach than a parking lot, warehouse, retail store or factory. Most indoor maintenance operations follow several basic steps, using a two-person team:

1. The electrical circuit leading to the bank of fixtures being worked on must be turned off ("locked" and "tagged") in accordance with Federal Occupational Safety and Health Administration (OSHA) regulations. Live sockets present a shock hazard. If this is an incandescent system, time should be allowed for the lamps and fixtures to cool.

2. Shielding material, such as a plastic lens, is removed by the technician at the fixture and passed to the technician who is on the floor.

3. Lamps are removed and passed to the technician on the floor.

4. The technician at the fixture cleans the fixture. Concentrated dirt deposits are removed first with a cloth or vacuum cleaner. Then he or she applies an appropriate solution to a brush, sponge or cloth and wipes the fixture. The fixture is then given another wipe to remove any dirt or solution left.

5. During step 4, the technician on the floor cleans the shielding

material with an appropriate damp material or an appropriate machine such as an ultrasonic cleaner.

6. If the shielding material is not glass, destaticizing fluid is applied to retard dust accumulation.

7. The lamps are replaced. The technician on the floor passes up fresh lamps and then the cleaned shielding material for replacement. Only lamps that are compatible with the system should replace the old lamps. Lamps must be properly seated in the lampholders (sockets).

8. When finished with a system, turn the system on to make sure all of the new lamps are operating properly. Then return any unused new lamps to storage for future spot relamping needs (if applicable). Spot relamping needs can also be satisfied by saving several of the newest lamps that were removed during group relamping. These will be the lamps that have little or no blackening or banding at the ends of the lamp.

Dispose of spent lamps (and failed ballasts) in accordance with state and/or Federal guidelines.

Notes and variations:

1. If this were only a cleaning procedure, then the lamps would also be cleaned by wiping with a sponge during step 5. In step 7, these same lamps would be returned to the fixture for further service.

2. Before and during a relamping and cleaning operation, the fixture should be inspected for defective components such as broken sockets, loose wires, failed ballasts and cracked/yellowed plastic lenses or louvers. After the operation is over and the technicians are at a safe distance from the fixtures, the lighting system should be turned on and given a visual inspection to ensure that all lamps are operational. The system should also be inspected for signs of abnormal operation such as swirling and flickering.

Use the right cleaning method

Typical interior fixture surfaces include porcelain enamel and synthetic enamel. If reflectors have been installed as the chosen upgrade option, the surface may be anodized aluminum or silver film. Typical shielding materials include plastic, aluminum or glass. Before attempting to clean a fixture, become familiar with manufacturer's recommendations on suitable cleaning solution, wiping material and drying methods.

Use the right equipment

Several factors will determine what type of equipment should be used, including the type of system to be maintained, the type of maintenance performed and the environmental characteristics of the building (obstacles in the area and mounting height of the fixtures). Typical equipment used for reaching fixtures includes ladders, scaffolding, aerial platforms and bucket trucks. Disconnecting hangers and lamp changers can be used for relamping. And vacuum cleaners and ultrasonic cleaners can be used for cleaning fixtures and shielding materials.

MANAGING A MAINTENANCE PROGRAM

To get the most out of a lighting system through planned maintenance, conduct a four-step procedure. First, survey existing conditions. Determine relamping and cleaning intervals that will make the system most effective and economical. Establish a permanent maintenance policy. Finally, develop a strategy of implementation.

SURVEY EXISTING CONDITIONS

The first step is to examine the characteristics of the building and the lighting system. This information can be obtained by reviewing the survey forms completed in Chapter 2.

DETERMINE MAINTENANCE INTERVAL

The economic maintenance interval indicates the period at which scheduled maintenance work will economically ensure recommended illumination. Performing excessive maintenance work may result in consistent light levels above what is recommended; however, this can prove costly in labor and equipment. Conversely, skimping on maintenance to "save" money may result in the consequences wrought by low light levels and can be costly in terms of resources.

Average Life in Hours of Fluorescent Lamps at Various Burning Cycles

	Hours Per Start					
Lamp Type	3	6	10	12	18	Continuous
40W Preheat	15,000	17,500	21,200	22,500	25,000	28,100
40W Rapid Start	20,000+	24,400	27,700	28,800	31,600	37,700
High Output (HO)	12,000	14,000	17,000	18,000	20,000	22,500
Very High Output (VHO)	10,000	12,500	14,900	15,900	17,900	24,900
Slimline (96T 12)	12,000	14,000	17,000	18,000	20,000	22,500

Figure 5-10. As can be seen the table above, the higher the hours per start, the longer the reated life. An rapid start F40/CW lamp can be rated at nearly 40,000 hours if operated continuously. An important step in scheduling group relamping and cleaning operations is to determine rated life based on the hours per start of the application.
Source: Osram Sylvania, Inc.

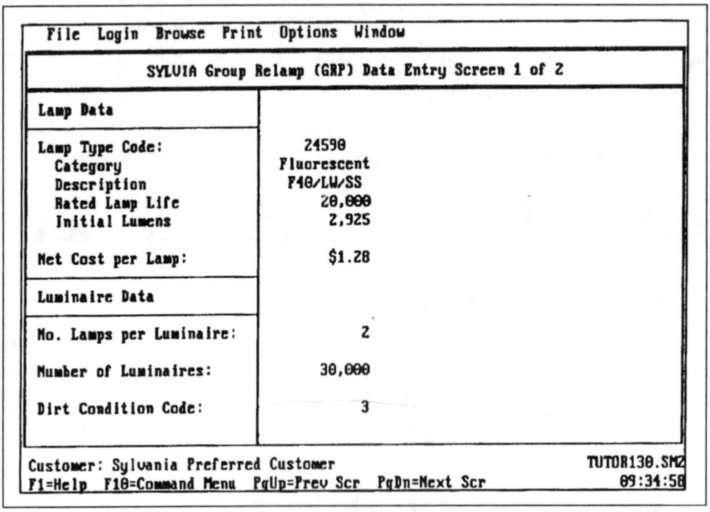

Figure 5-11. Some lighting upgrade analysis computer programs, such as Sylvia from Osram Sylvania, Inc., incorporate group relamping analyses. *Source: Osram Sylvania, Inc.*

To determine options, complete Maintenance Worksheet #1.

Complete Maintenance Worksheet #1

This worksheet allows the upgrade planner to determine up to three planned maintenance options based on group relamping and cleaning at 60%, 70% and 80% of rated lamp life. These intervals are expressed as years (and months) and the percentage values of life adjusted accordingly (e.g., 60% becomes 58%). Option 4 represents spot relamping and no cleaning for comparison purposes or actual selection if planned maintenance is not feasible.

Note that Maintenance Worksheet #1 does not incorporate room surface dirt depreciation. Typically, these surfaces are not cleaned as often as fixures, although some applications may warrant more frequent cleaning. The Worksheet assumes that room surface cleaning or repainting will be performed at 10-year intervals, which is a sensible practice for many applications.

Once this worksheet is completed, the upgrade planner can determine the lamp lumen depreciation and fixture dirt depreciation factors for the maintenance options (based on the operating characteristics of F40/CW lamps) and be prepared to determine the most economical option in Maintenance Worksheet #2.

Once the option is chosen, the appropriate light loss factors can be inserted into steps 10 and 11 in Chapter 3's Light Level Worksheet #2 on Page 76.

Maintenance Worksheet #1:
Determining Maintenance Interval, Page 1

Area description: WORD PROCESSING DEPT. Area No.: 4

Fixture type: FLUORESCENT RAPID START

IES maintenance category: V Dirt conditions: CLEAN

1. Rated life of lamps @ 15 hours per start	30,000	
2. Annual operating hours	÷ 3,900	
3. Years before rated life	= 7.69	
	x 0.6	
4. Maintenance interval at 60% of lamp life	= 4.61 years	
5. Round off to closest six-month interval	4.5 years x 12 =	54 months
6. Repeat maintenance interval at 60% of lamp life from **Step 4**	÷ 4.61 years	
7. Adjustment factor	= 0.98	
	x 60	
8. New maintenance interval as percentage of rated life	= 59 %	
9. Repeat years before rated life from **Step 3**	7.69	
	x 0.7	
10. Maintenance interval at 70% of lamp life	= 5.38 years	
11. Round off to closest six-month interval	5.5 years x 12 =	66 months
12. Repeat maintenance interval at 70% of lamp life from **Step 10**	÷ 5.38 years	
13. Adjustment factor	= 1.02	
	x 70	
14. New maintenance interval as percentage of rated life	= 71 %	

Maintenance Worksheet #1:
Determining Maintenance Interval, Page 2

Area description: WORD PROCESSING DEPT. Area No.: 4

15. Repeat years before rated life from **Step 3** 7.69

x 0.8

16. Maintenance interval at 80% of lamp life = 6.15 years

17. Round off to closest six-month interval 6 years x 12 = 72 months

18. Repeat maintenance interval at 80% of lamp life from **Step 16** ÷ 6.15 years

19. Adjustment factor = 0.98

x 80

20. New maintenance interval as percentage of rated life = 78 %

OPTION #1, "60%" of life: Group relamp and clean @ 59 % of life at 4.5 years (54 months); spot relamp early failures.

OPTION #2, "70%" of life: Group relamp and clean @ 71 % of life at 5.5 years (66 months); spot relamp early failures.

OPTION #3, "80%" of life: Group relamp and clean @ 78 % of life at 6 years (72 months); spot relamp early failures.

OPTION #4: Spot relamp all failures; no cleaning (assume 10-year fixture dirt depreciation factor).

	OPTION #1	OPTION #2	OPTION #3	OPTION #4
Lamp lumen depreciation factor	0.81	0.79	0.78	0.83
Fixture dirt depreciation factor	0.76	0.73	0.72	0.63

Complete Maintenance Worksheet #1 *(continued)*

Maintenance Worksheet #1 is a simplified method for determining viable planned maintenance intervals. First, write the fixture type, its maintenance category and the dirt conditions in the area.

1. Indicate the rated life of the lamps based on the hours-per-start interval.
2. Divide by the annual operating hours.
3. The result is the period of time before rated life, expressed as decimal years.
4. Multiply by 0.6 to determine the maintenance interval at 60% of rated lamp life, expressed in decimal years.
5. Round off to the nearest six-month interval in decimal years. Multiply by 12 to determine months, which will help in determining the fixture dirt depreciation factor.
6. Divide by the maintenance interval at 60% of lamp life from Step 4, expressed in decimal years.
7. The result is an adjustment factor.
8. Multiply by 60 to determine the new maintenance interval expressed as a percentage of rated lamp life, which will help determine an accurate lamp lumen depreciation factor.

Steps 9-20 repeat the process for 70% and 80% of rated lamp life. On Page 2, express the information obtained as options. On Page 3, determine the lamp lumen depreciation and fixture dirt depreciation factors for the options, adding a spot relamping/no cleaning option for comparison purposes.

Note that these are not the only options appropriate for every application. Extremely dirty environments such as factories with poor ventilation, for example, may require more frequent relampings and/or washings.

After the depreciation factors for the four options are determined, we are ready to determine the most economical option.

Choosing the most economical option

The formulas used to calculate annualized group relamping versus spot relamping costs are:

Group relamping with spot relamping of early failures:

A = (OPERATING HOURS ÷ OPERATING HOURS)
PER YEAR BETWEEN RELAMPINGS

B = (PERCENTAGE OF x NUMBER OF) x (LAMP COST + LABOR COST)
LAMPS FAILING LAMPS TO SPOT RELAMP
BEFORE GROUP 1 LAMP
RELAMPING

C = (LAMP COST, GROUP + LABOR COST TO) x NUMBER OF LAMPS
RELAMPING GROUP RELAMP
1 LAMP

ANNUALIZED COST ($) = A x (B + C)

Spot relamping:

AVERAGE
ANNUAL = (OPERATING HOURS ÷ RATED LAMP) x (LAMP + LABOR COST TO) x TOTAL NUMBER
COST ($) PER YEAR LIFE COST REPLACE 1 LAMP OF LAMPS

Cleaning. The basic formula used to calculate cleaning cost is:

CLEANING = TIME TO WASH x HOURLY LABOR x NUMBER OF FIXTURES
COST ($) 1 FIXTURE RATE ($) IN LIGHTED SPACE
(IN HOURS)

Now complete Maintenance Worksheet #2.

Maintenance Worksheet #2:
Maintenance Cost, Page 1

Area description: WORD PROCESSING DEPT. Area No.: 4

	Option #1	Option #2	Option #3	Option #4
SUPPORTING DATA				
1. Operating hours per year	3,900	3,900	3,900	3,900
2. Rated lamp life @ 15 per start				÷ 30,000
3. Variable A				= 0.13
4. Operating hours between group relamping intervals	÷ 17,700	÷ 21,300	÷ 23,400	
5. Variable B	= 0.22	= 0.18	= 0.17	
GROUP RELAMPING				
6. Labor cost to group relamp one lamp	$ 0.75	$ 0.75	$ 0.75	
7. Lamp cost to group relamp one lamp	+ 1.90	+ 1.90	+ 1.90	
8. Group relamping cost for one lamp	= 2.65	= 2.65	= 2.65	
9. Number of lamps	x 360	x 360	x 360	
10. Group relamping cost	=$ 954	=$ 954	=$ 954	
SPOT RELAMPING				
11. Percentage of lamps failing before group relamping	0.03	0.09	0.18	
12. Number of lamps	x 360	x 360	x 360	x 360
13. Number of lamps to be spot relamped	= 11	= 32	= 65	
14. Labor cost to spot relamp one lamp	$ 7.50	$ 7.50	$ 7.50	$ 7.50
15. Lamp cost to spot relamp one lamp	+$ 2.25	+$ 2.25	+$ 2.25	+$ 2.25
16. Spot relamping cost for one lamp	=$ 9.75	=$ 9.75	=$ 9.75	=$ 9.75
17. Number of lamps to be spot relamped from **Step 13**	x 11	x 32	x 65	x 360
18. Spot relamping cost				= $ 3,510
19. Variable A from **Step 3**				x 0.13
20. Final spot relamping cost	=$ 107.25	=$ 312	=$ 633.75	=$ 456.30

Maintenance Worksheet #2:
Maintenance Cost, Page 2

Area description: WORD PROCESSING DEPT. Area No.: 4

	Option #1	Option #2	Option #3	Option #4
CLEANING				
21. Labor cost to clean one fixture	$ 1.00	$ 1.00	$ 1.00	
22. Number of fixtures	x 90	x 90	x 90	
23. Total cleaning cost	= $ 90	= $ 90	= $ 90	
TOTAL COSTS				
24. Group relamping cost from **Step 10**	+ $954	+ $954	+ $954	
25. Final spot relamping cost from **Step 20**	+ $107.25	+ $ 312	+ $633.75	$456.30
26. Total cost	= $1,151.25	= $1,356	= $1,677.75	
27. **Variable B** from **Step 5**	x 0.22	x 0.18	x 0.17	
28. TOTAL ANNUALIZED COST	= $253.28	= $244.08	= $285.22	= $456.30

Completing Maintenance Worksheet #2

First, indicate the area's description and identification number.

To gain vital supporting data in the form of two variables — Variable A and Variable B, complete Steps 1-5:

1. Indicate the operating hours per year.

2. For the spot relamping option, divide by the rated lamp life at the actual hours per start.

3. The result is Variable A to be used in future calculations.

4. For the three planned maintenance options, divide the operating hours per year by the operating hours between the group relamping/cleaning intervals.

5. The result is Variable B.

To determine the total cost of group relamping over the life of the lamps, complete Steps 6-10:

6. Indicate the labor cost to group relamp one lamp by multiplying the hourly labor rate by the time to group relamp as a fraction of one hour. Suggest: 3 minutes, 1/20 of one hour.

7. Add this figure to the lamp cost to group relamp one lamp. Generally, the unit lamp cost for group relamping operations is relatively lower than for individual lamps purchased for spot relamping needs, due to volume discount. Obtain from the lamp supplier. Use the unit price for the lamp type in the existing system, although it may be replaced with a more-efficient lamp type; the most economical planned maintenance option can still be determined based on these prices, and after the retrofit option is chosen, new lamp prices can be applied and the form recalculated to determine actual maintenance savings.

8. The result is the group relamping cost for one lamp.

9. Multiply by the number of lamps in the system to be group replaced (number of fixtures x number of lamps per fixture).

10. The result is the total group relamping cost presented over the life of the lamps.

Steps 11-20 allow the maintenance planner to determine 1) the cost of spot relamping early failures for the group relamping options and 2) the total annualized cost of spot relamping the system for Option #4. The sequence for the spot relamping option is self-explanatory; here, we will briefly review the steps to determine the spot relamping cost component of the planned maintenance options:

11. Indicate the percentage of lamps failing before group relamping. Obtain from the mortality curve or from the lamp manufacturer.

12. Multiply by the number of lamps in the system.

13. The result is the number of lamps to be spot relamped as part of this option.

14. Indicate the labor cost to spot relamp one lamp by multiplying the hourly rate by the time to spot relamp expressed as a fraction of one hour. Suggest: 30 minutes, 1/2 of one hour.

15. Add this figure to the lamp cost to spot relamp. Obtain from the lamp supplier.

16. The result is the spot relamping cost for one lamp.

17. Multiply by the number of lamps to be spot relamped from Step 13.

Steps 18 and 19 are used for the spot relamping option (#4) to result in the annualized spot relamping cost.

20. The result of Steps 16-17 is the final spot relamping cost for the group relamping/cleaning options.

To determine the total cleaning cost, complete Steps 21-23:

21. Indicate the labor cost to clean one fixture by multiplying the hourly labor rate by the time to wash one fixture expressed as a fraction of an hour. When cleaning is combined with group relamping, suggest: 3-4 minutes, 1/15-1/20 of one hour.

22. Multiply by the number of fixtures to be cleaned.

23. The result is the total cleaning cost.

Complete Steps 24-28 to determine the total annualized costs for the group relamping/cleaning options:

24. Add the cleaning cost from Step 23 to the group relamping cost from Step 10.

25. Now add the result to the final spot relamping cost for this option from Step 20.

26. The result is the total maintenance cost.

27. Multiply by Variable B from Step 5.

28. The result is the total annualized cost. The lower the lamp cost, the higher the savings of group relamping/cleaning versus spot relamping will be. The economic benefit of group relamping is more apparent when replacing traditional F40/CWs than higher-priced lamps such as the T8 or T10 lamp. As these new lamps become more accepted by the market and supply increases, lamp prices are expected to drop. This will make group relamping F32T8s and other lamps even more economical in the future.

NEXT STEP: Now that the cost of each maintenance option has been determined, we can choose an option based on the light level to be maintained and economics. In our example, although group relamping and cleaning at 70% of life is more economical than at 60%, we will choose 60% to provide more flexibility in choosing a retrofit option. The cost difference of these two options is minor compared to the potential energy savings to be gained from a retrofit option that produces less light but also produces significant energy savings.

DETERMINE ENERGY MANAGEMENT OPPORTUNITY

In some situations, a good planned lighting maintenance program can be considered a low-cost energy conservation upgrade by itself. In a nutshell, planned lighting maintenance increases average maintained light levels, so these options are available: 1) energy-saving lamps can replace existing traditional lamps to reduce energy consumption while producing less lumen output, b) fewer fixtures/lamps can be designed into an original system, c) fewer fixtures/lamps need to be provided for in a redesigned system, d) lamps can be removed from a system that is being retrofitted, and e) equipment such as current limiters can be installed.

In terms of removing fixtures/lamps, planned lighting maintenance provides its most valuable support in dirtier environments and where redesign takes place.

Example: A retrofit of a factory's HID system results in the replacement of mercury lamps with high pressure sodium. Because high pressure sodium lamps produce more lumens than mercury lamps while requiring less energy, the factory's lighting system may have to be redesigned to account for fewer fixtures/lamps. If it were to be maintained on a scheduled periodic basis, even fewer fixtures/ lamps would be required to provide the necessary light levels, further saving installation and operating costs associated with those fixtures/lamps.

In other situations, however, planned maintenance may not provide a strong energy management opportunity beyond installing energy-saving lamps or current limiters. Fewer lamps may be required for an office lighting system, but if this quantity does not represent a uniform delamping, which lamps should be removed? If a system contains 100 fluorescent four-lamp fixtures and 20 lamps can be removed, which ones should go? Removing lamps from some fixtures but not others can affect appearance of the facility, occupant satisfaction with their lighting, and light distribution. It is generally desirable to consult with a professional before making a decision of which lamps, if any, should be removed.

ESTABLISH A PERMANENT MAINTENANCE POLICY

After the best maintenance option is chosen, a permanent maintenance policy should be established as part of corporate lighting energy policy. If a lighting system relies on planned maintenance to provide consistent proper light levels and later is denied the benefit of maintenance, it may continue to be energy-efficient but can become a failure as a lighting system.

DEVELOP A STRATEGY OF IMPLEMENTATION

Most companies attempt to economize on the cost of planned lighting maintenance by giving the responsibility to the current maintenance staff. However, like equipment specification and installation, maintenance should be performed by educated and trained personnel. For this reason and others, it is often more economical to contract a lighting management company for ongoing maintenance services.

FLUORESCENT LAMP AND BALLAST DISPOSAL

A lighting upgrade involving energy-efficient lamps and ballasts results in a large changeout of lamps and ballasts from the existing lighting system. Group relamping also results in scheduled large-scale replacement of lamps. Disposing of lamps and ballasts should be conducted in an environmentally responsible manner and in accordance with Federal and state guidelines.

FLUORESCENT LAMPS

Mercury has been targeted by the U.S. Environmental Protection Agency as a potential environmentally hazardous material. Mercury is an essential ingredient in fluorescent, mercury and metal halide lamp operation. No cost-effective or environmentally safe alternatives to mercury are currently available.

The electrical manufacturing industry has made progress toward reducing mercury content of fluorescent lamps. In 1985, the National Electrical Manufacturers Association (NEMA) reported that 48 milligrams were contained in an F40/CW lamp. By 1990, the content had been reduced to 41. NEMA estimates that by 1995, 27 milligrams will be the average content.

Disposal of lamps containing mercury became an issue in California in 1988, when the state banned land disposal of fluorescent lamps. As more states began examining the issue, the EPA undertook the question and found that fluorescent lamps passed the previously used Extraction Procedure (EP TOX) but failed the new Toxicity Characteristic Leaching Procedure (TCLP). Problems with the test procedure were ironed out by August 1992 and fluorescent lamps were found to consistently fail the TCLP — meaning that many types of fluorescent lamps are eligible to be classified as hazardous waste when disposed of in large quantities. This in turn means that large changeouts of lamps which typically accompany a retrofit can result in a company being designated a hazardous waste generator.

As of August 1992, the EPA had not issued a regulation. Currently, there are three options for lamp disposal. Landfilling and incineration are two, and recycling is the third. Recycling fluorescent lamps means separating the mercury from other components such as glass and aluminum endcaps. Glass can be remanufactured into non-food containers or used as filler in asphalt and cement. Aluminum endcaps can be recycled into other aluminum products. The mercury itself can be distilled and employed in thermometers and other products. According to the EPA, recycling a fluorescent lamp costs about $0.10 per foot, or $0.40 per four-foot lamp.

In August 1992, the EPA began considering requiring states to set up special collection programs for fluorescent lamps. These programs would be designed to reduce hazardous waste, regulatory paperwork and financial burdens on the generators (those disposing) of the lamps. They would also promote lamp recycling. The lamp generators would likely be charged a fee to use the collection services. The proposal was published in the Fall issue of the *Federal Register* and posted for public comment.

For more information, call the EPA's Hazardous Waste Hotline at (800) 424-9346.

Lamp recyclers include:

Lighting Resources, Inc.
Pomona, California
(800) 572-9253

Mercury Recovery Systems
Monrovia, California
(818) 301-1372

Mercury Refining Co. Inc.
Latham, New York
(518) 785-1703

Mercury Technologies Corporation
Benicia, California
(707) 745-5173

Mercury Technologies of Minnesota Inc.
White Bear Lake, Minnesota
(612) 426-2102

Quick Silver Products, Inc.
Brisbane, California
(415) 468-2000

Recyclights
Minneapolis, Minnesota
(612) 378-9571

Salesco Systems USA
Phoenix, Arizona
(800) 368-9095

Salesco Systems USA
Stoughton, Massachusetts
(617) 344-4074

The lamp recycling industry is growing along with the need for a suitable infrastructure to handle recycling of enormous quantities of fluorescent lamps.

FLUORESCENT BALLASTS

Polychlorinated biphenyls (PCBs) are a hazardous material which was once used widely as an insulator in electrical equipment such as fluorescent ballast capacitors. PCBs do not readily break down in the natural environment, but instead move up the food chain to humans. PCBs are a possible carcinogen and cause of developmental and reproductive defects in humans based on the results of tests with laboratory animals.

In 1979, the Federal Toxic Substances Control Act (TSCA) of 1976 went into effect. TSCA banned the manufacture, processing and distribution of PCB-containing ballasts. The Lighting Research Center estimates that 400 million fluorescent ballasts are installed across the U.S., although some ballast manufacturers offer estimates as high as 1.6 billion. According to FulCircle Ballast Recyclers, about half of these (200-800 million) are PCB-containing ballasts.

How to identify PCB ballasts

PCB ballasts are identifiable because of the lack of the notice "NO PCBs" on the label. (Non-PCB ballasts bear "NO PCBs" on their labels.) If PCB ballasts are encountered, they must be disposed of in accordance with Federal and state guidelines.

TSCA regulations for leaking ballasts

According to the EPA, if the ballast is leaking PCBs or has been punctured, disposal is governed by the TSCA. The ballast must be incinerated

Figure 5-12. Components of a magnetic ballast. In a PCB ballast, the capacitor and sometimes the asphalt is contaminated with PCBs. This means that the ballast must be properly disposed in an environmentally responsible manner in accordance with state and Federal guidelines. *Source: FulCircle Ballast Recyclers.*

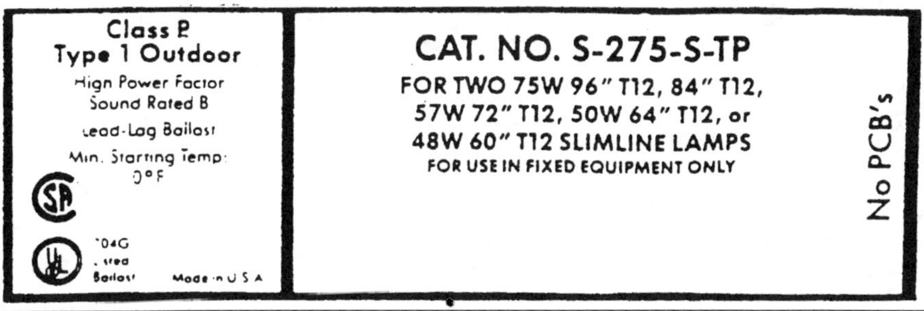

Figure 5-13. PCB ballasts are identifiable because of the lack of the notice "NO PCBs" on the label. In the generic ballast label shown above, notice is given that this ballast does not contain PCBs. *Source: FulCircle Ballast Recyclers.*

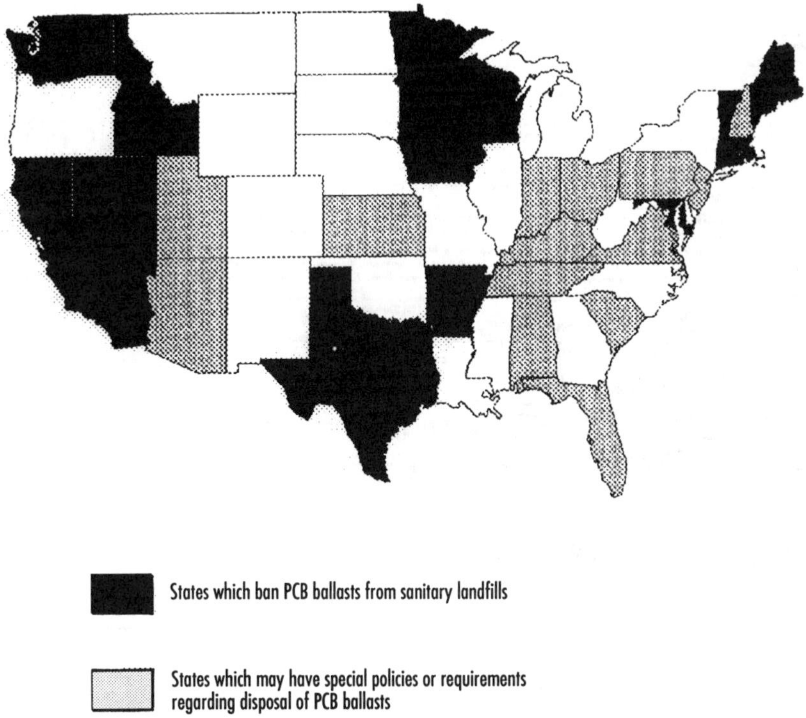

Figure 5-14. States which banned PCB ballasts from sanitary landfills as of 1991/1992. *Source: FulCircle Ballast Recyclers.*

UNIFORM HAZARDOUS WASTE MANIFEST

1. Generator's US EPA ID No. 40 C F R P A R T 7 6 1 | Manifest Document No. 0 0 0 1 4 | 2. Page 1 of 6 | Information in the shaded areas is not required by Federal law.

3. Generator's Name and Mailing Address: ABC Facility Owner, 123 Energy Street, Anytown, USA 99999

4. Generator's Phone (212) 580-5550

A. State Manifest Document Number

B. State Generator's ID

5. Transporter 1 Company Name: XYZ Trucking Company | 6. US EPA ID Number: NY D 9 7 7 8 2 2 1 1 1

C. State Transporter's ID

D. Transporter's Phone 212-267-4394

7. Transporter 2 Company Name: N/A | 8. US EPA ID Number

E. State Transporter's ID

F. Transporter's Phone

9. Designated Facility Name and Site Address: FulCircle Ballast Recyclers, 509 Manida St., Bronx, NY 10474 | 10. US EPA ID Number: NYD 9 8 6 9 8 0 2 3 3

G. State Facility's ID: 200000044

H. Facility's Phone: 212-328-4667

11. US DOT Description (Including Proper Shipping Name, Hazard Class, and ID Number)	HM	12. Containers No.	Type	13. Total Quantity	14. Unit Wt/Vol	I. Waste No.
a. RQ, Polychlorinated Biphenyls, 9, UN2315, PG II (EPA Waste Code PCB-2) (ERG-31)	X	2 5	D M	9 0 0 0	K	
b.						
c.						
d.						

J. Additional Descriptions for Materials Listed Above: old DOT description: RQ, Hazardous Substance, Solid, N.O.S. (Polychlorinated Biphenyls) NA 9188

K. Handling Codes for Wastes Listed Above

15. Special Handling Instructions and Additional Information: Used fluorescent light ballasts contain non-leaking, small capacitors, which are not regulated for disposal pursuant to 40 CFR 761.60 (b)(2)(ii). Ballasts are being shipped for capacitor removal, metals reclamation and recycling.

16. GENERATOR'S CERTIFICATION: I hereby declare that the contents of this consignment are fully and accurately described above by proper shipping name and are classified, packed, marked, and labeled, and are in all respects in proper condition for transport by highway according to applicable international and national governmental regulations.

If I am a large quantity generator, I certify that I have a program in place to reduce the volume and toxicity of waste generated to the degree I have determined to be economically practicable and that I have selected the practicable method of treatment, storage, or disposal currently available to me which minimizes the present and future threat to human health and the environment; OR, if I am a small quantity generator, I have made a good faith effort to minimize my waste generation and select the best waste management method that is available to me and that I can afford.

Printed/Typed Name: John Smith | Signature | Month Day Year

17. Transporter 1 Acknowledgement of Receipt of Materials

Printed/Typed Name: Robert Jones | Signature | Month Day Year

18. Transporter 2 Acknowledgement of Receipt of Materials

Printed/Typed Name: N/A | Signature | Month Day Year

19. Discrepancy Indication Space

20. Facility Owner or Operator: Certification of receipt of hazardous materials covered by this manifest except as noted in Item 19.

FulCircle Ballast Recyclers

Printed/Typed Name: Brin McCass | Signature | Month Day Year

Figure 5-15. A Uniform Hazardous Waste Manifest.
Source: FulCircle Ballast Recyclers.

or treated according to more stringent state guidelines if applicable. Call the EPA's TSCA Hotline at (202) 554-1404 for more information.

CERCLA regulations for non-leaking ballasts

According to the EPA, if 12-16 ballasts are not leaking but are to be disposed of together in a 24-hour period, the company is required by the

Comprehensive Environmental Response, Compensation and Liability Act (CERCLA) to report such disposal to the National Response Center at (800) 424-8802. Companies which throw away PCB ballasts expose themselves to CERCLA (Superfund) liability. The EPA encourages business and industry to dispose of PCB ballasts in a high-temperature incinerator facility or chemical waste landfill. Ballasts must be packed in an approved 55-gallon drum to be transported; one drum holds between 150 and 300 ballasts depending on their size and how tightly they are packed (tightly packed drums decrease total cost of disposal).

State guidelines

Some states, such as California, impose regulations more stringent than Federal requirements. As of June 1992, seven states treated PCB ballasts as hazardous waste and regulated them as such: California, Connecticut, Maryland, Maine, Minnesota, Vermont and Wisconsin. These states require companies disposing of PCB ballasts to notify the state environmental or other appropriate office of any PCB ballast disposal and to obtain identification numbers. These numbers are assigned to a uniform hazardous waste manifest, a form used by disposal firms and licensed waste carriers to record the chain of custody as the waste moves from the company to where it will finally be disposed. For information, call the state environmental office or regional EPA office.

Disposal options

There are six options for disposal of non-leaking PCB ballasts:

1. Landfill the ballasts. Many sanitary landfills will not accept PCB ballasts because they are usually not designed to accept this waste. Disposal in a sanitary landfill may expose both the landfill and the company to Superfund liability.

2. Incinerate in a municipal solid waste incinerating facility. Most municipal incinerating plants are not designed to handle ballasts well; the result usually is contaminated ash containing PCBs, dioxins and unburned metal fragments. This option can be expensive.

3. Dispose in a chemical waste landfill. Chemical waste landfills are designed to handle hazardous waste such as PCBs. This option is also relatively inexpensive (about $1 per ballast, not including packaging, transportation and profile fees). However, even the most highly secured landfills have been shown to suffer leaks of fluid, making this option expose the company to possible future Superfund liability.

4. Incinerate in a TSCA-approved facility. This option provides peace of mind that the PCBs are completely destroyed. It is, however, often the most expensive option (about $5-6 per ballast, not including

packaging, transportation and profile fees).

5. Recycle the ballasts. All PCB-containing components and material — the capacitor and asphalt potting — are removed and incinerated. The rest of the ballast, including steel and copper (about 80% of its weight) is recycled. This option has been estimated to be about 30-50% less expensive than whole-ballast incineration.

Here is a list of several ballast recyclers:

Eastern Environmental Technologies
Norwalk, Connecticut
(203) 856-2014

Ensquare, Inc.
Newton Upper Falls, Massachusetts
(617) 969-9238

Environmental Energy Group
Denton, Texas
(817) 383-3632

FulCircle Ballast Recyclers
Cambridge, Massachusetts
(617) 876-2229

Bronx, New York
(212) 328-4667

Salesco U.S.A.
Arizona
(800) 368-9095

It is generally desirable to keep permanent records of PCB ballast disposal activities.

Sources:

Dong, Mitchell and Brin McCagg, "When Retrofitting, Recycle PCB Ballasts: Part I," *MainLighter: The Lighting Management & Maintenance Magazine*, March 1992.

Dong, Mitchell and Brin McCagg, "When Retrofitting, Recycle PCB Ballasts: Part II," *MainLighter: The Lighting Management & Maintenance Magazine*, April 1992.

Foundations of Lighting Management. Somerset, New Jersey: Philips Lighting Company, Training program cosponsored with the interNational Association of Lighting Management Companies, January 28-30, 1991.

Fundamentals of Commercial and Industrial Lighting, Fifth Edition. Washington, DC: The Electrification Council, 1983.

Getting the Most From Your Lighting Dollar, Second Edition. Washington, DC: National Lighting Bureau.

interNational Association of Lighting Management Companies, Princeton Junction, New Jersey.

Lighting Handbook. New York, New York: Illuminating Engineering Society of North America, 1987.

Lighting Management Seminar. Danvers, Massachusetts: Sylvania Lighting Division, Training program cosponsored with the interNational Association of Lighting Management Companies, February 19-21, 1992.

Lighting Upgrade Manual, First Edition. Washington, DC: U.S. Environmental Protection Agency, Green Lights Program, November 1991.

Lighting Upgrade Waste Disposal. Washington, DC: U.S. Environmental Protection Agency, Green Lights Program, February 1992 (addendum to *Lighting Upgrade Manual*).

Lindsey, Jack. *Applied Illumination Engineering*. Lilburn, Georgia: The Fairmont Press, 1991.

Profiting from Lighting Modernization. Washington, DC: National Lighting Bureau, 1987.

Sales, Frank and Lynette Gehring, "A 'Hands-On' Guide to PCB Ballast Disposal," *MainLighter: The Lighting Management & Maintenance Magazine*, June 1992.

Victor, Jeff, "The Economics of Planned Maintenance," *Lighting Management & Maintenance*, July 1992.

Walitsky, Paul and Peter Bleasby, "Classification and Disposal of Lamps Containing Mercury: Tackling the Issue," *Lighting Management & Maintenance*, August 1992.

Further reading:

Controlling Electrical Hazards. Washington, DC: Occupational Safety and Health Administration. Order Number 029-016-00126-3.

Keck, Merle E., "Streetlighting Maintenance Changes Emphasis," *MainLighter: The Lighting Management & Maintenance Magazine*, September 1991.

Roth, Peter, "Repairing HID Fixtures," *MainLighter: The Lighting Management & Maintenance Magazine*, June 1992.

Strunk, Dorothy L, "Lockout/Tagout: A Sensible Combination to Save Lives," *Lighting Management & Maintenance*, July 1992.

Zarlenga, James W., "Reflector Retrofits: What You Should Know About Installation, Maintenance, Warranties," *MainLighter: The Lighting Management & Maintenance Magazine*, April 1992.

Source: National Lighting Bureau

CHAPTER 6

Retrofit Financing: Putting Money to Work

HOW TO PAY FOR YOUR LIGHTING UPGRADE

A lighting upgrade is a relatively low-risk investment typically bearing high financial rewards (which first pay for the cost of the new equipment). Some corporations, however, balk at directly purchasing the equipment upfront, preferring the status quo where they pay more — but in monthly increments — to the utility on the electric bill. At the same time, corporations require a minimum payback and internal rate of return on investment for energy efficiency projects. Meeting this figure (the Present Value Discount Rate) is no guarantee of approval of an upgrade, but not meeting it usually gets an automatic "no."

Corporations seeking to upgrade their lighting system with less risk and lower initial investment can turn to several sources for financial assistance: the local electric utility, conventional sources (the bank), a third party such as the installation contractor or an energy service company (ESCO), or a lighting manufacturer.

OVERVIEW OF UTILITY-SPONSORED DEMAND SIDE MANAGEMENT PROGRAMS

Utilities have positioned themselves as a major force promoting and accelerating the conversion of buildings to energy-efficient lighting systems. They are doing so through demand side management (DSM) programs.

WHAT IS DSM?

Thomas Morron, vice president of customer service and marketing for the Edison Electric Institute, describes DSM as "all electric utility customer service and marketing activities that influence how customers use electricity."

The Utility Data Institute indicates that customers are being influenced to reduce use of electricity: "Regardless of their size, service territory or corporate structure, American electric and gas utilities are embarking on massive programs to optimize and reduce the demand for their product."

Numerous forces are compelling many electric utilities to consider demand reduction as an energy resource. Encouraging corporations to reduce demand can take the form of direct cash rebates. These programs, highly successful, are growing in terms of number of utilities participating and reach of programs. The Electric Power Research Institute surveyed 267 electric utilities and their 628 commercial DSM programs in October 1991. The respondents were asked to provide information about DSM in their service territories:

Table 6-1. Focus of DSM program frequency by end-use.

HVAC	41%
All end uses	26%
Lighting	26%
Demand control	16%
Building envelope	13%
Water heating	13%
Motor drive	10%
Refrigeration	6%
Cooking	4%
Power conditioning	2%

Source: Electric Power Research Institute, 1991.

Table 6-2. Targeted customers in 1991 commercial DSM programs.

Lodging facilities	65%
Any commercial building type	63%
Large offices	19%
Retail establishments	18%
Small offices	15%
Elementary/Secondary schools	15%
Groceries	14%
Health care facilities	13%
Colleges	11%

Source: Electric Power Research Institute, 1991.

Table 6-3. Programs targeting retrofit, new construction or renovation markets.

Retrofit	90%
New construction	66%
Renovation	63%

Source: Electric Power Research Institute, 1991.

The Oak Ridge National Laboratory completed their own study to find out what common energy efficiency upgrades were encouraged by electric utility DSM programs:

Table 6-4. Measures commonly implemented in DSM programs.

COMMERCIAL CUSTOMERS
- Cooling with condensing water
- Energy-efficient lighting
- Refrigeration with heat recovery
- Cool storage
- Energy-efficient motors
- Workshops to promote energy efficiency
- HVAC maintenance
- Load reduction devices and actions

INDUSTRIAL CUSTOMERS
- Energy-efficient lighting
- Gaining "interruptible" customers
- Energy-efficient motors

Source: Oak Ridge National Laboratory, 1990.

Energy-efficient lighting, because it poses a faster payback and higher return on investment than most other building upgrades, is featured prominently in most DSM programs. And because the majority of large energy users are existing buildings, retrofit is the primary focus of DSM.

WHY UTILITIES ARE PROMOTING DSM

Electric utilities are promoting DSM primarily as a way to defer building new generating capacity. Rising energy demand in energy-strapped regions such as California and the Northeast has left many utilities hard-pressed to meet that demand. Due to the relatively high cost of new construction, environmental concerns over nuclear power/fossil fuel combustion, and Not In My Backyard (NIMBY) opposition to new power plants, hundreds of utilities have found it cheaper to encourage their customers to become permanently energy-efficient than to build new power plants. These programs initially took the form of advertising, but low response prompted more dramatic action. Electric utilities began offering financial incentives, sometimes in the form of a direct rebate on purchased energy-efficient equipment. Nearly 70% of current cash rebate DSM programs were not initiated until after 1987.

Utilities are finding other benefits of DSM, including load management — load shifting, peak clipping and valley filling. Additional benefits include reduced critical fuel usage, ability to evenly distribute system demands, reduced capital requirements and improved short-term financial performance.

In addition these benefits, utility-sponsored DSM is prompted by external forces acting on the marketplace, including state-mandated least-cost resource planning, the Clean Air Act amendments of 1990, and state-offered financial incentives.

Least-cost planning

According to the U.S. Environmental Protection Agency, least-cost planning, or integrated resource planning, refers to "least-cost energy conservation and electric power planning methodology employed by an electric utility that evaluates the full range of existing and incremental resources in order to meet expected future demand at lowest system cost."

Utilities are increasingly using least-cost planning for short-term (typically a period less than five years long) to long-term supply projections. Load growth is forecasted and all options considered for meeting this demand are ranked in order of cost-effectiveness. Based on this information, the utility develops a short-term and long-term plan for meeting future energy demand using reliable resources which pose the lowest cost. The short-term plan is submitted to the state public utility commission, while the long-term plan represents the utility's anticipations.

DSM programs are respected as a legitimate resource in least-cost planning. Strictly screened, they often rank high on the benefits/cost-effectiveness scale, making them attractive to utilities engaged in least-cost planning. And they are integrated into the planning process along with acquisition of new supply resources as a means to meet load growth forecasts.

Table 6-5. Comparative costs for electric power plant options versus demand side measures.

Gas turbine (peaker)	$1,500/kW
Oil-fired	$2,500/kW
Coal-fired	$4,000/kW
Nuclear	$3,000-6,000/kW
Demand side measures	$1,000/kW

Source: EUA Cogenex Corporation.

Environmental costing and the Clean Air Act

Many states are beginning to require that utilities internalize environmental costs — called "externalities" — into the least-cost planning process. By assigning a dollar cost to air pollution produced by new generating capacity, new supply-side options look less attractive compared to DSM measures, which prevent pollution rather than produce it. (Chapter 7 covers the relationship between energy production and the environment.)

In 1989, former U.S. Representative Richard Ottinger, now director of the Pace University for Environmental Legal Studies, attempted to quantify the environmental costs associated with electric generation. He found that each delivered kWh from existing coal-fired power plants presents an added environmental cost of 6.8 cents per kWh. For oil, he determined a cost of 3 to 7.9 cents. These figures are costs assigned to fish killed from acid rain, lost coastal area due to global warming, increased health care costs due to smog, etc. He even estimated a cost of 2.91 cents per kWh for nuclear power, which accounts for the costs of routine emissions, additional funds for decommissioning and the risk of accidents during operation.

New York State was the first state to formally include environmental costs to its utilities' resource-bidding process. An environmental cost of 1.4 cents per kWh was assigned to coal-fired power plants. The state levied $1,959 per ton for nitrogen oxides and $0.98 per ton for carbon dioxide. Massachusetts soon joined New York and levied the steeper value of $6,792.50 per ton of nitrogen oxides and $22.99 per ton of carbon dioxide. The State of Nevada Public Service Commission, the Long Island Lighting Company, Wisconsin Electric Power Company, the California Energy Commission, the California Public Utilities Commission, the South Coast Air Quality Management District and the Washington State Energy Office are also assigning environmental costs to electric generation. At the same time, those states are rewarding utilities with credits for deferred emissions.

As this trend develops, the Clean Air Act amendments of 1990 are beginning to take effect. The Clean Air Act seeks to cut sulphur dioxide (SO_2) emissions in half by the year 2000. The amendments set up an innovative system for levying a fine on sulphur dioxide emissions above EPA-prescribed allowances for each utility. They also allow utilities who produce below the limit to receive avoided-emissions credits which they can sell or trade auction-style to other utilities via commodity exchanges or other methods. Each ton of sulphur dioxide avoided gains one credit from the Conservation and Renewable Energy Reserve of 300,000 credits. In 1992, for example, the Tennessee Valley Authority purchased emissions credits for 10,000 tons of sulphur dioxide, at $250-$300 per ton or a total of $2.5-$3 million, from Wisconsin Electric.

DSM helps utilities reduce production of sulphur dioxide, keeping them on track to scale down their emissions according to the Clean Air Act schedule. It also gives them a commodity they can sell to other utilities which do not comply.

Financial incentive mechanisms set up by states

According to the Edison Electric Institute, annual net kWh sales have been reduced by 1.3% through DSM up until 1991. Annual net sales are expected to decrease by 3% by the year 2000.

Encouraging customers to buy less electricity makes poor economic sense for a utility in the long term. DSM presents several economic benefits to the utility, most notable of which is the deferred expense of new generating capacity. However, DSM presents a cost to the utility, albeit a smaller one, but nonetheless a cost that must be paid. In the long term, the utility (and ultimately the consumer) will have to pay for DSM.

This is why utilities, state regulatory commissions and public-interest groups are taking a hard look at incentives to make DSM more profitable for utilities. In 1992, utility commissions in 25 states allowed utilities to recover lost revenues for DSM expenses.

The several types of incentives include direct compensation for lost revenues due to reduced energy use (which may include some of the cost of the DSM program itself), and shared savings agreements.

CURRENT PROGRESS AND FUTURE OUTLOOK FOR DSM

More than 21,000 megawatts of new capacity additions have been deferred since 1977 through energy efficiency/conservation measures.

According to a Department of Energy study, utilities invested $840 million in DSM programs in 1989, saving 5,000 megawatts through energy efficiency/conservation projects, the equivalent of the capacity of some 100 peaking plants. (Other demand side measures, such as direct load control and interruptible load, cut peak load by an additional 6,000 megawatts, or 120 plants). Energy savings in 1989 totaled some 15 billion kWh.

In 1990, utilities invested $1.2 billion and saved 6,000 megawatts, or about 120 peaking plants, through energy efficiency/conservation projects. Another 11,000 megawatts, or 220 plants, were saved through other demand side measures. Energy savings in 1990 totaled some 17 billion kWh.

In 1991, total spending on DSM reached $2 billion. The average utility spent more than 2% of its revenues on DSM. The New England Electric System, representing a group of utilities, spends more than 5% of its revenues on DSM. In the future, more and more utilities are expected to invest 5% or more of their revenues in DSM programs.

The future outlook for DSM shows growth for the rest of the decade, according to the DOE survey and the Electric Power Research Institute (EPRI). EPRI estimates that utility-sponsored DSM programs are expected to reduce summer peak demand by 6.7%, or another 45,000 megawatts of load, from a DSM investment of $30 billion.

An Oak Ridge National Laboratory survey of 23 utilities indicates that by the end of the decade, DSM programs will represent 21% of resource additions to peak capacity (see Table 6-6). DSM will account for 7% of total resources (4% of total generation).

Table 6-6. Utility-planned resource additions by the year 2000.

Demand side management	21%
Purchases from utility and non-utility resources	44%
New utility-owned generation	35%

Source: Oak Ridge National Laboratory, 1990.

Other future DSM developments include coverage of more technologies and end uses, increased utility services and more emphasis on energy efficiency/conservation projects.

TYPES OF UTILITY-SPONSORED DSM PROGRAMS

Your utility may offer any number of types of DSM programs to help you finance your lighting upgrade, including cash rebates, direct assistance and low-interest loans. As stated, utility financing provides less investment risk for the customer to purchase new energy-efficient lighting equipment, as it carries some — or sometimes all — of the investment burden.

Besides the utility incentives listed here, the utility may also provide help with energy auditing, developing an energy management policy and plan, assisting with efficient lighting design, selling energy efficiency projects to top management and measuring and monitoring energy reduction (as well as other aspects of equipment performance).

REBATES

Receiving a utility rebate means that the company is reimbursed for part or all of the cost of the efficiency upgrade. The structure of the rebate may be based on purchasing qualifying technologies, reducing demand for power and/or reducing energy consumption. In the simple payback formula, simply deduct the total rebate from the total installation cost.

Technology-based

Rebates are often technology-based, meaning that if the company purchases qualifying technologies (a compact fluorescent lamp, for example), it will receive a reimbursement for part of the cost. This reimbursement can be 30 – 50% of the installed cost. Technologies typically covered include energy-efficient lamps, efficient ballasts, occupancy sensors, reflectors, compact fluorescents, HID systems and daylighting controls such as photocells/dimming systems.

Energy reduction

Many utilities are retreating from the equipment-specific rebate. The problem with this rebate is that customers are encouraged to purchase only certain equipment, which sometimes may not be beneficial to the application. In some instances, it also discourages creativity. The energy reduction rebate pays the customer a cash amount for every kW removed from load or kWh of energy consumption reduced. Other energy reduction rebates offer discount electric rates as the incentive.

Table 6-7 shows a list of utility rebate programs published by *Energy User News* in April 1992. Note that by the time of this book's printing, a sizeable number of these programs may have been altered; others may have been withdrawn to be reintroduced later.

Table 6-7. Utility rebate programs. Source: *Energy User News*, April 1992.

UTILITY	EQUIPMENT	CONDITIONS	INCENTIVES
Arizona Public Service Co. P.O. Box 53999 Phoenix, AZ 85072 (602) 250-3539, AZ	• Cool Thermal Storage • Efficient Lighting	Incentive based on average kw reduction 3:00-6:00 P.M. summer weekdays. Half incentive paid upon completion of storage tank, remainder when installation is operational.	COOL THERMAL STORAGE: $125/kw shifted for customers on non-time-of-day rate. $75/kw shifted for customers on time-of-day rate. EFFICIENT LIGHTING: Call for specific rebate amounts.
Alantic Gas Light Co. P.O. Box 4569 Atlanta, GA 30302 (404) 584-3846, GA	• Commercial Cooking Equipment	Pays dealer, who must be authorized sales agent who displays, stocks, and sells NG commercial cooking equipment and must sign the dealer annual agreement. End user must be our customer. Equipment must be used for commercial purposes. Installations must be verified by us.	Pays dealer sales rep $31/1,000 Btu input; if appliance replaces electric equipment, pays $1/1,000 Btu input for each appliance sold to the dealership. For counter equipment, pays $1/1,000 Btu input to both dealer and dealer sales rep. If the counter equipment replaces electric, the programs pay both the dealer and dealer sales rep each $2.00/1,000 Btu input sold.
Atlantic Electric P.O. Box 1264 Pleasantville, NJ 08201 (609) 645-4100 Southern 1/3 of N.J.	• Lighting • Air conditioning • Timelocks • Efficient Motors • Energy Management Systems • Other	Two programs. For commercial conservation program, CACS audit required, items required, items must have 7-year payback, max of $5,000 per metered account. For commercial/industrial lighting program, verification of replacement required by inspection.	COMMERCIAL CONSERVATION INCENTIVE PROGRAM: Up to 20 percent of cost of item, as recommended by audit. All bldgs. at least 2 years old with less than 4,5000 mBtu of energy use are eligible. COMMERCIALINDUSTRIAL ENERGY — EFFICIENT LIGHTING PROGRAM: 10¢ per watt saved with a life of 10,000 hours or greater. for equipment with life of 2,000-9,999 hours, 5¢ per watt saved. $4 per energy-efficient ballast installed.

UTILITY	EQUIPMENT	CONDITIONS	INCENTIVES
Austin, City of, Environmental & Conservation Services Dept. 206 E. 9th, 17-102 Austin, TX 78701 (512) 499-1859	• Lighting • Air Conditioning • Window, Roof Treatments • Refrigeration • Motors • Thermal Storage • Custom (large-scale) • Energy Management Systems	Minimum payment $25, maximum $150,000 per participant. Pre-and post-installation inspections required. Programs subject to change without notice.	LIGHTING: fluorescent relamping: 50¢-$1/lamp. Incandescent to fluorescent: $2-$15/fixture. Incandescent to HID fixture:$20/fixture. Ballast retrofits: $6-$11/high-performance ballast, depending on wattage. Reflectors: $7-$14/fixture. Occupancy sensors: $5-$32/unit, depending on wattage. WINDOW, ROOF-CEILING TREATMENTS: Payment per sq. ft., contact utility. AIR CONDITIONING: $200 or $300 per kw/reduced, depending on type of equipment, tonnage. MOTORS: Rates vary according to size efficiency. REFRIGERATION: Strip curtains: $1-$2.50/sq. ft., installed. Display Case Doors: $15-25/linear ft. Mechanical subcooling: $100-$150/ton. Evaporative condensors: $100-$200/ton. THERMAL STORAGE: $300/kw shifted off peak.
Baltimore Gas & Electric P.O. Box 1475 Baltimore, MD 21203-1475 (301) 234-5000 Baltimore, 5 adjacent counties	• Indoor Lighting • Thermal Storage • Commercial Construction • High efficiency Motors • High efficiency Compressors	10-year "Conserve 2000" conservation and load management program, also includes standby generation program, curtailable rates, financing for energy improvements, radio control of AC. New construction and major retrofit can qualify for design and construction incentives based on kwh saved. Variable rebates by size for incremental costs of high efficiency motors.	LIGHTING: BG&E pays up to 40% of installation cost of efficient lighting retrofit-will finance remaining 60% on shared savings. New construction, util. pays $200/deferred kw. THERMAL STORAGE: ("Cool Reserve"): $200/kw shifted off-peak. Also, BG&E pays up to $15,000 on feasibility study b P.E. COMPRESSORS: $41/scfm for 50-125 hp.
Benton County PUD P.O. Box 6270 Kennewick, WA 99336 (509) 582-2175 in a joint program with City of Richland P.O. Box 190 Richland, WA 99352	• Lighting • Insulation • Windows • HVAC	New construction or retrofit. Verification of payment required.	ENERGY SMART DESIGN: 70-100% of incremental cost of measures beyond code or current practices in new construction. 50-70% of installed cost for existing buildings. Equipment rebates available or custom payments based on energy audit.
Bonneville Power Administration (Commercial Programs) P.O. Box 3621-RMCC Portland, OR 97230-3621 (503) 230-5873, OR, WA, ID, and western MT	• Lighting • HVAC • Motors • Envelope Measures	New construction or retrofit. Equipment rebates available or custom payments based on energy audit. Verification of payment required.	ENERGY SMART DESIGN: 70-100% of incremental cost of measures beyond code or current practices in new construction. 50-70% of installed cost for existing buildings. Programs vary by member utility.
Bonneville Power Administration (Industrial Programs) P.O. Box 3621-RMIC Portland, OR 97208-3621 Call local offices: Puget	• Motors • Energy Review • Custom	New construction or retrofit. Clearance required before hiring consultant for energy review. All programs limited to customers of utilities that buy firm power from BPA.	MOTORS: 20% of purchase price on all motors 1 hp and up. ENERGY REVIEW: Pay all costs. CUSTOM: On measures with 15-year life, up to 80% of project cost and 15¢/kwh. Payments vary according to project life. Other incentives available. contact local area office for details.

Sound: (206) 553-1122 Snake River: (509) 353-2904 Upper Columbia (509) 353-2904; Lower Columbia (503) 230-4045			
Boston Edison Co. 800 Boylston St. Boston, MA 02199 (617) 424-3377 Greater Boston area	Energy Efficiency Partnership: • Lamps • Exit Sign Kits • HVAC Improvements • Refrigeration • Industrial Process • Motors • Other	For C&I retrofits with demand over 150 kw Edison provides technical review, assists customer. Audit proposal needed, contractor selected through bidding. For equipment replacement-remodeling program, Edison offers technical and design assistance, tabulates incentive based on established baseline, inspects installation. New construction, Edison contracts with outside firms to review designs, recommend measures. Also small C&I retrofits.	LARGE C&I RETROFITS: Maximum incentive equal to project cost less first year's savings. SMALL C&I RETROFITS: Cost-effective improvements provided at no cost to customer. EQUIPMENT REPLACEMENT-REMODELING: Incremental cost of efficiency upgrades. NEW CONSTRUCTION: Amounts up to incremental cost.
Boston Gas Co. 201 Rivermoor St. Boston, MA 02132 (617) 323-9210, ext. 344 Eastern MA	• Gas-fired Air Conditioning • Cogeneration • Heating • Controls • Water Heating • Insulation • Heat Recovery Equipment • High Efficiency Furnaces	Must be Boston Gas customer. Gas air conditioning and cogeneration equipment must be ordered before December 31, 1992. Cogen must expect to run minimum 3,000 full load hours. Gas conservation measures must be cost effective.	GAS-FIRED AIR CONDITIONING: $100/ton, financing of 50% of feasibility studies available at our discretion. Free feasibility studies performed by utility. $50,000 cap. COGENERATION: $75-$200/kw depending on expected full load hours. Financing of 50% of feasibility study available at our discretion. $210,000 cap. GAS CONSERVATION: 100% of cost measures for customers using 500,000 cf/yr or less. For customers using more than 500,000 cf/yr bring measures down to one-year payback. Additional subsidies available for non-profit and charitable organizations. Audits and engineering studies free if measures are implemented.
Burlington Electric 585 Pine St. Burlington, VT 05401 (802) 658-0300	• Lighting • Other Equipment	"Energy Advantage" lighting program can be either rebate or low-interest loan program where user pays back 75% of energy savings, max. of 5 years. Utility will write off any additional amount.	LIGHTING: Custom program rebate amount depends on kwh savings, cost of equipment, correlates to loan amount. Program will be expanded to include motors, HVAC, refrigeration. TOP 10 PROGRAM: For 15 largest customers only, any cost-effective custom.
Central Hudson Gas & Electric Corp. 284 South Avenue Poughkeepsie, NY 12601 (914) 486-5514 Mid Hudson Valley	• Lighting • Motors • Electric & Gas Air Conditioning	Preapproval required; call (800) 825-8669. For current program, all work must be completed and final applications received by December 1, 1992.	Call for rate information
Central Illinois Light Co. 300 Liberty Street Peoria, IL 61602 Central Illinois	• Outdoor Lighting • Commercial Heat Pumps • Process Heaters	Pre-, post inspection audits required. For lighting, dusk to dawn only. Heat pumps and process improvements, maximum rebates of $2,5000.	LIGHTING: up to $100 per kilowatt, minimum 1 kw. HEAT PUMPS: $50 per ton. ELEC. PROCESS; $50 for each 5kw. All prgrams for industrial, large commercial.

UTILITY	EQUIPMENT	CONDITIONS	INCENTIVES
Central Louisiana Electric Co. P.O. Box 5000 Pineville, LA 71361-5000 (318) 484-7689 14,000 sq. miles in LA	• Heat pumps • High EER Air Conditioning and Resistance Heat • Water Heating • Cooling • Infrared Heating • Microwave Processing • Industrial Process Resistance Heating	Applies to new and existing general service customers. Exclusions: public authority, municipal accounts, barns, hobby shops, expected load factors of 15% or less, and other resistance loads above 500 KW. Must be new installation or replacement of existing gas-fired equipment. Heat pumps 5 tons and below must have minimum SEER of 10 or greater and above 5 ton, SEER of 8 or greater. Air-conditioning or resistance heat must have minimum EER of 9 or greater.	HEAT PUMP: $50/ton HIGH EER AIR CONITIONING AND RESISTANCE HEAT: $15/kw resistance heat. Other resistance loads, $15/kw
Central Maine Power Co. Edison Drive Augusta, ME 04336 (800) 872-9937 Central/Southern ME	• Efficient Lighting • Efficient Motors • Custom	New construction and retrofit. Minimum purchase $50. User must agree to leave measures in place for useful life. Larger rebates may require contracts to ensure saving. Sample follow-up inspections.	EFFICIENT LIGHTING: Fixed rebate table for energy efficient lamps, hybrid ballasts, electronic ballasts, compact fluorescent lamps, occupancy sensor, and H.I.D. fixtures. Technologies not listed on Fixed rebate table may be calculated at $.01/kWh projected annual savings based on life of equipment up to 7 years. New construction calculated on watts.sq. ft. EFFICIENT MOTORS: Fixed rebate table for 5-75 hp mieeting minimum efficiency. Annual use greater than 1000 hours. CUSTOM: Energy-efficient electrical equipment other than lighting calculated at 1¢/kwh projected annual savings based on life of equipment. Maximum based on technology. Retrofit and new construction.
Central Vermont Public Service 77 Grove Street Rutland, VT 05701 (802) 773-2711	• Efficient Motors • Efficient Lighting • Efficient Industrial Processes • HVAC • Refrigeration • Energy Management Systems	Motors must meet or exceed established efficiency and hours of use. Customer must sign contract ensuring performance.	MOTORS: Rebates vary with motor size, designed to defray average incremental cost paid (tables available) ALL OTHER TABLES: Customized rebates negotiated with each customer.
Chelan County PUD No. 1 P.O. Box 1231 Wenatchee, WA 98801 (509) 663-8121 Chelan Co., WA	• Lighting • Motors • HVAC	New construction or retrofit. Verification of payment required.	ENERGY SMART DESIGN: 70-100% of incremental cost of measure beyond code or currrent practices in new construction. 50-70% of installed cost for existing buildings. Equipment rebates available or custom payments based on energy audit. Building commissioning option allow testing, balancing, and correction of building systems in operational mode. Payment to design team for additional time due to incorporation of recommended measures. Daylighting analysis including scale modeling.
Clark Public Utilities P.O. Box 8900 Vancouver, WA 98668 (206) 699-3000	• Lighting • HVAC	New construction or retrofit. Verification of payment required.	ENERGY SMART DESIGN: 70-100% of incremental cost of measure beyond code or currrent practices in new construction. 50-70% of installed cost for existing buildings. Equipment rebates available or custom payments based on energy audit. Building commissioning option allow testing, balancing, and correction of building systems in operational mode. Payment to design team for additional time due to incorporation of recommended measures. Daylighting analysis including scale modeling. Daylighting analysis including scale modleing.

Commonwealth Edison P.O. Box 767 Chicago, Illinois 60690 (312) 294-2871 (800) EDISON-1 Northern 1/3 of IL	• Outdoor Lighting • Efficient Motors • Heat Pumps (multifamily bldgs. only)	User must submit invoice.	OUTDOOR LIGTHING: \$100/kw. Must be photocell-controlled any type lamp. Commercial customers only. MOTORS: 5-200 hp, \$70-2810/motor. Must be high efficiency. Expires December 1992.
Connecticut Natural Gas Corp. 100 Columbia Blvd. Hartford, CT 06144-1500 (203) 727-3377	• Engine Driven (Reciprocating & Centrifugal) • Dessicant • Absorption Equipment	Program based on avoided cost of average unit being installed (COP) 7 installing a more efficient unit (COP). Maximum rebate \$400/ton, \$250,000/customer. Program dates July 1, 1992-July 1, 1993, but will probably be extended.	Rebates paid after installation is completed & inspected. In conjunction with the rebate program, we offer seasonal air conditioning rate established at the start of the season from May to September. we expect the rate to be approximately \$4.20/mcf.
Consolidated Edison 4 Irving Place New York, NY 10003 (800) 343-4646 New York City and most of Westchester County	• Lighting • Gas and Steam AC • Electric AC • Motors • Variable Frequency Drives • Cool Storage	Rebates are available to all non-residential electric customers. Equipment must meet minimu efficiency levels. Con Edison must visit customer and give written approval before equipment is ordered, purchased, or installed. Cool storage and some AC rebates require feasibility study by licensed PE>	LIGHTING: \$12-\$35 for qualifying ballasts. \$1.80-\$5/energy-efficient fluorescent lamp; \$12/compact fluorescent lamp; \$200/kw of lighting connected to qualifying controls; \$300kw of customized lighting reduction; \$1,000/kw of reduction achieved through fixture replacement. GAS AND STEAM AC: \$100-\$400/ton, 50% of cost of feasibility sutdy up to \$15,000. MOTORS: \$25/hp VARIABLE FREQUENCY DRIVES: \$100-\$160/hp ELECTRIC AC: \$40-\$160/ton COOL STORAGE: \$600/kw of demand reduction, 50% of cost of feasibility study up to \$15,000.
Consumers Power 212 W. Michigan Avenue Jackson, MI 49201-2277 (517) 788-1621 MI lower peninsula (except Detroit)	• HVAC • Ballasts • Lamps and Fixtures • Motors • Refrigeration • Commercial Cooking • Cooking	Retrofit only. Post-inspection may be required. No maximum payment.	REBATE AMOUNTS: Call for details
Delmarva Power & Light Co. 800 King Street P.O. Box 231 Wilmington, DE 19899-0231 (302) 454-4303 Delmarva Peninsula (DE,MD,VA)	• Commercial/Industrial Indoor Lighting • High-Efficiency Air Conditioning • Cool Storage	Challenge 2000 program. Rebates available for both new construction and retrofit. Air conditioning and new construction lighting rebates based on watts reduced versus standard efficiency equipment. Retrofit lighting and cool storage rebates based on watts reduced from equipment in place. Cistomer must submit invoices with rebate application. Post-installation review required. Lighting also requires a pre-installation on-sight survey by Delmarva Power rep. Cool storage requires a technical and economic feasibility study performed by a professional engineer.	COMMERCIAL/INDUSTRIAL INDOOR LIGHTING: High-efficiency fluorescent and compact fluorescent lamps: 8¢/watt reduction. High-efficiency fluorescent ballasts (exceeding federal standard): \$0.18/watt reduction. High-pressure sodium or metal halide fixturesL 10¢/watt reduction. Minimum rebate \$100, maximum \$10,000. AIR CONDITIONING: based on \$18/kW saved, minimum rebate \$100, maximum \$50,000. COOL STORAGE: \$140/kw shifted off-peak. Load shifts in excess of 50kw are eligible. Minimum rebate \$7,000, maximum \$100,000.

UTILITY	EQUIPMENT	CONDITIONS	INCENTIVES
Dale Power Co. P.O. Box 1006 Charlotte, NC 28201-1006 (800) 473-4000, Piedmont area of NC and SC	• High Efficiency Chillers	Air conditioning only, no CFC refrigerants. Preapproval required.	COOLING: $15/ton and up on sliding scale. For a 300 ton, non-CFC machine at .63 kw/ton, incentive would be $7,500. Call for full details.
Eastern Utilities EUA Service Corp. 750 West Center St. P.O. Box 543 West Bridgewater, MA 02379 (800) 253-9681 Southeastern MA, northern and southeastern RI	• Lighting • Motors (5-250 hp) • HVAC • Water Heating • Custom	Customers must receive energy analysis by contractor hired by EUA. Minimum hours of operation vary with equipment. Equipment must meet minimum criteria. Contractors must be preapproved. Invoices must be submitted as proof of purchased. Postinstallation inspection required.	PRESCRIPTIVE MEASURES: Amounts are predeteremined and designed to equal the lower of measure the cost minus one year's savings or 70% of cost. CUTOM MEASURES: Incentives calculated on a case-by-case basis to equal the lower of measure cost minus one year's savings or 70% of cost.
Elizabethtown Gas Elizabethtown Plaza Union, NJ 07083 (908) 289-5000	• Gas Air Conditioning	Approved on a yearly basis. Up to $30,000 per unit. Pre-, post-installation inspections required.	GAS-FIRED AIR CONDITIONING: $100/ton.
Equitable Gas Co. Allegheny Center Mall Suite 2000, Pittsburgh, PA 15212-5352 (412) 642-7900	• Gas Air Conditioning • Gas Cooking	Must be conversion to gas from other fuel. Program geared to assist customers in reducing total energy cost, but increasing more efficient and afforable use of gas.	Programs based on the total Btu input and resulting mcf of additional gas load. Finance and leasing options for equipment available.
Eugene Water and Electric Board P.O. Box 10148, Eugene, OR 97440, (503) 484-1125 City of Eugene	• Lighting • Motors	New construction or retrofit. Verification of payment required	ENERGY SMART DESIGN: 70-100% of incremental cost of measures beyond code or current practices in new construction. 50-70% of installed cost for existing buildings. Equipment rebates available or custom payments based on energy audit. Building commissioning option allows testing, balancing, and correction of building systems is operational mode. Payment to design team for additional time due to incorporation of recommended measures. Daylighting analysis including scale modeling.
Florida Power Corp P.O. Box 14042 St. Petersburg, FL 33733 (813) 866-5339 32 Gulf coast and central FL counties	• HVAC • Water Heating • Indoor Lighting • Motors • Custom	Florida Power energy audit required before initiating measure. Most programs require on-site inspection and verification. Dealer incentives apply to eqiupment meeting minimum efficiency standards; HVAC equipment 5 tons or greater; participants must permantently disconnect reheat. Lighting measures must maintain IES standards. Motor program applies to three-phase AC motors 1 hp or greater. New motors must meet minumum efficiencies. Custom measures will be reviewed and approved on individual basis; only projects with a payback over two years. will be considered.	HVAC: Maximum noncash incentive to dealer of $85/unit installed (based on earned points redeemable for merchandise). HEAT PIPES: Maximum rebate of $160/kw reduced ($100 summer; $60 winter). LIGHTING: Energy saving incandescent, $33/kw reduced; regular fluorescent to energy saving fluorescent, $83/kw reduced; incandescent to fluorescent, $83/kw reduced; HID fuxture replacment, $100/kw reduced: delamping through fixture removal, redesign or reflectors, $100/kw reduced; electronic ballasts, $100/kw reduced. Maximum allowable rebate for qualifying measure $100/kw reduced. MOTORS: $6/hp not to exceed half cost of motor. CUSTOM: Up to $150/kw reduced for energy-efficient equipment and thermal storage, both new construction and retrofit; not to exceed 25% of project cost. FIXUP: Half cost of minor energy repairs up to $100.

Florida Power & Light Co. P.O. Box 029100 Miami, FL 33102-9100 (305) 227-4301 Eastern and western FL	• Efficient Lighting • Water-cooled Chiller Retrofit • Thermal Energy Storage	Fluorescent lamp group replacement, minimum 20 lamps. Maximum incentive $10,000/account. Chillers must be ARi certified with a full load operating design performance of 0.63 kw/ton or better. Piping, insulation, and electrical work needed for installation are covered. Engineering feasibility study must be performed by a PE registered in Florida. Manufacturers must have aminumum of 10TES systems installed and operating for at least 3 years, unless otherwised approved by FPL.	LIGHTING: $1/lamp for 3-6 foot; $1.50/lamp for 8-foot energy efficient fluoresents. WATER-COOLED CHILLER RETROFIT: 11% of project cost. THERMAL ENERGY STORAGE: Up to $5,000 toward cost of professional engineering feasibility study (1/2 cost, up to $2,500, for study and remainder at project implementation). $250//ton of cooling load removed from summer on-peak.
Gas Co. of New Mexico 2444 Louisiana Blvd. NE Albuquerque, NM 87125 (505) 880-7485	• Gas Air Conditioning • Cogeneration Equipment	Gas AC program limited to 400 tons/project. Cogen: QF status required. In-person inspection required, both programs.	GAS AIR CONDITIONING: $200/ton, up to 400 tons maximum. COGENERATION: $65/kw.
GASCO Inc. P.O. Box 3379 Honolulu, HI 96842 (808) 547-3579 HI	• Cooking Equipment • Water Heating • Drying Equipment	Customer Installation Rebate limited to commercial cooking, water heating, and clothes drying equipment resulting in converted, new or additonal gas load. Rebate available to any customer for commercial equipment approved by AGA or Gasco that meets minimum Btuh ratings and is installed in Hawaii on Gsco's utility or non-utility systems between October 1. 1991, and June 15, 1992. Commercial eqiupment not eligible for rebate are those sold to customers with gas supply contracts with another gas supplier; relocation of existing business using same equipment; and for new projects that are specified for gas.	Rebates are $100-$250/piece of equipment depending on Btuh rating.
Grant County PUD West 312 Third Moses Lake, WA 98837 (509) 754-3541 Grant Co.; WA	• Lighting • HVAC	New construction or retrofit. Verification of payment required.	ENERGY SMART DESIGN: 70-100% of incremental cost of measures beyond code or current practices in new construction. 50-70% of installed cost for existing buildings. Equipment rebates available or custom payments based on energy audit. Payment to design team for additional time due to incorporation of recommended measures. Prescriptive path rebates based on square footage.
Houston Lighting & Power P.O. Box 1700 Houston, TX 77251 (713) 660-3862 (commercial) (713) 623-3663 (industrial) Greater Houston area	• Thermal (Cool) Storage • Custom	Utility must review and approve projects. Post-installation audit required before payment. Both new construction. and retrofits.	COMMERCIAL EFFICIENCY IMPROVEMENT PROGRAM: $50/kw of conservation that applies to peak load. $100/kw for load shifted from on- to off-peak. Peak hours are 1:00 P.M.-8:00 P.M. THERMAL STORAGE: $300/kw of avoided peak load, plus user qualifies for special thermal storage incentive rate. INDUSTRIAL PROGRAM: Custom program pays $50/kw for load reductions, $100/kw for load shifted to off-peak. $50,000 cap per customer.

UTILITY	EQUIPMENT	CONDITIONS	INCENTIVES
Inland Power & Light Co. P.O. Box 4429 Spokane, WA 99202 (509) 747-7151	• Lighting • HVAC	New construction or retrofit. Verification of payment required.	ENERGY SMART DESIGN: 70-100% of incremental cost of measures beyond code or current practices in new construction. 50-70% of installed cost for existing buildings. Equipemnt rebates avaiable or custom payments based on energy audit.
Jersey Central Power & Light 300 Madison Ave. Morristown, NJ 07960 (201) 455-8325	• Lighting • Motors • Cool storage • HVAC • Chillers • Adjustable Frequency Drives • Heat Pumps • Custom	Appropriate safety certification required. Rebate may not exceed 75% of installed cost. Equipment must meet minumum efficiency levels. Pre-and post-installation audits required. Lamp replacements except compact fluorescents are not eligible. Heat pipe rebate is only for centrifugal fans on VAV systems. Program implementation depends on regulatory hearings.	Proposed rates pending regulatory approval: LIGHTING: .30¢/watt reduced for permanent conversion, .15¢/watt reduced for non-permanent conversion, .10¢/watt controlled for occupancy sensors. MOTORS: Based on nameplate hp, \$3-\$29/hp. THERAML STORAGE: \$300/summer on-peak kw reduced. HVAC: Based on EER or SEER ratings, from \$3-\$15/1000 Btu. CHILLERS: \$400/kw reduced. DRIVES: \$77/motor hp on fan motors on variable air volume systems. HEAT PIPES: \$200/kw reduced. CUTOM: Energy-efficient measures that demonstrate overall energy savings to customer. Amounts based on demand and enrgy reductions.
Kansas City, Kan. Board of Public Utilities 700 Minnesota Ave. Kansas City, KN 66101 (913)573-9151	• Heat Pumps • Resistance Heating • Off-load Construction (Sept.-April) • Water Heater	Applicant must submit engineer's calculation for building heating requirement, must have engineers's approval of load requirements. Applicant must install a 52-gallon water tank with minimum insulation R-value of 12.	HEAT PUMPS: \$50/ton. RESISTANCE HEAT: (elec. furnaces, infrared heaters): \$1-/kw. NEW OFF-LOAD CONSTRUCTION: \$500,000 fund available WATER HEATERS: \$200.
Kentucky Utilities One Quality St. Lexington, KY 40507 (606) 255-2100 Parts of Kentucky	• Electric Heating Equipment Replacing Nonelectric (Mainly Heat Pumps) • Insulation • Geothermal Heating	Equipment must be installed in 1992. Building must experience heat loss of no more than 12 watts/square foot (new buldings also), to be determined by utility. Inspection to determine amount of heat loss required.	HEATING/COOLING EQUIPMENT: \$200-\$600, depending on size of area to be newly heated/cooled by electrical source. INSULATION: Up to \$700 total, also in increments based on size of area. GEOTHERMAL: \$1,000, to 6.5 tons. More than 6.5 tons: \$150/ton to \$5,000/bldg. For commercial users.
Long Island Lighting co. 170 Froelich Farm Blvd. Woodbury, NY 11797 (516) 364-7834 Nassau, Suffolk counties	• Lighting • HVAC • Window Film • Motors • Variable Speed Drives • Thermal Storage • Non-Elec. Cooling • Energy Mgmt. Systems • Refrigeration	"Dollars and Sense" program, a component of LILCO's Energywise commercial conservation programs. Pre-approval needed for rebates exceeding \$2,500, or rebates on customized HVAC or lighting, EMS, refrigeration, TES, non-electric cooling, or variable speed drives. Pre-and post-inspection required on above and including fixture replacements. For other items, paid invoice required.	LIGHTING: 34-watt fluorescents, 35¢/lamp; 60-watt fluorescent, 80¢/lalmp; screw-in fluorescents, \$6/lamp. Fluorescent fixtures, \$75/each. HID fixtures, \$50-100/each. Efficient ballast (magnetic), \$250/lamp powered. Electronic ballast, \$5/lamp powered. T-8 ballast: \$6/lamp powered. Reflectors: 20¢/lamp-watt removed. Occupancy sensors: \$24 each. HVAC: Unitary (air and water cooled), \$4 for every 0.1 above min. EER. WINDOW FILM: Shading coefficient less than .3, \$1/sq. ft: .3-.4, 75¢ sq. ft. THERMAL STORAGE AND NON-ELECTRIC COOLING: Call for more information and rebate levels. REFRIGERATION: Glass door, \$40/linear ft, also includes expansion valves and hot-gas defrost, \$45-\$145/rated compressor hp. EFFICIENT MOTORS: To 25 hp, \$12/hp, 26-200 hp, \$300 plus \$5/hp over 25hp. VARIABLE SPEED DRIVES: To 50 hp, \$110/hp, 51-200, \$85/hp, 201-1,000, \$65/hp. CUSTOM REBATES: Lighting, \$100/kw; HVAC, \$200/kw; general \$150/kw.
Los Angeles Dept. of Water and Power P.O. Box 111 Los Angeles, CA 90051 (800) U-ASK-DWP City of Los Angeles	• Lighting • HVAC • Building Envelope • Refrigeration • Motors • Controls	Customer pays 100% of project up to current practice/state standards baseline; then LADWP pays up to 100% of additional costs associated with more efficient designs and equipment. All measures subject to cost effectiveness screening by LADWP. Pre-approval, design review, and post-project inspection required.	COMPREHENSIVE PERFORMANCE-BASED DESIGN INCENTIVES: site-specific incentives tailored to cover up to 100% of incremental cost of energy efficiency improvement above current practice/state standards baseline. Commercial New Construction (CNC) program incentives available to cover incremental costs associated with system design, equipment purchase, commissioning, and O & M plan preparation. Energy Management Partnership (EMP) incentives available to large commercial customers (500 kw & up) for remodeling or replacing equipment. Costs include the above plus disposal.

Madison Gas and Electric P.O. Box 1231 Madison, WI 53701-1231 (608) 252-7222 Greater Madison area	• Lighting • Heating • Air Conditioning • Water Heating • Motors • Custom	For commercial, industrail, institutional facilities. Minimum performance and safety specifications for all projects. Preapproval required for rebates over $5,000 and for all custom projects.	LIGHTING: Compact fluorescent: $5/one piece bulb, $10/two piece bulb. Electronic ballasts: $7/lamp controlled or up to 50% project cost. HEATING: Boiler vent damper, $100-$300 each. Boiler hot water resets $150 each Boiler cutouts $50 each. Night setback thermostats $50 each. Pipe insulation 25% of cost. Infrared heating $4 per MBtuh. Furnace replacements to $250. Steam trap repair or replace: $30/trap. Boiler tune-up: $50-100/boiler. AIR CONDITIONING: Package and rooftop: $35-95/ton. MOTORS: $10-15/hp SERVICE HOT WATER: Contact Utility CUSTOM PROJECTS: Case-by-case basis. For commercial, industrial, institutional, multi-family housing.
Manitoba Hydro P.O. Box 815 Winnipeg, Manitoba R3C 2P4, (204) 474-3613 Province of Manitoba	• Motors • VSD • Industrial Process Electrotechnologies • Custom	Both commercial and industrial programs, new construction and retrofits. Incentives also paid to vendors.	HIGH EFFICIENCY MOTOR PROGRAM: $400/kw saved for high efficiency motor retrofits; aim is to buy down payback to 1.5 years. Vendors qualify for $100/kw saved rebate when qualifying motors are sold to end users. ELECTROTECHNOLOGIES: Incentives for industrial process electrotechnolgies, including microwaves, variable speed drives, high frequency heating, drying, infrared heat treatment. Call for details. CUSTOM: Call for details
Metropolitan Edison co. P.O. Box 16001 2800 Pottsville Pike Reading, PA 19640 (215) 921-6252 Cities of Reading, Lebanon, Easton, and New York	• Heat Pumps • Cool Storage • Heat Storage • Custom	Heat pumps: minimum SEER 11, units less than 65,000 Btuh; minimum EER 9.5, units over 65,000 Btuh. Thermal storage: Feasibility studies required. Custom: case-by-case analysis.	HEAT PUMP: Less than 65,000 Btuh, SEER 11.00-11.99, $70/ton, max, $5,000;over 65,000 Btuh, SEER over 12, $140/ton, max. $5,000; less than 65,000 Btuh, EER over 9.5, $70/ton, max. $5,000. COOLSTORAGE: $250/kw for shift to off peak, 50% feasibility study to $3,500, maximum rebate $50,000. HEAT STORAGE: $75/kw for shift to off peak, 50% feasibility study to $1,000, maximum rebate $50,000. CUSTOM: Benefits to company, level of feasibility study payment, and/or rebate will be determined on case-by-case basis. Not all projects qualify.
Metropolitan Utilities District 1723 Harney St. Omaha, NE 68102 (402) 449-8247 Omaha, surrounding area	• Space Heating Conversions • Water Heating Conversions • Gas Air Conditioning • Cogeneration • Desiccant Dehumidification	Expires end of 1992. User must have signed committment that equipment will be installed by program expiration date or another agreed upon date. Proof of purchase and installation rerquired.	SPACE HEATING CONVERSION: (Electric, oil, or propane to gas) $100/100,000 Btuh. Ma. $2,500 combined with water heating conversion. WATER HEATING CONVERSIONL (Electric, oil, or propane to gas) $100/100,000 Btuh. Max. $2,500 combined with space heating conversion. DIRECT OR INDIRECT NATURAL GAS A/C: (Includes desiccant and cogeneration) Captial and "flexed" gas price on project-by-project basis. OTHER: Economic development rate available to new commercial and industrail customers locating in service area.
Minnegasco, Inc. 201 South 7th St. Minneapolis, MN 55402 (612) 525-7755	• Conversions of Non-Gas Process or Heating Equipment to Gas • Gas Cooling	Conversions: Rebate cannot exceed total project cost. Rebate application must be approved before project starts; verification by inspection. Minimum size 50 mcf annual load. Cooling: Requires approval before purchase orders are signed. Inspection required to verify installation.	FIRM GAS CONVERSION: up to 75¢/mcf of added annual load, $15,000 maximum. INTERRUPTIBLE CONVERSION: up to 50¢/mcf of added annual load, $1,000 maximum. GAS COOLING: Up to $150/ton, amount varies according to tonnage. $50,000 maximum per facility.
Missouri Public Service 10700 East 350 Hwy. Kansas City, MO 64138 (816) 737-9380 Kansas City, west central MO	• Electric Space Heating Equipment	Special rates for electric space heating. Rate 120 — Commercial/Industrial Space Heating — allows separately-metered space heating equipment to be billed on a special flat rate September — April. Rate 250 — Commercial/Industrial Space Heating, single meter — allows all use to be billed year-round on load factor rate using a base of 100 use-hours.	ELECTRIC SPACE HEATING EQUIPMENT: Separately-metered space heating rate for eight winter months— Rate 120-3.64¢/kwh. Single-meter rate for electric space heating — Rate 250 — load factor rate using 100 base use-hours and excess use at 3.99¢/kwh year round.

UTILITY	EQUIPMENT	CONDITIONS	INCENTIVES
Montana Power Co. 40 E. Broadway Butte, MT 59701 (406) 723-5421 ext. 2338 60% of MT	• Lighting • HVAC • EMS • VFD • Energy Efficient Motors • Refrigeration • Other	New construction or retrofit. Detailed proposals required.	ELECTRIC CONSERVATION PURCHASE PROGRAM: New construction or retrofit. Utility assistance available for design, utility financial assistance for project implementation based on resource value. EFFICIENT LIGHTING PROGRAM: Electronic ballasts, high efficiency fluorescent tubes and fixtures. Halogen and HID technologies. Call for details. EFFICIENT MOTOR PROGRAM: Customer and dealer rebates for retrofit and new installation of energy-efficient motors. Fractional hp and rewinds do not qualify. Call for details.
Nevada Power Co. P.O. Box 230 Las Begas, NV 889151 (702) 367-5111 Southern Tip of NV	• Fluorescent Lamps • Compact FLuorescents • Reflectors • Other Lighting • Replacement A/C	For lighting, Nevada Power does audit and economic analysis, user provides invoice. For AC, units must be 2-20 tons with minimum EER of 9.2 (higher ratings for increased tonnage). Utility will do spot check, post-installation inspection, user provides invoice. New construction, user must submit blue-prints.	LIGHTING: Based on demand reduction during daylight peak hours, equaling $50-$125/kw saved. NEW CONSTRUCTION: Lighting: Custom rebate, $100-$200/kw reduced beyond energy-efficient standard. COMMERCIAL AC: $25-$150/ton (for 2-5 ton units), depending on EER. CUSTOM REBATE: For any electric equipment, amount based on life of measure, value to customer. MOTOR: Based on hp, efficiency, contact utility.
New England Electric (Includes Mass. Elec., Granite State ELec, Narragansett Elec.) 25 Research Drive Westborough, MA 01582 (508) 366-9011 Areas in MA, RI, NH	• Motors • Variable Speed Drives • Process Refrigeration • Custom Measures • Food Service • HVAC	Preapproval, postinstallation inspection required. Energy conservation analysis, technical assistance commissioning for operation and management and design assistance available.	ENERGY INITIATIVE: (for existing facilities): Unit pricing for incentive (per motor, per lighting fixture) based on total installation cost, $500-$800/kw, simple incentive application process. DESIGN 2000 (New Construction): Based on incremental cost of measures having efficiency that exceeds building codes or standard practice.
New York Power Authority 1633 Broadway New York, NY 10019 (212) 468-6930	• T-8 lamps • Reflectors • Compact FLuorescents • Electronic Ballasts • Other Lighting	"Help" initiative for gov't buildings (state,local) in New York City and Westchester County, for Long Island School Districts, and for state buildings in rest of state.	LIGHTING: NYPA will buy down costs to a two-year payback, and provide funding to finance up-front costs. Lighting incentives for the Upstate and Long Island School prgrams coordinated through local utility rebateprograms.
New York State Electric and Gas (Electric Utility) Corporate Dr., Kirkwood Industrial Park Binghamton, NY 13902-5229, Areas of upstate NY	• Lighting • HVAC • Elec. Thermal Storage • Refrigeration • Motors, AC drives • Gas Cooling • Elec. Water Heaters • Process • Energy Analysis	Thermal storage must be used off-peak. Rebates over $100,000 need corporate approval. Installations subject to verification. For A/C, process, and refrigeration, minimum efficiencies apply.	LIGHTING: Fluorescent fixtures with electronic ballasts, $10-$36. HID fixtures, $45-$68. Occupancy sensors, $22-$27. Reflectors, $8-$10. Compact fluorescents, $7-$24. HVAC, Process, Refrigeration: Air conditioners, $50-$250/ton depending on size, efficiency. Heat pumps, $50-$150/ton. Motors, $20/hp. Chillers, $30-$120/ton. Variable speed drives, $70-$140/hp. Gas A/C, $125/kw deferred. Electric water heaters, $40-$75. COOL STORAGE: $113/kw. CUSTOM: Available for any cost-effective efficiency measure not covered in fixed rebate schedule.
New York State Electric and Gas (Gas Utility) P.O. Box 3607 Binghamton, NY 13902 35% of NY	• Space Heating • Water heaters	Rebates apply to conversion from another energy source to gas in existing facilities. Verification inspection required.	SPACE HEATING: Under 150,000 Btu/h: $300, over 150,000 Btu/h: $2.00/annual decatherm. WATER: $100.

Utility	Programs	Requirements	Rebates
Niagara Mohawk Power 300Erie Blvd. Syracuse, NY 13202 (315) 474-1511 Upstate NY	• Lighting • HVAC • Heat Pumps • Chillers • Motors, Drives • Custom	All items need pre-approval. No upper limits to rebates. Minimum EER's, SEER's exist for all HVAC rebates. Minimum motor efficiency 84%. Call us for specific terms and conditions.	RETROFIT LIGHTING: T-8 lamps: three-foot, four-foot, $1; eight-foot, $1.50. T-12 electronic ballasts: $7.5-$15. T-12 hubrid ballasts: $15. T-8 dedicated ballasts, $7.50-$25. Compact fluorescents: single piece, $4; modular, $8; hard-wired, $10. Fixture reflectors: four-foot, $25; eight-foot, $30. HPS or MH HID lamps: 100-250W, $75/fixture; 200W and up, $100/fixture. Occupancy sensors: wall unit, $30; ceiling unit, $40. Exit signs: hard-wired fluorescent, $15; LED, $75. NEW CONSTRUCTION LIGHTING: T-8 energy-saving fluorescent lamps, 50¢. Solid state electronic T-8 lamp ballasts: $15 if less than 20% THD; $5 if 20-30% THD. Hard-wired fluorescent fixture: $10. Exit signs: fluorescent, $10; LED, $40. Occupancy sensors: wall-mounted, $15; ceiling mounted, $20. MOTORS, DRIVES: Motors range from $50 for 1 hp to $4,000 to 400 hp. Drives range from $1,100 fro 5 hp to $24,000 for 400 hp. consult utility literature for further information. CUSTOM: Case-by-case review of customer incentive proposals outside of specified rebate programs. Consult utility representative to praticipate. Subject to terms and conditions.
Northeast Utilities (includes Connecticut Light & Power, Western Mass. Elec.) P.O. Box 270 Hartford, CT 06141 (203) 665-5000 Parts of CT and MA	• Lamps • Ballasts • Lighting fixtures • Reflectors • Occupancy Sensors • HVAC • Energy Mgmt. Systems • Efficient motors	Lighting rebates: Preapproval for amounts over $1,000, postinstallation inspection required, mas. $35,000/account/year. Energy Action Program (EAP): Assessment by utility contractor-arranger who recommends measures, assists customer. EAP and Customer-Initiated Program participants must agree not to selfgenerate more than 250 kw/year.	LIGHTING REBATE PROGRAM: Fixtures: T-8 systems, $10-$15. Exit signs: LED, $80; electroluminescent, $85; retrofit kits, $10. Ballasts: hybrid, $7.50 elec; elec. controllable, $45. Compact fluors., $7-$10. Hardwired compacts, $20. Reflectors, $12.50 (T-12 standard 930 ma plus slimlines only) per lamp removed. Occupancy sensors, $40. ENERGY ACTION PROGRAM: Utility pays 50% of energy assessment, second 50% remibursed if measures are installed. CUSTOMER-INITIATED PROGRAM: Process-related measures receive buydown in installed equip. cost to one-year payback. Non-process measures receive lesser of 50% of equip. costs or amount to buy down to three-year payback. ENERGY CONSCIOUS CONSTRUCTION: To fill incremental cost. EAP and CIP for large commercial and industrial; others for industrail, commercial, institutional.
Northern Illinois Gas 1844 Ferry Raod Naperville, IL 60563 (708) 983-8676 Northern IL	• Gas Air Conditioning	Industrial facilities and commercial or multifamily buildings must order and install at least 10 tons of natural gas air conditioning equipment in each qualified building.	GAS-FIRED AIR CONDITIONING: Pays up to $100/ton for installation of natural gas air conditioning. Available through 12/31/92. Contact company for details.
Northern Minnesota Utilities 910 Cloquet Ave. Cloquet, MN 55720 (218) 879-1571, 50 communities in N. 1/2 of MN	• Commercial Food Service • Custom	Natural gas food service equipment must replace electric or propane equipment.	COMMERCIAL FOOD SERVICE: $1 per 1,000 Btu/h input. CUSTOM: Other equipment evaluated on individual basis.
Northern States Power Co. — Wisconsin, Marketing — Commercial & Industrial Divison, P.O. Box 8 Eau Claire, WI 54702 (800) 428-5223 Parts of WI and Upper MI	• Lighting • Cooling • Motors and AC Drives • Boiler Efficiency • Steam Trap Maintenance • Electric Thermal Storage • New Construction • Feasability Study Grants • New Construction	Site inspection or invoices required. Some programs require preapproval. Some programs and study grants have incentive caps.	MOTORS: $30 to over $500 for motors that meet efficiency requirements; ASD rebates are $20 per qualifying hp. THERMAL STORAGE: $5-$20 per installed kw of heating, includes slab heat, room units, and furnaces. COOLING: $7-$35 per ton, based on type; Ballasts: $3-$8 per fixture. NSP's Lighting Program also includes rebates for conversions, controls and reflectors, call for details. LIGHTING: 25¢-$3 per lamp, based on size and type; Ballasts: $3-$8 per fixture. NSP's Lighting program also includes rebates for conversions, controls and reflectors, call for details. GAS CONSERVATION: Up to $100 for each boiler tune-up, up to $30 per steam trap repaired or replaced. Call for more information on multifamily, gas air-conditioning, and customized gas rebates. NEW CONSTRUCTION: Call for details and literature on this and all NSP programs.

UTILITY	EQUIPMENT	CONDITIONS	INCENTIVES
Ontario Hydro 700 University Ave. Toronto, Ontario M5G 1X6 (800) 263-9000 Province of Ontario	• Lighting • Streetlighting • Motors • Pumps, Fans and Blower Systems • Occupancy Sensors • Refrigeration	Lighting conversions must be completed within 18 months of approval. Payment upon completion. Motors must meet specified efficiency levels and be connected at least one year.	LIGHTING: incentives for energy saving fluorescent lamps, compact fluorescents, reflectors, power reducers, T-8 systems, efficient electromagnetic ballasts, exit signs, photocell lighting controls. Payment ceiling of 50% of total project cost. Apply per energy-efficient lighting product; or up to $500 per kilowatt of demand reduction. Options to finance through interest free loan or lease. STREET LIGHTING: Communities must commit to convert 100% of existing street lighting, and complete conversions by December 31, 1994. Cash rebate on 25% of the total project cost, and equivalent interest-free loan, or a seven-year low-interest loan. MOTORS: T-frame, AC three-phase 1-200 hp motors. Over 200 hp all induction motors apply. $12 per hp, 1-500 hp high efficiency motors. Maximum incentive $2,400, minimum $50. PUMPS, FANS, AND BLOWER SYSTEMS: Engineering and incentives for industrial customers. Up to $50,000 per project towards engineering feasibility study to identify and assess opportunities. Up to $300,000 per project applied against capital equipment, engineering, and installation costs. Incentives above $300,00 evaluated on a project-specific basis. Flexible financing also available for capital projects. OCCUPANCY SENSORS: incentives apply to new, renovation, and retrofit applications. $35-$150 per unit installed. REFRIGERATION: For industrial, ammonia-based refrigeration systems in Canada. Up to 100% funding for plant audits; incentive of 10¢/kwh of annual energy savings, or the amount required to reduce the project payback to 1.5 years, whichever is less, to a maximum of $300,00.
Orange and Rockland Utilities One Blue Hill Plaza Pearl River, NY 10965 (800) 752-8861 Orange and Rockland counties in southern NY	• Efficient Lighting • Efficient Motors • Electric A/C • Cool Storage • Nonelectric Cooling	Preapproval required for projects over $1,000. Equipment must meet minimum efficiency levels. Proof of purchase required for all rebates. O & R reserves the right to perform inspections.	LIGHTING: efficient fluorescents, $1-$2/lamp; compact fluorescents, $8 each; electronic ballasts, $8 each; custom projects $200/kw saved; outdoor lighting, $200/kw saved; efficient exit sign kits, $7.50 each. NEW CONSTRUCTION PROJECTS: $200/kw saved. CONTROLS: $25-$50/kw connected MOTORS: $20-$6,100 depending on hp. AIR-COOLED CHILLERS: $45-$105/ton. WATER-COOLED CHILLERS: $15-45/ton. CENTRAL A/C AND HEAT PUMPS: $40-$125/hp depending on size & efficiency. COOL STORAGE: $250/kw shifted to off-peak. NON-ELECTRIC COOLING: $250/kw displaced with a nonelectric cooling system.
Pacific Gas & Electric 123 Mission, Room 2458 San Francisco, CA 94106 (415) 973-8575 Northern & central CA	• Retrofit Program • New Construction • Custom	Custom program requires savings calculations with applications. Retrofit and custom programs both require pre-and postinstallation inspections. In retrofit program, user must submit invoices with applications. Most energy-saving technologies qualify.	RETROFIT PROGRAM: Pays according to equipment installed, from 50¢ to $100,000. Maximum $100,000 per account. CUSTOM PROGRAM: Pays according to energy savings. 6¢/kwh or 20¢/therm based on first-year energy savings. Up to 50 percent of project cost to a maximum of $300,000 per account. NEW CONSTRUCTION: Similar to retrofit programs. Call for Details.
Pacific Power & Light (includes Utah Power & Light) 920 SW 6th Ave. Portland, OR 97204 (503) 464-6866 Parts of CA, ID, MT, OR, UT, WA, & WY	• New Construction • Custom	Proposals are modeled with DOE-2 software. Utility performs postinstallation commissioning and saving check. Call for further details.	ENERGY FIN-ANSWER: For new construction and major remodeling, pays entire up-front cost of measures above cold; user repays through savings at interest rate of prime or prime-plus-three, depending on measure. User selects loan term up to 20 years. RETROFITS: (pilot program only available in Albany, OR) Financing incentives similar to Energy FinAnswer. Energy audit required.
Pasadena, City of, Water and Power Dept. 150 S. Los Robles, Suite 20[illegible] Pasadena, CA 91101	• Heat Pumps • Thermal Energy Storage • Lighting • Electronic Ballasts	Commercial cool storage: Feasibility study: $5,000 CAP/Building rebate-$60,000 CAP/Building. Heat Pump: Rebate to participant-$5,000 per metered account. Rebate to installer=$500 per	COMMERCIAL COOL STORAGE: Feasibility study, $100/ton of cooling capacity, $5,000 cap. Installation, $300/kw of demand shifted to off-peak; $60,000 cap. HEAT PUMP: $60-$80/ton for SEER of 9.0 or above. $5,000 cap per metered acct. for users, $500 cap per metered cap for installers.

(818) 792-POWER City of Pasadena	• Commercial Energy Audits	metered account. Lighting: $20,000 per facility cap.	LIGHTING: $25-$100kw saved depending on installation. $20,000/facility cap.
Pennsylvania Electric Co. 1001 Broad Street Johnstown, PA 15907 (800) 652-3232 30 Counties in central and northern PA	• Industrial Electrotechnologies • Heat Pumps • Heat Storage • Cool Storage • Water Heating • Food Service • Lighting	Projects must be preapproved. Industrial technologies cannot exceed 5% of project capital costs. Heat pump efficiency requirements: split systems, HSPF 6.8, SEER 10.0; packaged system, COP 3.0, EER 9.0; ground source, COP 3.0, EER-12.5. Installed size must be within 135 of design cooling load for A/C. Cool storage systems must have a designed feasibility study performed. Food service equipment must be greater than 5 kw load.	INDUSTRIAL ELECTROTECHNOLOGIES: 5% of the capital costs up to $25.00 per KW. HEAT PUMPS: air source, $100/ton. Ground source: $150/ton. HEAT STORAGE: $100/kw up to 50 kw installed. Individual revenue analysis exceeding 50 kw. COOL STORAGE: Feasibility study funding of 2/3 of study costs. $250/kw shifted to storage. WATER HEATING: $200/ton for heat pump water heaters. $25/kw for other installations FOOD SERVICE: $30/kw new or replacement. $50/kw replacing fossil fuel with electric. Dealer incentive. LIGHTING: 5¢/watt of new outdoor, security, area, and architectural lighting.
Pennsylvania Power & Light Co., 2 North 9th Street Allentown, PA 18101-1179 (215) 774-5694 Central Eastern PA	• High Efficiency Heat Pump Systems • Off-peak Electric Thermal Storage Space Conditioning & Water Heating Systems	Limited to U.L.-listed equipment and all-electric heating and cooling systems. Equipment must be connected within three years of agreement. Heat pump minimum size 1 ton. available for new commercial buildings, replacement of fossil-fueled system or upgrade of electric resistance system in existing building. Thermal storage minimum size 1 kw.	HEAT PUMPS: Rebate not to exceed equipment cost. Pay $8-$50/ton depending on COP, to maximum $50,000. THERMAL STORAGE: Off-peak system grant limited to system size not exceeding 15 watts/square foot of conditioned space. Grant will not exceed cost of equipment. Electric underfloor storage, $50/kw to maximum $50,000. Central storage system: heating, $100/kw installed; cooling, $100avoided ton; both to maximum $100,000. Off-peak storage water heating. $50/kw installed to maximum $ 50,000.
Peoples Gas Light & Coke Co., 122 South Michigan Avenue Chicago, IL 60603 (312) 431-4861 Chicago	• Gas Fired Cogeneration • Gas Air Conditioning	Preapproval and postinspection required. Installations must be permanent. Premises must be heated and cooled with gas-fired equipment to qualify for air conditioning allowance. Programs expire February 1994.	COGENERATION: $75/kw of installed capacity up to a maximum of $100,000 per facility. 75% of cost of an economic feasibility study up to maximum $15,000. GAS AIR CONDITIONING: $150/ton for the first 200 tons of installed capacity and $50/ton for remainder up to a maximum of 1,000 tons.
People's Gas System, Inc. 111 Madison Street Tampa, FL 33602 (813) 273-0074 21 counties in FL	• Water Heaters and boilers • Pulse Combustion and Other Advanced Technology Furnaces. • Pulse Combustion and Other Advanced Technology Fryers and Cooking Equipment. • Gas Space Conditioning • Other	Available to commercial customers replacing electric appliances with input of 6,000 watts or more. Replacement equipment must be an energy efficient natural gas appliance. Gas space conditioning equipment must replace electric equipment with a COP of .8 or above.	REPLACING ELECTRIC APPLIANCES: $4/kw to maximum of 100 kw. Space Cooling: $150/ton, maximum 100 tons/project.
People's Natural Gas Co. 1815 Capitol Avenue Omaha, NE 68102 (402) 221-2438 Parts of MN, IA, NE, KN, & CO.	• Space Heating • Air Conditioning • Water Heating • Food Service Equipment • Process Equipment	Applications due by Dec. 1, 1992. Retrofit and new construction.	Rebates available on retrofit or conversion to natural gas on project-by-project basis. Rebates available on new construction on a project-by-project basis. Call for details.

UTILITY	EQUIPMENT	CONDITIONS	INCENTIVES
Plains Electric Generation and Transmission Cooperative Inc., 2401 Axtec Road, Albuquerque, NN 87107 (505) 884-1881 FAX: (505) 889-7636	• Motors • Electric Thermal Storage • HVAC	Thermal storage must be off-peak. All installations subject to verification. Minimum efficiencies required for all items. Energy conservation analysis and design assistance available.	MOTORS: $25/hp as used in irrigation applications, other custom rebates based on application. HVAC: air and ground source heat pumps, $50/ton THERMAL STORAGE: for central and room heating, $25/kw.
Port Angeles City Light P.O. box 1150 Port Angeles, WA 98362 (206) 457-0411 in conjunction with Clallum County PUD #1, P.O. Box 1090 Port Angeles, WA 98362 (206) 452-9724	• Lighting	New construction or retrofit. Verification of payment required.	ENERGY SMART DESIGN: 70-100% of incremental cost of measures beyond code or current practices in new construction. 50-70% of installed cost for existing buildings. Equipment rebates available or custom payment based on energy audit. Prescriptive path rebates based on square footage. Operations and maintenance services offered to ensure measures are maintained for optimum savings.
Portland General Electric 7881 SW Mohawk St. Tualatin, OR 97062 (503) 691-3800, Northern OR	• Lighting • Motors • Adjustable Speed Drives	Lighting incentives paid on equipment identified by PGE, provided electricity savings are at least 10%. Incentives paid to vendors and installing contractors. Motors must total at least 10 hp and meet nominal efficiency standards. ASD's must provide at least 10% electricity savings. Participant must document savings by energy study prepared by qualified PE; output from ASD computer analysis package; or competent hand-calculated analysis.	FLUORESCENT LAMPS: 34-watt T-12, 15¢ each; F-96 T-12, 20¢ each; T-8, 20¢ each. COMPACT FLUORESCENTS: Screwin, $1 each; fixtures, $3 each. ELECTRONIC BALLASTS: $3 each HPS fixture, 47.50 each; metal halide fixture, $5 each. LIGHTING CONTROLS:occupancy sensors, $5 each. MOTORS $3/hp paid to dealer to maximum $150. $3/hp paid to customer. ADJUSTABLE SPEED DRIVES: $5/hp paid to dealer to maximum $250. $25/hp paid to customer.
Potomac Electric Power co. 1900 Pennsylvania Ave, NW Washington, DC 20068 (202) 872-4630, DC, Parts of MD	• Lamps • Reflectors • Motors • Packaged AC • Chillers • Other • Daylighting • Occupancy sensors • Efficient ballasts • Thermal Storage	Lighting: Customer must complete program application, Pepco engineer preinspects. Cap of 50% material and labor costs on some technologies. Building Design: Prescriptive (limited to specific technologies) and comprehensive (design study, demand building simulation) options. All demand reduction technologies qualify.	LIGHTING: Daylighting, $30 per control unit. Occupancy sensors $60 each. THERMAL STORAGE: Call for details. Also pay 50 percent of cost of feasibility study up to $5,000.
PSI Energy 1000 E. Main St. Plainfield, IN 46168 (317)838-1939 Southern 2/3 of IN (excluding Indianapolis)	• Custom Equipment, Includes Lighting, Motors, ASD's HVAC, Building Envelope, Ice Storage, etc.	Total incentives cannot exceed PSI avoided cost or the amount necessary to provide a two-year payback. Supporting engineering estimate required.	EFFICIENCY IMPROVEMENT MEASURES: for commercial and industrial customers, the target is a two-year payback on incremental costs.
Public Service Electric & Gas P.O. Box 1258 Newark, NJ 07101 (800) 844-4444 Northwest to southwest NJ	• Thermal Storage • Fluorescent Lamps • Gas Air Conditioning • Energy Mgmt. Systems • Insulation • Other	For thermal storage, audit required. For "Cash Rebate" program, CACS audit required. Only items with 7-year payback or less are eligible, with a $5,000 limit per facility. Lighting rebate for fluorescent lighting only. $5,000 limit. Gas rebate for replacing or installing gas-fired air conditioning equipment. Expires 12/31/92.	THERMAL STORAGE: $250 per kw shifted off-peak up to 500 kw, $125 over 500. CASH REBATE PROGRAM: Up to 20% of cost of items PSE&G has recommended. Examples are lighting and ballast retrofits, reflectors, pipe insulation, weatherstripping, ventilation dampers. LIGHTING PROGRAM: To replace 40W bulbs to 34W or less (4 feet): $1.25 per bulb. To replace 75W bulbs to 60W or less (8 feet): 42.50 per bulb. GAS AC REBATE: For replacing existing gas-fired absorption chillers, $75 per ton. For new construction or replacing electric chillers with gas, $1.25 per ton. Minimum of 100 tons. Cash rebate program only for small commercial, multi-family housing. Other programs for commercial, industrial, institutional.

Public Service of Oklahoma 212 E. Sixth St., P.O. Box 201 Tulsa, Oklahoma 74102-0201 (918)599-2203, Northeast, southeast, southwest OK	• Heat Pumps	Available to commercial and institutional customers.	GOOD CENTS INCENTIVE: Pays $75-$120/ton for Good Cents-certified heat pumps, depending on EER/SEER. Non-certified. pays $50-$80/ton. Other options for building evaluation and technical assistance. Contact utility.
Puget Sound Power & Light Co., P.O. Box 9034 Bellevue, WA 98009-9734 (206) 462-3688, (800) 562-1482, Western WA	• Lighting • Motors • Occupancy sensors • HVAC • New construction	Must replace incandescent lighting with compact fluorescent lamps. Motors must be new AC/induction, normal torque, general purpose. Must replace existing light switches with commercial quality wall-mounted occupancy sensors. New constructions or retrofit. Verification of payment required.	LIGHTING: $8 each for nonreplaceable lamp; $10 each for replaceable lamp without reflector: $14 each for replaceable lamp with reflector. MOTORS: Instant rebate, save up to $2,250. OCCUPANCY SENSORS: $15/300 watts or less, $35/300 watts or more. ENERGY SMART DESIGN: 70-100% of incremental cost of measures beyond code or current practices in new construction. 50-70% of installed cost for existing buildings. Custom payments based on energy audit. Building commissioning option allows testing, balancing, and correction of building systems. in operational mode. Payment to design team for additional time due to incorporation of recommended measures. Daylighting analysis including scale modeling. Operations and maintenance services offered to ensure measures are maintained for optimum savings.
City of Riverside Utility District, 3900 Main Street Riverside, CA 92522 (714) 782-5485 FAX: (714) 369-0548	• Thermal Energy Storage	New construction or retrofit.	THERMAL ENERGY STORAGE: 50% of cost of feasibility study, up to $5,000; $200/kw shifted to off-peak.
Rochester Gas & Electric Corp., 89 East Avenue Rochester, NY 14649-0001 (716) 546-2700 Western NY	• Lamps • Ballasts • Custom Lighting • Motors & Drives • Water-Cooled AC • Electric Chillers • Gas Chillers	Rebates for equipment generally used during business hours (lighting must be indoors). Motors must meet minimum NEMA efficiency, minimum efficiencies exist for all hp. Air conditioning equipment: ratings must be at standard ARI conditions.	STANDARD LIGHTING: 8ft. fluorescents. $1.10/lamp; compact fluorescents, $8-$12; exit sign retrofit, $10; electronic ballasts, $10-$15; occupancy sensors, $23 wall switch, $64 remote sensor. T-8 systems, $20. CALCULATED (CUSTOM) LIGHTING: 50% of cost up to $200/kw of daytime demand saved. Preapproval required. MOTORS: $12 (1hp) to $1,425 (250 hp), graduated scale based on efficiency, hp. WATER-COOLED A/C: Packaged AC, $19-$200/ton, based on tonnage, EER. Water-source heat pumps, based on EER, tonnage. ELECTRIC WATER CHILLERS: Water-cooled, $15-$63/ton, based on kw/ton, based on kw/ton, EER. Air-cooled, $20-$80/ton, based on EER, tonnage. VARIABLE SPPED DRIVES: $75/hp on fans and pumps.
Rockland Electric co. One Blue Hill Plaza Pearl river, NY 10965 (800) 752-8861 Bergen county, NJ	• Efficient Lighting • Cool Storage • Controls • Custom • New Construction	Preapproval required for projects over $1,000. Equipment must meet minimum efficiency levels. Proof of purchase required for all rebates. Inspection may be performed.	LIGHTING: Fluorescent lamps, $1-$2/lamp; compact fluorescents, $4/each; electronic ballasts, $5 each. CUSTOM PROJECTS: $150/kw saved. NEW CONSTRUCTION: $100/kw saved. CONTROLS: $25-$50/kw connected. COOL STORAGE: $250/kw shifted to off-peak.
Sacramento Municipal Utilities District P.O.Box 15830, Mail Stop 73A, Sacramento, CA 95852-1830, (916) 732-6166	• Refrigeration • Lighting • Motors • Process • Daylighting • HVAC • Thermal storage	Preapproval and postinstallation required on all projects. Customer must show invoices.	RETROFIT PROGRAM: Rebates based on kwh saved over life of project. Peak bonus may be added for depending on technology. Rebates could vary from 1.5-3¢/kwh. Limited to term of lease on 70% of SMUD's estimate of project cost. NEW CONSTRUCTION: Commercial program includes a range of prescriptive measure incentives for projects under $50,000 square feet and a customized performance approach including design and construction incentives for projects 50,000 square feet or large. Hook-up fee discounts available for specific savings levels. Both prescriptive and customized approaches provide incentives at three tiers set at energy saving of 20%, 25%, and 35% over Title 24/current practice.
Salt River Project P.O. Box 52025 Phoenix, AZ 85072-2025 (602) 236-8586	• Thermal Energy Storage • Heat pumps • Reflectors • Refrigeration • Fluorescent Lamps • Hardwire Compact Fluorescents • HID Fixture Conversions • Occupancy Sensors	Thermal storage: engineering report with documentation and utility audit required. Refrigeration and heat pumps (PurchasePower program): 10-ton limit. Lighting Edge Program: preapproval necessary: both new construction and retrofit.	THERMAL STORAGE: $260-$60/kw, depending on kw shifted. PURCHASE POWER PROGRAM. $30-$135/ton, depending on SEER rating, or financing is available. 10 ton limit. LIGHTING EDGE PROGRAM: Lamps, 40¢-$2; efficient ballasts, $3-$7. New fluorescent fixtures with electronic ballasts: T-8, $3-$16; T-10, $3-$6; T-12 (eight-foot), $4-$20; occupancy sensors, $10-$20, based on number of fixtures.

UTILITY	EQUIPMENT	CONDITIONS	INCENTIVES
San Diego Gas & Electric P.O. Box 1831 San Diego, CA 92112 (619) 696-4610, San Diego Co., parts of Orange Co.	• Thermal Storage • Lighting • Gas A/C • Custom • Other	Lighting must be contractor-installed, new construction paid upon installation. All programs have free utility audit.	COMMERCIAL COOLING: $50-$200/kw shifted off-peak, amount determined by utility. CUSTOM: No set range. Efficient measures must pass utility payback and cost-effectiveness test. LIGHTING: 4-foot fluorescent lamps, electronic ballasts, hybrid ballasts, reflectors — contact utility. TITLE 24: Incentives for changes in building plans to achieve net energy savings of 10% or greater in one or more categories.
Seattle City Light 1015 Third Avenue Seattle, WA 98104 (206)684-3254 or 3534	• Lighting • Lighting Controls • Motors • HVAC • Air Conditioning	New construction or retrofit. Verification of pre-existing condition and payment required. Rebates based on rebate list, which is limited by incremental cost. Custom incentives based on kwh savings.	ENERGY SMART DESIGN: 70-100% of incremental cost of measures beyond code or current practices in new construction. 50-70% of installed cost for existing buildings. Equipment rebates available or custom payments based on audit. Building commissioning option allows testing, balancing, and correction of building systems in operational mode. Payment to design team for additional time due to incorporation of recommended measures. Daylighting analysis including scale modeling. Operations and maintenance services offered to ensure measures are maintained for optimum savings.
Sierra Pacific Power Energy Services Dept. P.O. Box 10100 Reno, NV 89520 (702) 689-4242	• Lighting • Window film • A/C • Refrigeration • Economizers • Time Clocks • Motors • Strip Curtains and Doors • Motion Detectors • Photo Cells • Custom	For Peak Performance Program. user must submit plans signed off on by engineer. For "Light Brigade," audit required. Programs apply to both new construction and retrofits. Payments up to 100 percent of project cost.	LIGHT BRIGADE: For business operating 8:00 A.M.-5:00 P.M. or from dusk to dawn with exterior lighting, $125/kw reduced through lighting measures. For operations 10:00 A.M.-10:00 P.M. or more hours, $150/kw. PEAK PERFORMANCE PROGRAM: Incentive negotiated depending on benefit to utility. SPECIAL INCENTIVE PROGRAM: Window film, solor shades;50¢/square foot. Clean A/C condenser coils: $2.50/ton. Economizers: $50/ton. Time clocks: $40 each. Motion detectors: $15 each. Photo cells: $10 each. Strip curtains & doors: $1.50/square foot. Refrigerator door gaskets:; $1/linear foot. Energy-efficient motors 1/2-1/4 hp, $7.50 per motor; 1-10 hp, $7.50/hp: $5/hp. clean refrigeration condensor coils, $10/ton.
Southern California Edison P.O. Box 800 2244 Walnut Grove Ave. Rosemead, CA 91770 (800) 952-5062 Southern CA	• Lighting • Upgrades and Medification • Daylighting Controls • Heat Pump Water Heaters • Air Conditioning • Heat Pumps • Chillers • Evaporative Coolers • Window Treatment • Insulation • Motors • Adjustable Speed Drives • Low Temperature Chemical Dishwashers • Custom	Audit required. Preapproval and postproject inspections required. Installations must be permanent. Proof of purchase and installation required, utility must approve calculations. Rebates cannot exceed 30% of installed cost unless otherwise indicated.	INDOOR LIGHTING SYSTEM REPLACEMENT: $140/kw reduced LIGHTING EFFICIENCY MODIFICATION: $100/kw reduced LIGHTING MODIFICATION: 3¢/annualized kwh saved PHOTO-ELECTRIC DAYLIGHTING CONTROLS: $90/kw controlled HEAT PUMP WATER HEATERS: $15/KBtu capacity LOW TEMP CHEMICAL DISHWASHERS: $50/kw reduced A.C. HEAT PUMPS: $9-$20/ton capacity EVAPORATIVE COOLERS: $70/ton of electric A.C. replaced WINDOW TREATMENT: 80¢/sq. ft. ROOF.WALL INSULATION: 16¢/sq. ft. R-11/R-19 HIGH EFFICIENCY ELECTRIC MOTORS: Per motor, based on hp HIGH EFFICIENCY CHILLERS: $15-$18/ton, based on max kw/ton efficiencies ELECTRONIC ADJUSTABLE SPEED DRIVES: 5¢/annualized kwh saved CUSTOMIZED REBATE: 5¢/annualized kwh saved or $65/kw reduced as determined by Edison.
Southern California Gas Co. P.O. Box 3249, Los Angeles, CA 90051-1249 (213) 244 3729 (800) GAS 2000 FAX: (213) 244-8261, -8262 Most of southern CA excluding San Diego & Long Beach	• Space Heating • Water Heating • Cooking • Boilers • Double-Effect A/C • Process • Other	Installations must be recommended by utility after audit. Preapproval, postproject inspection required. Installations must be permanent. Proof of purchase and installation required. Audits and applications due by Dec. 1.Incentives based on equipment input ratings or estimated energy savings not to exceed 20% of total installed cost unless otherwise noted.	COMMERCIAL NEW CONSTRUCTION: Space heating. $2.50-$15/M Btuh depending on efficiency. Water heating, $2.50-$38.25/MBtuh depending on efficiency. Cooking: $2.50-$10 MBtuh depending on efficiency. Combination hot water-space heating. $3.75/MBtuh. Program will be adjusted July 1. COMMERCIAL RETROFIT: Space heating, $5-$15/MBtuh depending on efficiency. Cooking, $5-$15 MBtuh depending on efficiency. Dryers, $2.50/MBtuh. Steam, water boilers, $1.50/MBtuh depending on efficiency. Gas engine, $50/hp. Weatherization: contact utility. INDUSTRIAL RETROFIT: Space heating, $7/MBtuh. Boiler, $70/hp. Dryer, $4/MBtuh, Furnace/kiln/oven, $3/MBtuh. Process cooking equipment, $2/MBtuh. Misc. equipment (eg., burner replacement), $1.50/MBtuh. Industrial heat recovery, 50% of project cost or 50¢/estimated therm saved, whichever is less. COMMERCIAL/INDUSTRIAL ENERGY MANAGEMENT CONSULTANT PROGRAM: 50¢ x estimated annual therms saved or 50% of consultant study, whichever is less.

South Carolina Electric & Gas Co. P.O. Box 764 Columbia, SC 29218 (803) 748-3220 Southern half of SC	• Thermal Storage • HVAC • Lighting • Standby Generators • Motors and Drives	Pre-and postinstallation inspections required. Most programs both new construction and retrofit; some are retrofit only. Call for details.	THERMAL STORAGE: $100/kw deferred plus demand plus demand rate of $2/kw for additional load added by the system. HVAC: High-efficiency chillers: $100/kw deferred from baseline efficiency. Commercial HVAC: $50/ton for 9 EER or higher for rooftop packaged equipment. GREAT APPLIANCE TRADE-UP (GATU):Variable rates on 11 SEER or higher equipment. GATU piggy back: $200/system for HVAC systems that meet GATU efficiencies and use gas for supplemental heat. GATU Financing: Up to $10,000 at 9% for 48 months for 13 or better SEER. LIGHTING: Relamping: 10¢/6 watts removed from peak load. Ballasts: high frequency units, $165/kw or $5/unit, new construction; $247/kw or $7.50/unit, retrofit. STANDBY GENERATORS: $2/kw contracted plus 7¢/kwh generated. HIGH EFFICIENCY MOTORS: $100/kw deferred. VARIABLE SPEED DRIVES: $235/kw deferred.
Southern Connecticut Gas 880 Broad St. Bridgeport, CT 06604 (203) 786-6117 S. Central CT	• Direct or Indirect air Conditioning • Cogeneration • Gas AC • Non-Emergency Power Generation • Gas engine/Gas or Steam Turbine Mechanical Drives • Conversion Burners	Payments made on a custom basis related to consumption levels. Alternate financing is available to interruptible rate customers committing to minimum consumption for up to five years.	CONVERSION BURNER PROGRAM: $2/mcf to a maximum of $4000/burner. Margin on consumption is used to offset loan principal. OTHER PROGRAMS: Call for information.
Southwest Gas Corporation P.O. Box 98510; Las Vegas, NV 89193-8510; (702) 364-3192; AZ, northern & southern NV, southern CA	• Cogeneration • air Conditioning • Agricultural Water Pumping	Incentive rates available to encourage use of gas for cogeneration, air conditioning, and agricultural water pumping.	Call Scotty Chisholm for information.
Snohomish County PUD #1 9930 Evergreen Way P.O. Box 1107 Everett, WA 98206 (206) 347-1703	• Lighting • Motors • HVAC • Chillers • Insulation • Glazing	A payment cap is set on capital expenditures for plant expansion. Cost-effectiveness criteria are based on predicted savings, expected life of measures and value of saved energy. Projects must be evaluated and approved by utility prior to installation. Current program includes a commercial incentive of up to 50% of installed cost for cost-effective measures. Rebates are established for energy efficient motors. Industrial incentive is $.15 per kWh saved during first year up to 80% of installed cost. Program enhancements are ongoing.	ENERGY SMART DESIGN: 70-100% of incremental cost of measures beyond code or current practices in new construction. 50-70% of installed cost for existing buildings. Equipment rebates available or custom payments based on aud000030ilding commissioning option allows testing, balancing, and correction of building systems in operational mode. Payment to design team for additional time due to incorporation of recommended measures. Daylighting analysis including sale modeling. Operations and maintenance services offered to ensure measures are maintained for optimum savings.
Springfield City Water, Light, and Power 109 N. 7th St. Springfield, IL 62702 (217) 789-2070	• Air Conditioning • Heat Pumps • Insulation • Lighting	Audit not required but utility performs inspection before and after installation. Invoice needed. Rebates apply to retrofits only.	WINDOW A/C, CENTRAL A/C, HEAT PUMPS: Rebate calculated by formula involving savings utility receives over base value (if no incentives), approximate $/kw saved value. Different with each technology. Example: central AC,3-ton unit with 36,000 Btu/h of cooling gets $142.20 rebate; 3-ton heat pump with 11 SEER, rebate is $309.96 (heating value for winter added in). INSULATION: 10% cost of materials up to $1,000 of work, covers installation by qualified contractor. LIGHTING: Based on $220/kw of reduced demand during peak hours weekdays. Payments made to businesses closed part of the week are calculated with proportional fractional multiplier.
Springfield Utility Board P.O. Box 300 Springfield, OR 97477 (503) 746-8451	• Lighting • HVAC	New construction or retrofit. Verification of payment required.	ENERGY SMART DESIGN: 70-100% of incremental cost of measures beyond code or current practices in new construction. 50-70% of installed cost for existing buildings. Equipment rebates available or custom payments based on audit. Building commissioning option allows testing, balancing, and correction of building systems in operational mode. Payment to design team for additional time due to incorporation of recommended measures. Daylighting analysis including scale modeling. Operations and maintenance services offered to ensure measures are maintained for optimum savings.

UTILITY	EQUIPMENT	CONDITIONS	INCENTIVES
Superior Water, Light and Power Co. 1230 Tower Avenue P.O. Box 519 Superior, WI 54880 (715) 394-2226 (715) 394-2224 City of Superior	• Gas Space Heaters • Water Heaters • Boilers • Steam Traps • Air conditioners • Electric Lighting Systems • Variable Speed Drives • KW Reduction Incentives • Energy Efficient Motors	Pre-and post-installation inspections required. Project must be pre-approved by SWL&P representative. Customer must provide proof of purchase. SWL&P must approve or develop the energy savings calculations and must verify installation before rebate can be issued.	COMMERCIAL GAS SPACE HEATING HIGH EFFICIENCY EQUIPMENT: $250 COMMERCIAL GAS WATER HEATING HIGH EFFICIENCY EQUIPMENT: $125 COMMERCIAL GAS COOKING HIGH EFFICIENCY EQIUPMENT: $125 CUSTOM: $150 rebate for each kw reduction up to 50% of total purchase cost.
Tacoma Public Utilities 3315 South 23rd St. P.O. Box 11007 Tacoma, WA 98411 (206) 383-9652 Tacoma, adjacent areas	• Lighting • Motors (1-200 hp) • HVAC	Must save electricity without fuel switching. Current rebate offer good through September 1992. Resulting lighting systems must be more efficient than required by energy code. Motors and HVAC must meet certain minimum efficiencies. New construction or retrofit. Verification of payment required.	LIGHTING: New fixtures, $10-$80 depending on type. Retrofit fixtures with electronic ballasts, $10 (T-12); $15 (T-8), compact fluorescents, $7-$35. OCCUPANCY SENSOR LIGHTING CONTROL: Wall mount $15-$25. MOTORS: $70 to $2,250 depending on size and enclosure. CHILLERS: $300/ton per kw/ton improvement. A/C UNITS: $90/ton/EER improvement. HEAT PUMPS: $250/ton/COP improvement ENERGY SMART DESIGN: 70-100% of incremental cost of measures beyond code or current practices in new construction. 50-70% of installed cost for existing buildings. Equipment rebates available or custom payments based on audit. Building commissioning option allows testing, balancing, and correction of building systems in operational mode. Payment to design team for additional time due to incorporation of recommended measures. Daylighting analysis including scale modeling. Operations and maintenance services offered to ensure measures are maintained for optimum savings. Other measures on a case-by-case basis. Call for specifics on all incentives.
Tampa Electric Co. P.O. Box 111 Tampa, FL 33601-0111 (813) 228-4111 Southwest FL	• Lighting • Standby Generators • Custom	Lighting: Minimum 1 kw load reduction. Ballasts must operate only energy-efficient lamps. Inspection required. Standby generators: Pays for use of existing equipment to reduce peak load. Minimum 50 kw, 50% load factor during peak to qualify, one hour advance notice provided. Peak Value Program: Performance bond and inspection required. new construction and retrofit. Must cut minimum 100 kw from summer peak or 50 kw from winter peak.	
Taunton Municipal Lighting Plant 55 Weir St., P.O. Box 870 Taunton, MA 02780 (508) 824-5844 Taunton, Raunham, Berkeley, North Diahton	• Lighting	"Lightwaves" program. Utility performs on-site assessment, customized plan, installation, quality control check. Customer receives the benefit of and immediate reduction in the electric bill without any up-front costs. Monthly savings are guaranteed. Customer is charged a small monthly fee for five years that equals about 20% of savings.	LIGHTING SYSTEMS: 13¢/watt removed. Program designed for T-8 lamps with 265 milliamp ballast. Savings from lamp/fixture removal, decrease in lumen levels, add-on energy-saving devices not eligible. STANDBY GENERATOR PROGRAM: Base credit of $2/kw/month of generation normally available during our peak. When we call on you to generate, pay 25¢/kw and 10¢/kwh. PEAK VALUE PROGRAM: Any conservation measure is eligible, such as thermal storage, heat pipes, variable speed drives. Pay $250/kw for measures that reduce both summer and winter peaks; $200/kw for reducing summer peak; $150/kw for reducing winter peak. 50% of rebate paid upon completion, balance paid out over five years, during which we meter the project at no cost to the customer.
Texas Utilities, P.O. Box 660268, Dallas, TX 75266-0268, (214) 954-5164 Northern 1/3 of TX	• Thermal Storage • Custom	Programs require capital investment by customer.	ENERGY EFFICIENCY IMPROVEMENT PROGRAM: $50/kw for on-peak efficiency improvements. $100/kw for load shifted from On- to off-peak. THERMAL STORAGE: $250/kw for first 500 kw, $125/kw for additional kw.

TransAlta Utilities P.O. Box 1900, Station M Calgary, Alberta T2P 2M1 (403) 267-7533	• High Efficiency Electric-Motors (HEM) • Energy-Saving Fluorescent Lamps • High Pressure Sodium (HPS) Yard Lights • Refrigerators	Products must meet qualifying requirements. Proof of purchase required for rebate. On-site inspection in some cases. Motors: must be a T-frame (NEMA specifications), 3-phase, AC, Constant speed, new induction motor, 1hp or larger. Must meet or exceed qualifying efficiency, as tested at full-load or 3.4 load (whichever is higher). Motors 500 hp and above considered on an individual basis. Program ends June 1992. Reduced wattage lamps must be purchased before September 30, 1992. Refrigerators must meet qualifications and be purchased before December 31, 1992.	MOTORS: $400/kw saved up to 25% purchase price. Formula used to calculate rebate. $500,000 payment cap. FLUORESCENT LAMPS: 25¢ for three-and four-foot lamps, 50¢ for three-and four-foot lamps. Minimum rebate claim $7.50. HPS YARD LIGHTS: $20/HPS yard light fixture or conversion kit. REFRIGERATORS: $40 each.
Tucson Electric Power Co. P.O. Box 711 Tuscon, AZ 85702 (602) 745-3539 Tucson, AZ	• Efficient Lighting	Minimum of $100 rebate. Pre-and post-construction inspections and original invoice required. Ballasts must meet harmonics requirements.	LIGHTING: Fluorescent-fluorescent, 38¢-$1.88/lamp based on wattage. Incandescent-compact fluorescent, $2.13-$8.40/lamp, varies with wattage. Standard ballast-hybrid, electronic, or dimmable electronic, $9.87-$18.35. Delamping, $5.70-$7.50/tube removed based on wattage. Unused ballasts must be given to TEP for disposal. Specular reflectors, $15-$20. New T-8 fixtures, $14.26-$38.25 depending on number of lamps. Exit signs, $11.07-$17.11 depending on wattage. Timers and occupancy sensors, $10.33.
UGI Corp. Gas Utility Division P.O. Box 13009 Reading, PA 19612-3009 (215) 375-4441 Harrisburg, Lancaster, Lehigh, Reading areas	• Cogeneration • Gas A/C • Dessicant Dehumidification • New Technology • NGV's	Actual rebate amount determined by UGI rep based on nominal cooling capacity, project economics.	EFFICIENT GAS AC: Engine-driven or two-stage absorption systems: $50/ton. Dessicant dehumidification systems evaluated on an individual basis.
United Cities Gas Co. 5300 Maryland Way Brentwood, TN 37027 (615) 373 0104 Portions of TN, GA, VA, IL, MO, IA, KS, SC	• Gas A/C	Projects evaluated individually. Gas air conditioning may qualify for special air conditioning rate.	GAS AIR CONDITIONING: Determined on project-by-project basis. Our pipeline suppliers may also provide rebates. All rebated paid to equipment supplier. Customers and energy decision makers should contact United Cities Gas for more information.
The United Illuminating Co. P.O. Box 1564 New Haven, CT 06506 (203) 787-7585 Bridgeport/New Haven	• Cool Storage • Lighting • Motors • HVAC • Thermal Envelope Measures • Efficient Chillers • Variable Speed Drives • Industrial Electric Technologies	Energy Opportunities pays a percentage of installed cost of electric efficiency improvements. Required pre-, post- inspections. Incentives include costs for design, engineering materials, labor taxes, demolition, and disposal.	COMMERCIAL: 15¢/kwh cap. MANUFACTURING: 30¢/kwh cap. Call for details.

UTILITY	EQUIPMENT	CONDITIONS	INCENTIVES
Washington Gas — District of Columbia Division 6801 Industrial Road. Springfield, VA 22151 DC	• Multifamily • Cogeneration • Gas Cooling • Space Heating • Water Heating • New Construction • Peak-shaving Operators • Commercial Cooking	New construction and retrofits. Efficiency ratings apply in some cases.	MULTIFAMILY: Up to $500/unit for gas heating retrofits; max, 30 units/project, 800 total units. $100/unit for individual unit gas clothes dryer and/or cooking equipment; max, 50 incentives/housing projects. COGENERATION: Up to $1,000/kw for packaged systems 60-100 kw. Requires shared savings agreement. GAS COOLING: Up to $400 for systems 7.5-1,000 tons. Payments vary with system size. SPACE HEATING: $150-$540 for commercial, multifamily, or government installations. depending on system size and efficiency. WATER HEATING: $85/10,000 Btuh input for units at least 80% efficient. NEW CONSTRUCTION: Up to $200,000 for building owners or designers who submit new building plans that use efficient gas technologies for space and water heating and cooling. Payments based on square footage. PEAK-SHAVING GENERATORS: Up to $250/kw for systems 100-1,000 kw to provide peak-shaving or dedicated electricity for building. COMMERCIAL COOKING: No-cost survey plus incentives of 10% of cost for recommended purchases for broilers, convection ovens, fryers at least 65% efficient, infrared griddles, and ranges with power burners.
Washington Gas —Maryland Division, 6801 Industrial Rd Springfield, VA 22151 Montgomery, Prince George's, St. Mary's, Charles, Calvert counties	• Gas Chillers • Furnaces • Boilers	Programs expire Oct. 31, 1991, but may be renewed pending regulatory approval. Must agree to allow us to inspect and monitor equipment twice a year. Installation verification required.	GAS CHILLERS: Up to $400/ton for gas-fired double-effect absorption or engine-driven equipment up to 1,000 tons. Up to $200/ton for steam-fired double-effect absorption process equipment up to 1,000 tons supplied by gas or dual-fuel boilers. Five chiller rebates available. FURNACES: $250-$540 depending on efficiency. BOILERS: Small boilers, $250-$540 depending on efficiency. Large boilers, $150. Rebates are per dwelling unit or per 50,000 Btuh input. total 5,000 boiler/furnace rebates available.
West Koolenay Power P.O. Box 130, Trail, BC VIR4L4 (604) 368-3321 (800) 663-7799, S. Central BC	• Lighting • Motors	All commercial and industrial customers qualify. Programs valid until December 31, 1992. Payments are rebates for certain products or rebates based on energy saved.	LIGHTING: Fluorescent tubes, 25¢ for 4-inch tubes; 50¢ for 8-inch tubes. Efficient magnetic ballasts, $2 per ballast. electronic ballasts 2¢/kwh saved annually for 5 years. Metal halide and HPS lamps, $400/kw saved. Reflectors, $25 for 2-lamp, $35 for 4-lamp fixture. Compact fluorescents, $1-$7.50. Exit sign lamps, $4 each. Controllers, 2¢/kwh saved annually over 5 years. MOTORS: $400/kw saved.
West Texas Utilities Co. P.O. Box 841 Abilene, TX 79604 (915) 674-7294, West TX	• Unitary Air Conditioning • Heat Pump Equipment	Structure must meet minimum insulation requirements. Rebate based on efficiency rating of equipment. On-site inspection of insulation and HVAC equipment required for payment.	UNITARY HEAT PUMPS: $110/ton, 8.7-10.9 EER. $140/ton, 11 or higher EER. UNITARY AIR CONDITIONING: $50/ton, 9.7 EER and above. PACKAGED TERMINAL HEAT PUMPS: &75/unit, 8.7 EER and above.
Wisconsin Electric Power Co. P.O. Box 2046, Milwaukee, WI 53201, (414) 221-4900 Southeast WI, Appleton area, upper peninsula of MI	• Lighting • Motors • Air Conditioning • Insulation • Low-temp. Dishwashers • Controls • Design Incentive • Process Equipment • Refrigeration • Custom	"Smart Money" program. Both retrofit and new construction. Rebates shown are typical; actual amount may vary. Inspections generally necessary before, after installation. Utility will negotiate custom rebates for any improvement that reduces elec. use. Pre-approval required. Contact WEPCo early in the project for best results.	LIGHTING: $4-41/ballast; 50¢-$2/efficient fluorescent lamp; $7-$14/compact fluorescent; $10/current limiter; $10-$50/exit sign; $10/reflector; $1.20-$3.75/incandescent lamp converted to halogen or tungsten halogen; $25-$50/occupancy sensor; $200/kw or lighting load connected to daylighting controls. MOTORS: $5-$20/hp depending on size. AIR CONDITIONING: $20-$35/ton for minimum EER (varies according to size), plus $4/ton per 1 EER above minimum. High efficiency chillers; $10-$40/ton plus $3 for each .01 kw/ton below maximum. Thermal energy storage up to $550/kw and 8¢/kwh shifted off-peak. SOLAR FILM: 80¢/square foot. WATER HEATER INSULATION: $10 each LOW-TEMPERATURE DISHWASHERS: $30-$50/kw removed. REFRIGERATION: $90/lin. ft. for glass doors, $5/lin. ft. for strip curtains. CUSTOM: Available for process, EMS, variable speed drive, refrigerations, air compressors.
Wisconsin Gas 626 E. Wisconsin Ave.	• Industrial Process • Heating Systems	WISE Choice Program. Industrial: standard process equipment must increase to high-efficiency type.	FURNACES, INFRARED, UNIT HEATERS: $100-$1,000 SETBACK THERMOSTATS: 50% of cost up to $75 BOILER CONTROLS: Up to $420. EFFICIENT BOILERS: 10% of installed cost.

Milwaukee, WI 53202 (414) 291-6985 Southeast WI	• Food Service • Water Heating Conversions • Desiccant Dehumidification • Gas A/C • Large Chillers • Custom	Maximum payment $30,000 with cap of 35% of incremental cost. For industrial HVAC systems, maximum $15,000/item with 10% cap. Preapproval required. Commercial: both new construction and retrofits. New multi-family must meet total building energy rating of 3 Btu/Sq. ft./degree day. Boiler tune-ups require combustion test form. Free energy surveys available.	STEAM SYSTEM CONVERSIONS: Up to $15,000 WATER HEATERS: $75-$1,000 depending on usage. DESICCANT DEHUMIDIFICATION: Up to $4,000/unit. GAS COOLING: Up to half cost premium over comparable electric unit. WEATHERIZATION: (existing bldgs) 40% of cost up to $3,000. CUSTOM: Other measures considered on case-by-case basis.
Wisconsin Natural Gas Co. 233 Lake Ave. Racine, WI 53407-0007 (414) 635-2313	• Space Heating • Custom • Water Heating • Air Conditioning • Food Service Equipment • Process Equipment • Boiler Tune-ups • Steam Trap Repair/Replacement • Natural Gas Vehicles	Program runs through December 1992. Available to new or existing facilities in our service territory. Technologies must run on natural gas. Pre-purchase approval required. Maximum rebate per customer, $15,000/year. High efficiency criteria must be met.	SPACE HEATING, WATER HEATING, FOOD SERVICE EQUIPMENT: 15% of project cost. AIR CONDITIONING: Steam absorption chiller $100/ton; engine-driven chiller $125/ton; direct-fired absorption chiller $150/ton. BOILER TUNE-UP: $100 each STEAM TRAP SURVEY: $3 each up to $100/year. STEAM TRAP REPAIR/REPLACE: $30 each up to 50/year. CUSTOM TECHNOLOGIES: 15% of project cost. Also provide submeters free of charge for up to 24 months and free energy surveys of the facility.
Wisconsin Power & Light P.O. Box 192 Madison, WI 53701 (608) 252-3311 (800) 327-2708	• Lighting • Motors • Gas Technologies • Custom • Shared Savings	Inspection, invoice required. both new construction and retrofit.	LIGHTING: Lamps: 4-foot, 50¢: 4-foot slimline, 41; 8-foot, $1.25; screw-in, $5; fluorescent exit sign kits, $10. Ballasts: electronic, $7/lamp controlled; hybrid, 45 lamp controlled. MOTORS: 1-10 hp, $15/hp; 11-50 hp, $10/hp; 51-150 hp, $6/hp; over 150 hp, custom. GAS TECHNOLOGIES: Furnaces: under 60 MBtuh, $200; 61-120 MBtuh, $250; 121-19=80 MBtuh, $300. Steam Traps, $3/trap. Water heater flue dampers, $40. Boiler hot water reset, $150. Boiler temperature cutout, $100. Night setback, $50. Combination reset, setback $250. CUSTOM: Call for information SHARED SAVINGS: If project meets our criteria we will provide upfront financing and guarantee positive cash flow.
Wisconsin Public Service Corp. P.O. Box 19001 Green Bay, WI 54307-9001 (414) 433-4952 Northeast, North Central WI	• Motors • Lighting & Controls • Air Conditioning • Refrigeration • Gas Heating • Insulation • Food Service/Ice Makers • Steam Traps/Boilers • Design Studios • Custom	No limit on total rebate amounts. All rebates apply to retrofits and many to new construction. Custom rebates require pre-approval.	DESIGN STUDIES: 50% of cost up to $5,000 for in-depth study by professional engineer. WISE BUYS COMMERCIAL & INDUSTRIAL REBATES: Motors: 1-200 hp, $15-$2,900: Lighting: 60¢ - $1.50/fluorescent lamp; exit signs: $10-$25; electronic ballasts replaced: $11-$19; delamping: $10/ballast removed; occupancy sensors: $25; fixture conversions: $9-$150; rooftop AC: $14-$34/ton; chillers: $12-$40/ton; refrigeration curtains and doors: $5-$50/linear ft; ice makers: $50-$350; natural gas hot water boilers, furnaces, infrared and unit heaters, gas water heats: $1-$8/1000 Btuh; boiler controls: $150; setback thermostats: $25-$50; vent dampers: $75; insulation: 1¢R-Value added per sq. ft.; solar film: 70¢/sq. ft.; gas boiler tuneup: $125-$175. Contact utility for details or for other programs.
Wisconsin Southern Gas Co., Inc. 120 E. Sheridan Springs Rd. Lake Geneva, WI 53147 (414) 248-8861 Southeast WI	• Gas-fired Equipment • Insulation • Controls • Steam Traps • Other	Payback of 10 years or less required according to our calculations. Installations must be in existing facilities. Rebate program dates October 1 - September 30.	COMMERCIAL: (less than 24,000 therms annual consumption) 10% of installed cost. Maximum rebate $1,500, minimum $50. INDUSTRIAL: (greater than 24,000 therms annual consumption) 10% of installed cost. Maximum rebate $3,000, minimum $50. STEAM TRAPS:. (greater than 24,000 therms annual consumption) $3/trap tested by contractor, $30/trap repaired or replaced. Maximum rebate 100 traps/customer/year.

DIRECT ASSISTANCE

Under this type of DSM program, the utility pays a portion — or all — of the upgrade cost to the installing contractor of the company's choice. Another direct-assistance option is for the utility to provide lighting equipment and/or services through utility engineers or through a utility-chosen contractor.

LOW–INTEREST LOANS

Some utilities offer low-interest loans to help finance lighting efficiency projects. Payments can be added to the electric bill.

GETTING INVOLVED

To take full advantage of a possible DSM program offered by your utility, contact the local utility customer representative and request information. If the utility is in the planning stage of a DSM program, offer to participate in a focus group; this is an opportunity for an organization which is planning an upgrade to tell the utility about its energy policy, its concerns, the factors that influence decisionmaking, and if company personnel would like to see equipment demonstrations.

To maximize the effectiveness of limited DSM dollars, the utility may set requirements for participation. The upgrading organization, for example, may be required to perform a comprehensive energy analysis and purchase only qualifying technologies.

Make sure all deadlines for the rebate application and other paperwork are met. This and the review process may take some time before an application is approved. If the application is approved, complete the installation by the set deadline. Doing so will guarantee continued qualification for the rebate. Adhere to all other requests such as measuring and monitoring of energy savings via special metering equipment.

BE PREPARED TO STAY ENERGY–EFFICIENT PERMANENTLY

DSM's maximum benefit to the utility is derived over the long-term (a permanent reduction in energy use). This type of energy savings is called "persistent savings," and makes the utility's DSM program a real investment in reduced load. If, for example, energy-efficient lamps are replaced with the previously used types that use more energy — but are priced lower — the utility is given "vaporwatts" (demand for power and energy consumption go back to previous levels) and its money is wasted.

Gaining persistent savings is why many equipment-specific programs are being withdrawn and replaced by energy-reduction rebates. Under these agreements, the company's reduced energy use may be measured and monitored for 10 years or more. That means the equipment, such as energy-efficient lamps and ballasts, must be used during that period.

Sometimes, this can result in the customer picking up the cost of replacing lamps. In some third-party financing agreements, the third party will take care of this. If the customer ends up with the bill, it should continue to use energy-efficient lighting, as it has high merit, economy and value on its own without a rebate. Preferably, the customer will have incorporated a commitment to energy-efficient lighting into an energy policy.

MANUFACTURER REBATE

Sometimes, the manufacturer of energy-efficient lighting will offer a direct cash rebate on its equipment. This rebate can be combined with utility rebates to further reduce initial investment.

The manufacturer may also be willing to enter into a lease purchase agreement or long-term financing of all costs (see next section).

THIRD-PARTY FINANCING

Third parties can provide various forms of financial assistance for your lighting upgrade. These institutions generally offer four types of financing agreements: conventional financing, lease-purchase, lease-purchase with guaranteed savings, and shared savings. These options should be compared to the pros and cons of a direct purchase of the services and equipment needed for the upgrade.

If rebates are offered in conjunction with the customer's pursuit of any of these options, these rebates will commonly be paid to the owner of the upgrade. In some cases, this is the customer. But in some lease-purchase and shared-savings agreements, this will be a third party.

CONVENTIONAL FINANCING

Under a conventional financing agreement, the customer makes a downpayment of up to 30% of the project cost and agrees to fixed loan payments until the debt to the lending institution (bank, credit corporation or lighting supplier) is fulfilled. The customer assumes full ownership of the equipment and 100% of the risk (meaning if actual energy savings do not meet the anticipated amount because of a change in building usage

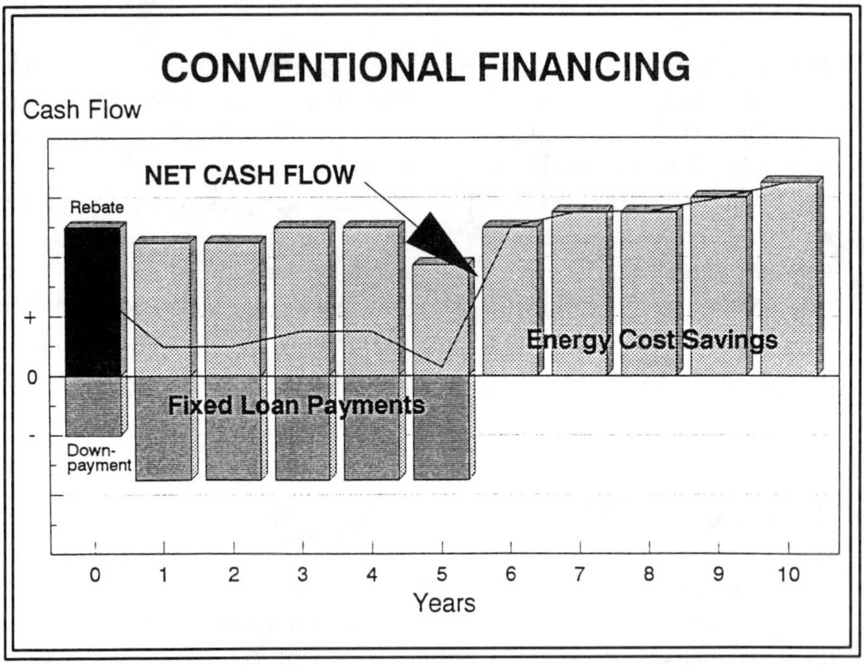

Figure 6-1. Conventional financing.
Source: U.S. Environmental Protection Agency, Green Lights Program.

or other event, the customer suffers the consequences). Depending on current tax laws, depreciation and the interest can be considered tax-deductible. Conventional loans are generally considered the simplest method of financing an upgrade.

LEASE–PURCHASE

Lease-purchasing is offered by banks, leasing companies, contractors, equipment suppliers and some utilities. The features of a lease-purchase is no customer downpayment, tax-deductible payments and customer-assumed performance risk.

The customer makes no downpayment, although the leasing company may require a processing fee and/or the first and last monthly lease payments. The leasing company holds actual ownership of the equipment throughout the duration of the contract (generally five to 10 years). At the end of the contract, the customer can purchase the equipment at a predetermined, mutually agreeable amount, renegotiate the contract or give up the equipment to the leasing company.

The periodic payments on the lease are tax-deductible as a business operating expense. Additionally, the customer "owns" (has in legal possession) the equipment during the contract period, so it can derive tax

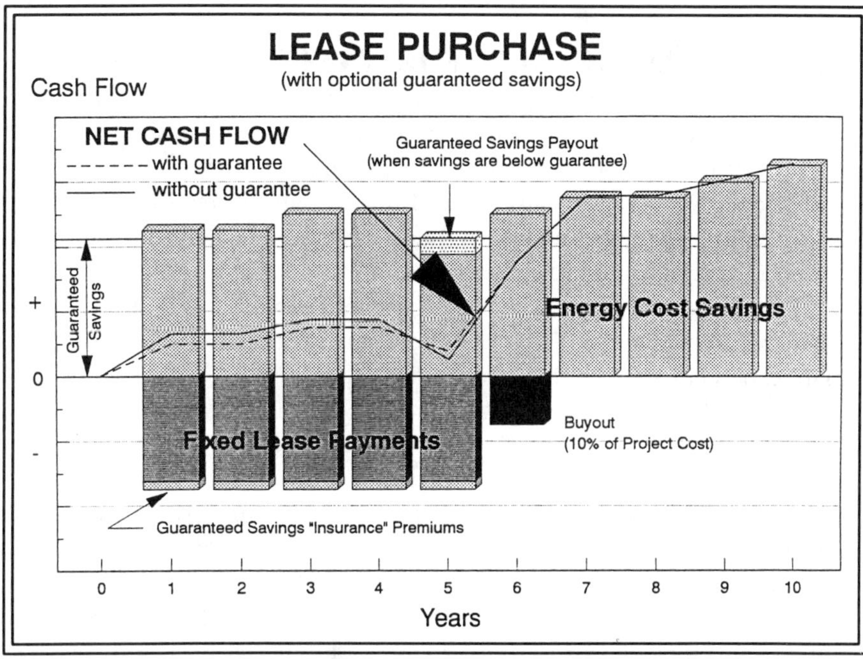

Figure 6-2. Lease-purchasing agreements.
Source: U.S. Environmental Protection Agency.

benefits from depreciation. The equipment offered under these agreements are generally removable types, such as outdoor fixtures. This means the equipment may not be considered an improvement to the customer's real property, and therefore may be depreciable on a relatively fast schedule. (Consult with your chief financial officer and accountant for more information.)

In lease purchase agreements, the customer assumes all of the equipment's performance risk. The customer is also responsible for system maintenance.

LEASE–PURCHASE WITH GUARANTEED SAVINGS

The lease-purchase agreement — as well as conventional financing agreements in some cases — can be amended with a guaranteed-savings clause. This clause is typically offered by the same organization which leased out the equipment. The customer pays for the protection of this clause in relatively small increments as if it were purchasing an insurance policy on the projected energy savings. Essentially, the clause guarantees that energy savings will be greater than an agreed-upon minimum amount. This amount typically equals the fixed lease payment for the same period. This guarantees that the customer will experience positive cash flow.

SHARED SAVINGS (PERFORMANCE CONTRACT)

The shared savings agreement, or performance contract, is an innovative approach to financing a lighting upgrade. It is usually offered by an energy service company (ESCO), which commonly represents a group of professionals involved in business and finance, and in engineering as well. Other contractors may also offer a shared savings agreement directly, or may be subcontractors of the ESCO.

Features of the shared savings agreement include no customer downpayment, immediate positive cash flow, payments based solely on performance, no performance risk, and third-party ownership of the equipment.

Here is how such an agreement works. The utility sends out a request for resource bids to meet new energy demand. The ESCO bids on satisfying the resource need by reducing demand in the service territory by a number of megawatts. If agreed, the utility awards the ESCO these megawatts to remove from the utility's load. To fulfill this pledge, the ESCO approaches various utility customers with optimum energy savings potential and proposes building upgrades, which typically includes lighting but can also include HVAC improvement, thermal storage, controls, energy management systems and other improvements.

The customer pays nothing upfront for the cost of the upgrade. In a shared savings agreement, the customer enjoys installation of a new lighting system by the ESCO — or a contractor connected with the ESCO — and

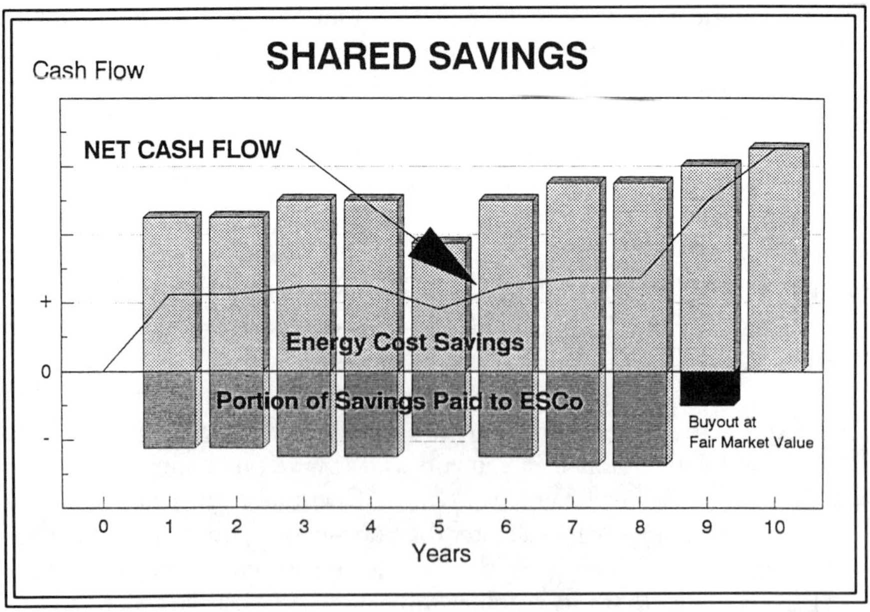

Figure 6-3. Shared savings.
Source: U.S. Environmental Protection Agency, Green Lights Program.

OVERVIEW OF FINANCING OPTIONS

	DIRECT PURCHASE	CONVENTIONAL FINANCING	LEASE PURCHASE	SHARED SAVINGS
Initial Payment	100% of project cost	0-30% of project cost	$0 or deposit	$0
Periodic Payments	none	fixed	fixed	% of energy cost savings
Performance Risk	owner 100%	owner 100%	variable: based on guarantee	investor 100%
Contract Termination Options*	n/a	principal payoff	fixed buyout; renew; return	fair market value buyout; renew; return
Ownership	building owner	building owner	lessor	investor
Tax Deductions**	depreciation	depreciation and interest	lease payments	shared savings payments

* at end of term

** subject to change in tax laws; consult with tax advisor regarding eligibility

Figure 6-4. Comparison of financing options.
Source: U.S. Environmental Protection Agency, Green Lights Program.

faces no initial investment. The ESCO will reap its profit from the energy savings, making it a true "performance contractor." The burden of risk of the new lighting system's performance is placed squarely on the ESCO.

The customer makes immediate positive cash flow. It pays the ESCO a percentage of the resulting energy savings (a payment which, depending on tax laws, may be tax-deductible as a business operating expense) and keeps the rest. Since the customer did not pay anything upfront, it enjoys immediate positive cash flow. The actual amounts paid to the ESCO and kept by the customer will vary year to year as energy use fluctuates. To determine accurate payment, and demonstrate to the utility that the savings are real and persistent, lighting energy use is often measured and monitored with electronic submetering equipment for the duration of the contract (typically 10 to 15 years). Generally, the percentage allocated to the ESCO is higher in the first few years (such as 80%) to allow the ESCO to recover a payback for the initial investment. After this point, the split gradually becomes more even.

The ESCO may receive payment from the utility as well as the customer. In these cases where the ESCO receives the utility incentives, the customer can negotiate a reduced payment and keep more for himself.

Because the ESCO is paid by saving energy, it has incentive to maintain the lighting system, replace defective components, and relamp with energy-efficient lamps. All equipment is owned by the ESCO. At the end of the contract period, the customer can purchase the improvements of a mutually agreeable value, renegotiate the contract or terminate the contract, in which case the ESCO can recover the equipment.

ROLLING FINANCING

Facility managers are usually interested in seeing a trial installation or at least equipment demonstrations before committing large resources to an upgrade. This process assures them, at least initially, that the equipment works reliably for the occupants and their needs.

The trial installation idea can be taken one step further to encourage the company to upgrade its buildings one at a time. The concept is called "rolling financing."

Here is how it works. The company chooses the building which presents the greatest opportunity. Factors influencing opportunity include installation cost, availability of utility assistance, electric utility rates and amount of energy to be saved through the proposed upgrade.

After the lighting system in this building is upgraded, the energy savings are put into an "account." The company waits until this account grows large enough to fund the initial cost of the next building's upgrade — and so on down the line, until all of the buildings are upgraded. This process can take years, but there are creative ways to use rolling financing to see all of the buildings upgraded within a reasonable time frame.

Sources:

Barrett, Larry B. "What Customers Should Know About Utility DSM Programs," *Proceedings of GLOBALCON 1992*, March 23-27, 1992.

Costello, Richard J. and David Blazon, "ESCOs Bring Opportunities for Lighting Management Companies," *The MainLighter: The Lighting Management & Maintenance Magazine*, March 1992.

Energy User News. "Embracing DSM, Utilities Plan to Spend Billions in the '90s," April 1992.

Flanigan, Ted, "Environmental Costing for Utilities Brings Opportunity for

Lighting Management Companies," *The MainLighter: The Lighting Management & Maintenance Magazine*, November 1991.

Hines, Virginia, "Demand-Side Management Rebate Programs: Which Way Are They Headed?" *Strategic Planning for Energy and the Environment*, Winter 1992.

Lucy, Jim, "The Energy Opportunity: Market Drivers," *Electrical Wholesaling*, December 1992.

Meade, William and Elliot Roseman, "Make Room for DSM," *Independent Energy*, January 1992.

Morron, Thomas D. "Electric Utility DSM Programs: Current Status and Future Direction," *The MainLighter: The Lighting Management & Maintenance Magazine*, April 1992.

Payne, William F. "DSM Needs Redefinition," *Strategic Planning for Energy and the Environment*, Fall 1991.

U.S. Environmental Protection Agency, Green Lights Program, *Lighting Upgrade Manual*, First Edition, November 1991.

Further Reading:

Demand-Side Management, Volume 3: Technology Alternatives and Market Implemention Methods. Palo Alto, California: Electric Power Research Institute, December 1984. EPRI EA/EM-3597 v3.

Demand-Side Management, Volume 4: Commercial Markets and Programs. Palo Alto, California: Electric Power Research Institute, September 1987. EPRI EA/EM-3597, v4.

DSM Regulatory Impacts. Palo Alto, California: Electric Power Research Institute, March 1988. EPRI EM-5631.

Destribats, Alan F. "DSM: Customer Value and a Profit Opportunity," *Marketing Executives Conference*, October 15, 1991.

Gellings, Clark W. and John H. Chamberlin. *Demand-Side Management: Concepts and Methods*. Lilburn, GA: The Fairmont Press, 1988.

Reid, Michael. "Analyst Predicts $30 Billion Market Over Next Decade, 20% of Expected Growth," *Electric Power Alert*, October 15, 1991.

Schweitzer, Martin, et al. *Key Issues in Electric Utility Integrated Resource Planning: Findings from a Nationwide Study*, Oak Ridge National Laboratory, April 1990.

Utility Energy Strategies: The Role of Efficiency, Productivity and Conservation. Palo Alto, California: Electric Power Research Institute, February 1989. CU-6272.

Source: U.S. Environmental Protection Agency, Green Lights Progam.

CHAPTER 7

Green Engineering: Assessing Environmental Impact

LIGHTING UPGRADES BRING GREEN BENEFITS

Studies and polls of American consumers in recent years indicate a developing strong awareness of environmental degradation brought about by air and water pollution. They also indicate a strong desire for government and private industry to take stronger action to both prevent pollution and conserve natural resources. Practical engineering solutions and advancing technology can make a significant impact on these environmental problems. Upgrading the lighting system to reduce energy consumption, for example, can measurably reduce air pollution and the amount of non-renewable natural resources consumed during electric generation at utility power plants. And it can do so without the sense of painful sacrifice generally tagged to corporate environmental protection. In fact, corporate commitment to environment protection through lighting upgrade can be powerful ammunition in a marketing campaign. Such a campaign can target environmentally aware and sensitive consumers, possibly influencing choices during a period when these consumers are starting to "vote with their wallets."

ENERGY'S RELATIONSHIP WITH THE ENVIRONMENT

When electric energy is generated at a fossil-fuel-fired utility, byproduct gases including carbon dioxide, sulphur dioxide and nitrogen oxides are emitted into the atmosphere from the power plant's smokestacks. The utility industry has made great strides in improving efficiency, switching to low-sulphur fuels, and installing flue gas desulphurization equipment (scrubbers) and other pollution control devices. Coal used by U.S. power plants in 1985, for example, produced 44% less sulphur dioxide than in 1975. Despite these advances by the utility industry, substantial pollution is nonetheless produced as demand for electricity continues to grow and we continue to rely primarily on fossil fuels for the needs of our "electric society."

CARBON DIOXIDE (CO_2) AND GLOBAL WARMING

Carbon is considered one of the building blocks of life and is found in almost every material, including food, wood and fossil fuels (coal, oil and natural gas). Whenever any substance containing carbon is burned, carbon dioxide is produced. Earth's atmosphere contains much of this naturally produced gas. Carbon dioxide does not hurt property, plants or trees. Large amounts are absorbed by oceans and by plants and trees, which through photosynthesis convert the gas to oxygen, a process which aids in plant growth and productivity.

The Industrial Revolution brought dependence on fossil fuels to generate electricity and the high standard of living we enjoy today. Because the combustion of these fuels has added large amounts of carbon dioxide into the atmosphere, and this gas has a lifespan of some 500 years, Earth has not been able to assimilate it naturally. Since 1860, worldwide carbon dioxide production (all sources) has increased from a negligible quantity to six billion tons, with two-thirds of this increase occurring in the past 30 years. In 1990, carbon dioxide levels were measured at about 350 parts per million by volume of air — a peak untouched for 130,000 years. Enough carbon dioxide has accumulated to cause the evolution of the so-called "Greenhouse Effect."

According to the Greenhouse theory, the sun's rays freely penetrate carbon dioxide and strike the earth, reflecting upward as infrared radiation (heat). Carbon dioxide absorbs infrared radiation at certain wavelengths and traps it within the atmosphere. This trapped energy raises worldwide mean temperature, the process called global warming.

Consequences of global warming may include:

A *warmer climate*. Earth warmed one degree Fahrenheit since 1890. The 1980s were the warmest decade in over a century, with 1987 and 1988 the warmest years on record.

A *more hostile climate*. As temperatures rise, heat waves may occur in regions in high and middle latitudes. In drier and warmer areas, forest fires will be more likely to occur. Reduced snowpack in mountain regions may lead to more flooding in the winter and less moisture during the summer.

A *drop in food production*. The cultivated land marked for agriculture in the American West may shrink by nearly one-sixth (15%) to as much as one-third (31%) due to drought. U.S. corn production dropped 17% in 1980, 28% in 1983 and one-third (34%) in 1988. The 1988 drought marked the first time in recent U.S. history that grain production fell short of consumption.

Forest migration. Healthy forests now account for 33% of U.S. land area. In Greenhouse scenarios, many species of American trees attempt to migrate north because they live in a narrow temperature range optimum for healthy life. A rapid increase in temperature may cause forests to shrink during migration. Many species of trees in the eastern United States will be replaced by shorter, hardier shrublike trees that will be more adaptable to the new climate.

A *rise in sea level*. Sea level is measured as mean (worldwide) and relative (local). Generally, land subsidence caused by compacting soil and other phenomenon causes sea level to rise four to eight inches every 100 years. According to the Intergovernmental Panel on Climate Change (IPCC), however, global warming expands seawater and will erode (melt) the polar ice caps, causing sea level to rise at a greater rate: a mean rise of about four inches to one foot, with a "best guess" of about eight inches, by the year 2030.

The best-guess rise of about eight inches by the year 2030 is expected to raise local sea level by 1 foot-1 inch along the East Coast and over three feet in sensitive areas in California, Texas and Louisiana, particularly the ecologically important wetlands. These increases are expected to erode beaches, disperse sewage and other wastes and damage coastal infrastructure such as roads and bridges. They may also increase flooding and saltwater intrusion into groundwater supplies.

The best guess estimate of three feet by 2100, while a seemingly modest figure, is expected to severely erode shore front. According to the U.S. Environmental Protection Agency, the cost of holding the coast against a 3-foot, 4-inch rise in sea level would cost the U.S. between $73 and $111 billion.

SULPHUR DIOXIDE (SO_2) AND ACID RAIN

Like carbon dioxide, sulphur dioxide is produced both naturally and artificially. In the case of electric generation, this gas is the byproduct of the combustion of coal or oil. Electric generation in 1987 accounted for 66.2% of sulphur dioxide emissions. Once in the atmosphere, sulphur dioxide mixes with moisture and forms sulphuric acid, the primary ingredient in the phenomenon known as acid rain.

Consequences of acid rain include:

Threat to aquatic life. Thousands of lakes across the North American continent no longer support any form of aquatic life. The National Acid Precipitation Assessment Program (NAPAP), set up in 1979 by then-President Jimmy Carter to measure acid rain's effect on the environment, concluded in 1990 that acid rain is responsible for a large share of the acidification of sensitive lakes and streams in the eastern U.S. (As a lake's pH drops to a level of 6 to 6.5, sensitive fish and other organisms begin to lose their ability to reproduce. Below a pH of 5, the entire ecosystem begins to collapse.) Although the NAPAP study showed the damage was not as widespread as previously believed, many fish populations in the Adirondacks have vanished because of the acidity's damage to their reproductive cycles. In a survey sponsored by New York State and a group of electric utilities, 23.5% of the 1,469 lakes were found to be fishless.

Threat to vegetation. Acid rain also damages trees and vegetation. The NAPAP study, again finding that the damage to forests was not as widespread as many believed, nonetheless concluded that populations of high-elevation red spruce in the northern Appalachians are likely to be weakened by contaminated soil. In the future, accumulating long-term acidification of soil may pose a deeper threat to the trees as the acid displaces nutrients and absorbs aluminum.

NITROGEN OXIDES (NO_X) AND OZONE POLLUTION (SMOG)

Nitrogen oxides are produced naturally and by manmade processes. During electric generation, it is produced when small amounts of nitrogen in fossil fuels mix with oxygen during the fuels' combustion. In the atmosphere, nitrogen oxides mix with other elements and forms of air pollution (oxygen, carbon monoxide and hydrocarbons other than methane). Sunlight triggers a chemical reaction and produces ozone (O_3) and other photochemical oxidants in the troposphere (lower atmosphere). In the stratosphere (upper atmosphere), ozone shields the earth from the sun's harmful ultraviolet radiation, but in the troposphere this same molecule is ozone pollution, or smog.

In the 1980s, ozone pollution appeared to be decreasing modestly. The U.S. keeps tabs on tropospheric ozone pollution through a network of 42 monitoring stations. They reported a decline in peak ozone levels by 13% between 1979 and 1986. Some five to 10 stations per day, however, reported levels that exceed the U.S. ozone standard.

Consequences of smog include:

Threat to plant life. Ozone damages plant and leaf tissues and interferes with photosynthesis. It also weakens plants and trees to disease, drought and other pollutants. Even an ozone concentration considered "average" is strong enough to reduce corn crop yield by up to 1%, soybeans and cotton by about 7%, and alfalfa by 30% or more. The NAPAP estimates that aggregate crop yields are reduced by 5% to 10%. This represents, according to the EPA, a loss of $1 billion to $2 billion annually to U.S. agriculture.

Threat to humans. Ozone can damage lungs and the respiratory tract. In 1986, more than 75 million Americans lived in areas which exceeded the national ozone standard. In 1990, a U.S. Department of Energy survey found that 121 million Americans lived in areas where ozone smog and other air pollution exceeded Federal health limits. Breathing unhealthy air is believed responsible for up to 50,000 premature deaths among Americans every year. Ozone pollution is also generally attributed with irritation of the eyes, decreased visibility and objectionable smells.

CORPORATIONS CAN HELP THE ENVIRONMENT BY HELPING THEMSELVES

For years, energy conservation (and environmental protection) have often been equated with depressed economic growth. Jimmy Carter told America to "sacrifice and suffer" to save energy during the Energy Crisis. Ronald Reagan equated energy conservation with "just being too hot in the summer and too cold in the winter."

Many companies followed this lead, and believing that environmental protection occurs at the expense of economic growth, placed themselves in a confrontational position with environmentalists. Today, a majority of consumers consider themselves environmentalists and allow corporate environmental reputation to affect their purchasing decisions.

Reducing energy consumption via strategic lighting management presents the opportunity to commit to environmental protection while saving money. With environmentalism as high on the consumer's agenda as it is now, that fosters public good will and makes good marketing.

MARKETING GREEN BENEFITS

Good marketing strategies identify consumer needs and appeal to them through marketing tools such as advertising, public relations, brochures, annual reports and other communications. Excellent marketing strategies do all of that *and* tell consumers that the company is committed to their long-term needs. A company which reduces energy use and thereby reduces air pollution is contributing to a quality of life that most Americans want to see sustained or improved for their own future and for future generations. And unlike some products and companies which are merely "dressed up" as environmentally sensitive through marketing, this company's impact on the environment is sincere. Consumers generally respond both to perceived sincerity and to environmental commitment.

In fact, three out of four are likely to.

THREE OUT OF FOUR AMERICANS ARE ENVIRONMENTALISTS

Many people believed America's environmental craze would peak with the first Earth Day back in the 1970s and wane. The continued swing toward environmental concern, however, has outlasted the notion of it being a fad for two decades. Today, environmental awareness has become part of mainstream values. Here are the trends:

ITEM #1: More than three-fourths (77%) of Americans say a company's environmental reputation affects their purchasing decisions, according to a July 1989 survey conducted by the Michael Peters Group. This was confirmed by a Gallup poll ("over 75%").

ITEM #2: Nearly all Americans (97%) want natural resources preserved for future generations (*American Demographics*, March 1989), a strong sentiment that went on in 1991 to stall efforts to drill for oil in the Arctic National Wildlife Refuge.

ITEM #3: Three out of four (75%) Americans consider themselves "environmentalists," according to a later Gallup poll.

ITEM #4: Four out of five (80%) believe air and water pollution to be the greatest problems facing America in the next quarter-century, a Roper poll indicates.

ITEM #5: World membership in environmental groups jumped from

13.3 million in 1988 to 15.9 million in 1989, according to *Time* magazine.

ITEM #6: Four out of five Americans (80%) agree that "protecting the environment in the United States is so important that requirements and standards cannot be too high," a *New York Times* and *CBS News* public opinion poll indicates.

ITEM #7: In 1990, the fourth goal on a list of nine set by the National Earth Day Committee stated, "A transition to renewable energy resources and increased residential and industrial energy efficiency."

ITEM #8: In September of 1989, the Coalition of Environmentally Responsible Economies (CERES) — a group of corporate investors who control $150 billion in assets — set forth the Valdez Principles as a reaction to the *Valdez* oil spill and other environmental concerns. The document was a collaborative effort between these investors and environmental groups; it urges corporations to "reduce waste, use resources prudently, market safe products and take responsibility for past harm." One of the 10 environmental guidelines for American companies is "wise use of energy." CERES has even promoted the idea of a competitive rating system which could illustrate levels of corporate environmental concern to insurance brokers, lenders, employees and consumers. The document suggests that how far companies go to implement the Principles may affect investment.

Because some Americans do not consider that the business community has done enough to protect the environment, companies are seeking ways to benefit the environment in order to satisfy consumer demands for environmental responsibility. A proven method of fostering a consumer's good will is to demonstrate a commitment to his or her long-term benefit.

SELLING GREEN BENEFITS TO SENIOR MANAGEMENT

Selling a lighting upgrade to top management generally requires a satisfying payback and return on investment based on projected energy savings. As recent years have produced more environmentally aware and sensitive consumers, incorporating an environmental commitment into marketing strategy can provide an additional incentive to upgrade to senior management. This is an excellent opportunity for the facility manager, energy manager and/or energy engineer to work with the marketing communications department and develop a plan to promote the corporation's commitment to the environment through efficient use of energy. Many companies today are committing to lighting upgrades primarily for the environmental benefits, particularly in California. Overall, green benefits provide additional incentive for management to say "yes" to lighting efficiency.

FEDERAL GREEN LIGHTS PROGRAM

In January 1991, the U.S. Environmental Protection Agency (EPA) initiated Green Lights, its voluntary, non-regulatory program which encourages major U.S. corporations to upgrade lighting systems to reduce air pollution associated with electric generation — but only where profitable and where lighting quality is maintained or improved. Green Lights represents an opportunity to receive support, from choosing equipment to locating financing resources, on to marketing the environmental benefits of a corporate commitment to energy-efficient lighting.

Participating corporations, called Green Lights Partners, start the process by making a firm commitment to the lighting upgrade at the senior management level. Partners sign a Memorandum of Understanding which stipulates their commitments to Green Lights and the EPA's commitments to Partners. Partners agree to:

1. Survey lighting systems in all of their U.S. facilities.
2. Upgrade the lighting system in 90% of all qualifying building space within five years. It is mutually understood by the Partner and Green Lights that no upgrade need be undertaken unless it first passes the Profitability Test. The Profitability Test stipulates that the chosen upgrade should generate an internal rate of return of the prime interest rate plus 6%.
3. Educate employees about the company's commitment to Green Lights energy efficiency and pollution prevention goals.
4. Participate in the Green Lights public recognition program to increase awareness of the environmental benefits of energy-efficient lighting.
5. Meet Green Lights goals in new construction.

Green Lights, in return, commits to helping the Partner complete each step. The Partner can take advantage of several programs from Green Lights:

1. The National Lighting Product Information Program has been developed by Green Lights and the Lighting Research Center to help Partners bypass the process of sorting through competing vendor claims and objectively assess the performance of lighting equipment. The Program publishes specifier reports for different equipment types such as electronic ballasts, current limiters, specular reflectors, occupancy sensors and compact fluorescent lamps.
2. Lists of manufacturers, contractors and utilities, respectively called Green Lights Manufacturer Allies, Lighting Management Company Allies and Utility Allies, are available so that Partners can quickly access professional assistance. Green Lights Allies sign their own Memorandum of Understanding with the EPA and have committed to the same pollution prevention goals as Partners. Green Lights does not endorse any

Figure 7-1. Green Lights was launched by the U.S. Environmental Protection Agency in 1991 to encourage U.S. corporations to upgrade their lighting systems to protect the environment and reduce their electric bill. *Source: U.S. Environmental Protection Agency, Green Lights Progam.*

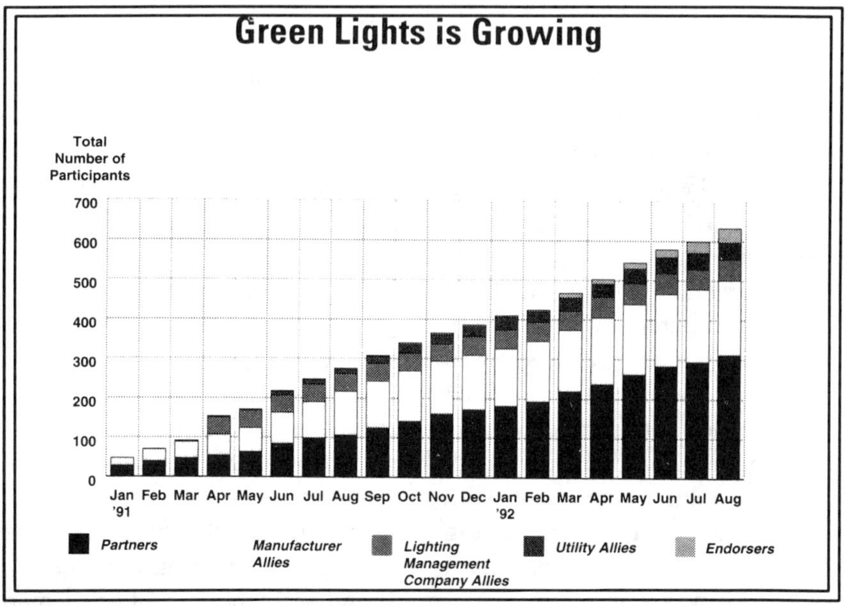

Figure 7-2. Green Lights has built strong momentum in recruiting companies to participate. As of January 1993, more than 350 corporate "partners" and 300 trade and utility "allies" agreed to Green Lights goals. Some 58 organizations, including the Association of Energy Engineers, have endorsed the program.
Source: U.S. Environmental Protection Agency, Green Lights Program.

Ally, but provides the standards Allies agree to meet in their Memorandum of Understanding.

3. The Decision Support System package can help Partners survey their facilities, assess equipment options and check for their qualification in the Profitability Test. The package contains a computer software program which conducts many of the calculations and checks the results against required return on investment, IESNA recommendations and other specifications. The package also contains the *Lighting Upgrade Manual*, which takes the Partner step by step through the process of a lighting upgrade. Green Lights holds workshops around the country to educate Partner staff about how to use the Decision Support System.

Additionally, the computer software allows Partners to tap into the Green Lights Bulletin Board, an electronic bulletin board which posts information about how Partners can provide progress reports on the status of upgrades. It also provides information regarding disposal of fluorescent lamps and electronic ballasts, national and regional pollution prevention figures, *Light Briefs* about equipment technologies, a lighting glossary, and a summary of ASHRAE/IES 90.1 building standards.

4. Resources to educate employees about the benefits of energy-efficient lighting are available. Green Lights was designed with the philosophy that no upgrade should be taken unless full support is received by the Partner from both senior management and their personnel.

5. A public recognition program has been launched which calls attention to the pollution prevention efforts of Green Lights Partners. Advertisements featuring Partner corporate logos are placed in magazines. A special Green Lights Partner logo is available for use in the Partner's own marketing communications as a badge signifying its commitment to Green Lights goals.

6. Green Lights can supply a financing database containing a comprehensive computerized list of electric utility demand side management programs which specify rebates for lighting and other forms of support such as a free audit of the lighting system.

7. A newsletter, called the *Green Lights Update*, is distributed monthly to all program participants, highlighting Partner successes and providing additional information about Allies and programs.

The designers of Green Lights acknowledge the Electric Power Research Institute's estimate that if energy-efficient lighting were installed nationwide wherever technically feasible and profitable, total U.S. demand for electricity would be cut by about 10%. A 10% reduction translates to $18.6 billion in electric bill savings. Regarding the environment, it translates to a 4% reduction in total U.S. carbon dioxide emissions (the equivalent of 43 million cars), a 7% reduction in total sulphur dioxide emissions and a 4% reduction in total nitrogen oxides emissions. Green Lights has ambitiously accepted these estimates as its goals.

Combined, Green Lights Partners — and Allies, who also agree to comply with Green Lights goals — represent some 2.9 billion square feet, or 3% of all U.S. industrial/commercial building space.

Green Lights Participants

(733 as of January 11, 1993)

Green Lights Partners (364)
Green Lights Allies (311)
Green Lights Endorsers (58)

CORPORATE PARTNERS (339 total)

Parent Organizations:

A & C Enercom
Abbott Laboratories
AES Corporation
Alamance County Schools
Albany General Hospital
ALCOA
Allergan
Alliance to Save Energy
Alliance for Environmental Education
Alta Bates Medical Center
American Express Company
American Public Power Association
American Council for an Energy Efficient Economy
American Standard, Inc.
Amoco
ARCO
Ashland Oil, Inc.
Automatic Data Processing, Inc.
Aveda Corporation
Baldor Electric Company
Bancorp Hawaii, Inc.
Bank of America
Baxter Healthcare Corporation
Bay Area Hospital
Bear Stearns Companies, Inc.
Bechtel
Bell Atlantic
Bellcore
Blue Cross & Blue Shield Mutual of Ohio
BMG/RCA Music
Boeing
Boulder Valley Public School District
Brandeis University
Brooklyn Union Gas Company
Broward Community College
Brown University
Browning Ferris, Inc.
The Bruce Company
California State University System
Camp Dresser & McKee, Inc.
Canyon Ranch
• Cape Canaveral Marine Services, Inc.
Cardolite Corporation
Career Track
Carnegie Mellon University

The Oliver Carr Company
The Catalyst Group
Caterair International Corporation
Center for Applied Engineering
Central Florida Community College
Central Carolina Bank
Charming Shoppes, Inc.
Chemical Bank
Cherry Hill Board of Education
Chevron
Childhelp USA
Cibola County Schools
Citicorp/Citibank
Citizen's Photo
City University of New York
Cleveland State University
Club Corporation, International
Colonial Pipeline
Colorado Sate University
Columbia University
COMPAQ Computer Corporation
Continental Insurance
Coolfont Resort
Cosmair, Inc.
Cracker Barrel Corporation
Crestar Bank
CTEC Corporation
Data General Corporation
Defender Services, Inc.
The Dexter Corporation
Digital Equipment Corporation
DMB Associates, Inc.
Downtown Plaza Towers Associates
Dresser Rand
Dura Pharmaceuticals
Duracell U.S.A.
E.J. Brach Corporation
Earth Care Paper
Earth Share
Eaton Corporation
Electric Power Research Institute
Eli Lilly and Company
Elkhart General Hospital
Enron Property Company
Environmental Defense Fund
Environmental Law Institute
Facility Management Consultants
First Data Corporation
First Union National Bank
Flynn Lumber & Supply Co., Inc.
Fred Meyer, Inc.
Friend Public Schools
G.M. Popkey Company, Inc.
General Dynamics Corporation
Genovese Drug Stores, Inc.
Georgia Institute of Technology
Gerber Products Company
Gibson Speno Companies
Gillette Company
Godfrey Realty Company, Inc.
Goodyear Tire & Rubber Company
Government Development Bank of Puerto Rico
Grainger
Graphics Technology International
Gross Enterprises
Grumman Corporation
GTE
Harris Corporation
Hasbro, Inc.
Haworth, Inc.
Herman Miller, Inc.
Hewlett-Packard Company
Highlands Regional Medical Center
Hilton at the Circle
Hoechst Celanese
Hofstra University
The Home Depot
Honeywell, Inc.
Hope Network
Horton Memorial Hospital
Hotel San Remo-Ramada
Humana, Inc.
Huntsville City Schools
ICF International
INOVA Health Systems
Intergraph Corporation
International Technology Corporation
International Institute for Energy Conservation
Interstate Hotels Corporation
IPS Electric and Midwest Gas
Irish Boat Shop
ITT Corporation
Izaak Walton League
J.E. Seagram Corporation
Jaakko Poyry
Jackson State University
Jay Peak Ski and Summer Resort
The Jewish Home for the Elderly
Joan Fabric Corporation
Johns Hopkins/SAIS
Johnson Controls World Services
Johnson & Johnson
Kerr-McGee Corporation
Kolar Management, Inc.
L.L. Bean, Inc.
Leon County School Board

Liz Claiborne, Inc.
Lockheed Corporation
Louisville & Jefferson County Metropolitan Sewer District
Louisville Resource Conservation Council
Lyondell Petrochemical Company
M-Tac Enterprises
MagneTek, Inc.
Marriott Corporation
Martin Marietta Corporation
Maryland Science Center
Massachusetts Institute of Technology
Mattel, Inc.
Maytag
McKeesport Hospital
McNeil Real Estate Management
The Meade Group, Inc.
Meadowcreek
Medcenter
Medical College of Ohio
The Melville Corporation
Memorex Telex
Memorial Hospital of Lafayette County
Mercer University
Methodist Hospital of Lafayette County
Metropolitan Water Reclamation District of Greater Chicago
3M
Mobil Corporation
Monsanto Company
Motorola Incorporated
Mt. Bachelor Ski & Summer Resort
National Westminster Bancorp
National Service Industries
National Wildlife Federation
National Semiconductor Corporation
Natural Resources Defense Council
The Nature Conservancy
NBD Bank, N.A.
NBS Consultants
Nestlé USA
New Canaan YMCA
Nike, Inc.
North Ottawa Community Hospital
North Carolina Alternative Energy Corporation
North American Philips Corporation
Northeast Bancorp, Inc./Union Trust Corporation
Northern Arizona University
Northern Illinois Medical Center
Northwest Georgia Regional Hospital
Northwood School District
NYNEX Corporation
Odyssey of America
OECO Corporation
Okaloosa-Walton Community College
The Old North Church
One Marconi Place, Inc.
Oxford Properties Florida, Inc.
Palmer Bellevue Corporation
Pasadena City College
Perry Drug Stores
Phillips Petroleum Company
Phoenix Home Life Mutual Insurance Company
Pima Community College
Planned Parenthood of South Central Michigan
Polaroid Corporation
Proper Management, Inc.
Provident Life & Accident Insurance Company
Quad Graphics
Ravenswood Hospital Medical Center
Redlands Federal Bank
Richman Gordman, Inc.
Ricoh Electronics, Inc.
The Ritz-Carlton Hotel Company
Rochester Institute of Technology
SAIC
Santa Cruz Valley Union High School
School Administrative Unit #51, New Hampshire
The School Board of Sarasota County, Florida
SCT Yarns, Inc.
Sealed Air Corporation
Service Merchandise Company
Shell Oil Company
Siemens Corporation
Sisters of St. Francis of Sylvania, Ohio
Skeff Distributing Company, Inc.
Sony Corporation of America
St. Paul Fire and Marine Insurance
St. Elizabeth Medical Center
St. Michael Hospital
Stamats Communications, Inc.
Standard Microsystems Corporation
State University of New York at Stony Brook

State Farm Mutual Automobile Insurance Company
Steelcase, Inc.
Straub Clinic and Hospital
Student Loan Marketing Association
Sun Company, Inc.
Super Valu Stores, Inc.
Supermarkets General Corporation
Tampa General Hospital
Texaco, Inc.
Texas Air Control Board
The Timberland Company
Toshiba America, Inc.
Trade Press Publishing Corporation
Transamerica Corporation
Tufts University
Turner Broadcasting System
• Tucson Solar Village
Underwriters Laboratories, Inc.
Union of Concerned Scientists
Union College
Union Camp Corporation
Unity College
University of Redlands
University of Southern Maine
University of Illinois at Chicago
University of Miami
University of Rochester
University of Florida
University Corporation for Atmospheric Research (NCAR)
University of Georgia
US West, Inc.
US Bancorp
USF & G
USX
Van der Hosrt USA
Wachovia Corporation
Walton Monroe Mills
Warner-Lambert Company
The Washington Times
Waste Management, Inc.
Western Digital Corporation
Westin Hotels & Resorts
Westinghouse Electric Corporation
Westminster College
Whirlpool Corporation
White Castle System, Inc.
Wolverine World Wide
Woodloch Pines
World Resources Institute
Xerox Corporation
Yellow Freight System, Inc.
Yosemite Community College District
Zurn Industries, Inc.

Subsidiary Organizations:

Alaska Airlines, Inc.
American & Efird, Inc.
ANR Pipeline Company
Aristech Chemical Corporation
B.P. Exploration
B.P. Exploration-Alaska
Bath Iron Works
BellSouth Telecommunications
Carolina Freight Carriers Corporation
Carrier Corporation North America
Chem-Nuclear Geotech
Colonial Pacific Leasing
Dean Witter Realty
Domino's Pizza Corporation
Energy User News
First Data Resources, Inc.
First National Bank of Chicago
Geneva Pharmaceuticals, Inc.
GPU Services Corporation
Home Box Office
Horizon Air Industries, Inc.
IMS America Ltd.
Jantzen, Inc.
Jewel Food Stores
Karastan Bigelow
Kenyon Oil Company, Inc.
Lone Star Steel
Powell Electrical Manufacturing Company
Preston Trucking
Reliance Standard Life Insurance
Solar Turbines, Inc.
Solvay Minerals, Inc.
Southern California Gas Company
Southwire Company
Thrift Drug, Inc.
Viskase Corporation
Walt Disney Studios

Divisions:

CIBA-GEIGY Pharmaceuticals
Cox Newspapers
Macwhyte Company

GOVERNEMENT PARTNERS
(25 total)

The State of California
The State of Florida
The State of Hawaii
The State of Idaho
The State of Maine
The State of Maryland
The Commonwealth of Massachusetts

The State of Missouri
The State of Nebraska
The State of Oregon
The Commonwealth of Pennsylvania
The State of South Dakota
Virgin Islands, Government of the United States
The City of Azusa, California
The City of Houston, Texas
The City of Naperville, Illinois
The City of Oxnard, California
The City of Portland, Oregon
The City of Tallahassee, Florida
The Town of Northwood, New Hampshire
Broward County, Florida
Dade County, Florida
Douglas County, Oregon
Hillsborough County, Florida
Leon County, Florida

MANUFACTURER ALLIES
(203 total)

A Weatherization Co/AWXCO
A.L.P. Lighting + Ceiling Products
Advance Control Technologies, Inc.
Advance Transformer Company
ALCOA Aerospace/Commercial Rolled Products
Amalco Metals, Inc.
American Energy Management
American Illuminetics, Inc.
American Lighting Corporation
American Lighting Systems
American Louver Company
American Systems and Services
Amerlux, Inc.
Appliance Control Technology, Inc.
Area Lighting Research
Art Directions, Inc.
Badger USA, Inc.
Brayer Lighting, Inc.
Bright Side Lighting
Brownlee Lighting
Bryant Electric
Canterra Electronics International
CCR Lighting Technologies
C.E.W. Lighting, Inc.
• Chloride Systems, Manufacturer of Exide Lightguard Products
CMB Associates, Inc.
Columbia Lighting, Inc.
Control Systems International
Cooper Lighting
CSL Lighting Mfg., Inc.
Dark To Light, Inc.
Davis Controls Corporation
Dazor Manufacturing Corporation
Dielectric Coating Industries
Digecon
DuraLux Industries
Duray Fluorescent Manufacturing
Duro-Test Corporation
Dynamic Energy Products, Inc.
East Rock Manufacturing and Technologies
Edison Price Lighting
Elba USA, Inc.
Electronic Ballast Technology, Inc.
Emergency Safety Products, Inc.
Energy & Environmental Lighting Services
Energy Dezign Corporation
Energy Saving Products, Inc.
Enersave Company
Enerton Technologies
Enterprise Lighting, Inc.
Environmental Energy Group
ESCO International
Etta Industries
Exitronix Division of Barron Manufacturing
Fail-Safe Lighting Systems
Feit Electric Company
Finelite
First Lighting
Flexiwatt Corporation
Flexlite, Inc.
FTI
FulCirlce Ballast Recyclers
GE Lighting
The Genlyte Group
Good Earth Lighting, Inc.
Guardian Lighting Controls, Inc.
Harris Manufacturing, Inc.
Heath Company
Hetherington Industries
Holphane Company, Inc.
Honeywell, Inc.
House O'Lite
Hubbell Incorporated, Lighting Division
Illumination Control Systems

INCON Industries
Industrial Energy Systems, Inc.
Indy Lighting
Integrated Power & Lites
International Energy Conservation Systems
Isolite Corporation
Janmar Lighting
Jedcor Energy Management Company, Inc.
Johnson Controls, Inc.
Juno Lighting
K-Lite Division of ICI Acrylics/K-S-H Inc.
Kenall
Kilowatt Saver, Inc.
Kim Lighting
King Technology, Inc.
The Kirlin Company
Lamar Lighting Company
Legion Lighting
LexaLite International
Light Energy Corporation
Lighting & Lowering System
Lighting Resources, Inc.
LightMedia Corporation
Lightron of Cornwall, Inc.
LiteTronics International
Lights of America
Lightway Industries
Litecontrol
Lithonia Lighting
Lorin Industries
LSI Industries
Lumatech Corporation
Lumax Industries, Inc.
Lumen-Tronics
Magnaray International
MagnaTek, Inc.
Marvel Lighting Corporation
Megalite Corporation, Inc.
Mercury Recovery Services
MetalOptics, Inc.
3M
MirrorLight, Inc.
ML Systems
Moldcast, Division of USI Lighting
Mor-Lit
Motorola Lighting, Inc.
Mule Emergency Lighting
MyTech Corporation
National Lighting Company
Neonix
Norbert Belfer Lighting
NOVA Conservation and Load Management
Novitas, Inc.
NRG Lighting, Inc.
Optical Coating Laboratory, Inc.
Optilight, Inc.
OrEqual, Inc.
Osram Sylvania, Inc.
Paragon Electric Company, Inc.
Paramount Industries
Parke Industries, Inc.
Parrish Lighting and Engineering
Pearless Lighting Corp.
Peschel Energy Inc.
Philips Lighting Company
PLC-Multipoint
Pleamonn Products
Powerline Communication
Pre Finish Metals, Inc.
Prescolite, Division of USI Lighting
Prescolite Controls, Inc.
Prime Ballast
Pritchett Wilson Group
Progress Lighting, Inc.
RAB Electric Manufacturing
Reflect-A-Light
Reflective Light Technologies
Remtec Systems
Robertson Transformer Company
Roth Bros., Inc.
Ruud Lighting, Inc.
Salesco Systems USA
Save-A-Watt, Inc.
Scientific Component Systems
Sea Gull Lighting Products
Sensor Switch
Sharlin-Lite
Silverlight Corporation
Simkar Lighting Fixture Company
Solar Electric Systems of Kansas City, Inc.
Solar Kinetics, Inc.
Southco Metal Services
Spaulding Lighting Inc.
SPI Lighting, Inc.
Sportlite, Inc.
Standard Enterprises, Inc.
Steelcase, Inc.
Sterling, RMC
Stocker & Yale
Systematix, Inc.
Tamarack Corporation
Tek-Tron Enterprises
Teron Lighting
Terralux, Inc.
The Robert Group
Thomas & Betts Commercial and Industrial Lighting
Thomas Industries, Inc.
Topaz Energy Systems
Toshiba America Consumer Products
Triad Technologies
TrimbleHouse Corporation
TSAO Designs
Ulster Precision, Inc.
UNENCO
Valmont Electric

- Videssence

Venture Lighting International
Vision Impact Corporation
Visual Images
Waldmann Lighting Company
Warner Technologies
The Watt Stopper, Inc.
Wellmade Metal Products Company
H.E. Williams, Inc.
Wismarq Light Company, Inc.
Xtra Light
X-Tra Light Systems, Inc.
Zumtobel Lighting, Inc.

UTILITY ALLIES
(49 total)

American Electric Power Service Company
Arizona Public Service Company
Atlantic Energy
Bangor Hydro Electric
Boston Edison Company
Central Maine Power
City of Georgetown, Texas
City Utilities of Springfield
Consolidated Edison of New York
Duke Power Company
Florida Power Company
Grant County Public Utility District
Green Mountain Power Corporation
Greenville Utilities Commission
Idaho Power Company
Jersey Central Power & Light Company
Kansas City Power & Light Company
Los Angeles Department of Water and Power
New England Electric System
New York Power Authority
Northern States Power Company
O & A Electric Cooperative
Oklahoma Gas and Electric Company
Omaha Public Power District
Orange and Rockland Utilities
Orlando Utilities Commission
Pacific Gas & Electric Company

- Potomac Electric Power Company

Pike County Light and Power Company
Port Angeles Light Department
Portland General Electric Company
PSI Energy, Inc.
Puerto Rico Electric Power Authority
P.U.D. #1 of Grays Harbor County
Public Service Electric and Gas Company
Puget Sound Power & Light Company
Rockland Electric
Sacramento Municipal Utility District
Salt River Project
San Diego Gas & Electric
South Carolina Electric & Gas Company
South Carolina Public Service Authority
Southern California Edison Company
Springfield Utility Board
Tampa Electric
Taunton Municipal Lighting Plant
The UNITIL System of Companies
Wisconsin Electric Power Company
Wisconsin Power & Light Company

LIGHTING MANAGEMENT COMPANY ALLIES (59 total)

A-1 Lighting Service Company
ABD Lighting Management Company
Aetna Corporation
American Lighting, Inc.
Amtech Lighting Services

- Applied Energy Management, Inc.

Approved Lighting Corporation
Barney Roth Company
BK Engineering
Broadway Maintenance Company
Cherry City Electric
Chicago-Edison Corporation
Colorado Lighting
Conserve Electric Company, Inc.
Continental Lighting Services, Inc.
Creative Lighting Maintenance
Efficient Lighting and Maintenance
Energy Controls & Concepts
Energy Matrix
Eveready Electric Company
Fluorescent Maintenance Company (CO)

Fluorescent Maintenance Service, Inc. (FL)
Fluorescent Maintenance Service, Inc. (MS)
Fravert Services, Inc.
General Lighting and Sign Service
IllumElex Corporation
Imperial Lighting Maintenance
Innovative Lighting Services
Kenetech Energy Management, Inc.
Light Source
LighTec, Inc.
Lighten Up, Inc.
Lighting Consultants International
Lighting Maintenance, Inc.
Lighting Maintenance and Service
Lighting Management Corporation
Lighting Solutions
Lighting Systems Too!
Luminaire Service, Inc.
M E Energy Resources
Master Lighting Service
Mira Lighting and Electric Service
Murphy Electric Maintenance Company
New Mexico Energy Consultants
Planned Lighting, Inc.
Primo Lighting Management
Professional Lighting, Inc.
ProLite Lighting and Sign Maintenance
Reflections, Inc.
Sica Electrical & Maintenance
Stay-Lite Lighting
Suburban Lighting, Inc.
Superior Light and Sign Maintenance Co.
Sylvania Lighting Services
United Electrical Maintenance Corporation
Universal Lighting Services
USA Energy Corporation
Vista Universal, Inc.
Xenergy, Inc.

ENDORSERS
(58 total)

Aerospace Industries Association of America
• Airport Council International — North America
Alliance for Environmental Education
American Advertising Federation
American Gas Association
American Hotel & Motel Association
American Public Power Associates
American Society for Hospital Engineering for the American Hospital Association
American Society of Interior Designers
Associated Industries of Massachusetts
Association of Demand-Side Management Professionals
Association of Energy Engineers
Association of Professional Energy Managers
Atlanta Regional Commission
CommonSense
Connecticut Business & Industry Association
Consulting Engineers Council of Metropolitan Washington
Consumer's Counsel Governing Board, State of Ohio
Council of State Governments
Edison Electric Institute
Electric Ideas Clearinghouse
Energy Efficiency Trade Alliance
Environmental Action
The Environmental Exchange
Environmental Law Institute
Florida Institute of Government
Georgia Municipal Association
Greater Atlanta Chamber of Commerce
Illuminating Engineering Society of North America
Institute For Alternative Futures
interNational Association of Lighting Management Companies
international Facility Management Association
International Institute for Energy Conservation
Iowa Association of Business & Industry
Izaak Walton League
Maine Chamber of

Commerce and Industry
The Maryland Chamber of Commerce
Minnesota Chamber of Commerce
National Association of Chain Drug Stores
National Association of Regulatory Utility Commissioners (NARUC)
National Energy Management Institute
National Restaurant Association
National Rural Electric Cooperative Association
Nevada Professional Facility Managers
Association
New Hampshire Business & Industry Association
New Jersey Hospital Association
North Carolina Consumers Council
North Carolina Solar Energy Association
Northeast Public Power Association
Northwood, New Hampshire Conservation Commission
Ohio Pollution Prevention Network
Pacific Northwest Pollution Prevention Research Center
Saddleback Mountain
Lion's Club, New Hampshire
Smaller Business Association of New England
Union of Concerned Scientists
University of Tennessee Center for Industrial Services
Virgin Island Retailers Association
Wisconsin Center for Demand-Side Research

• Denotes new participant the week of January 11, 1993.

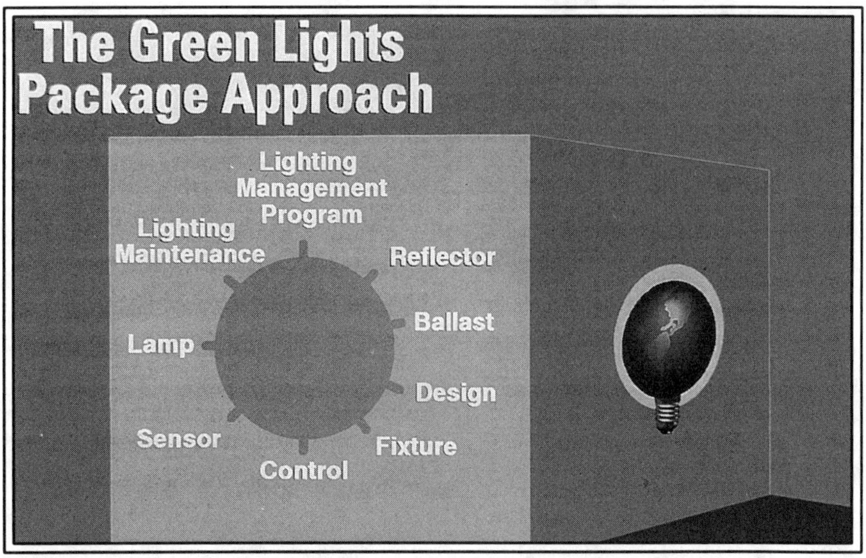

Figure 7-3. The Green Lights program is comprehensive, incorporating lamps, ballasts, occupancy sensors, reflectors and other technologies in addition to quality concerns and planned maintenance. *Source: U.S. Environmental Protection Agency, Green Lights Program.*

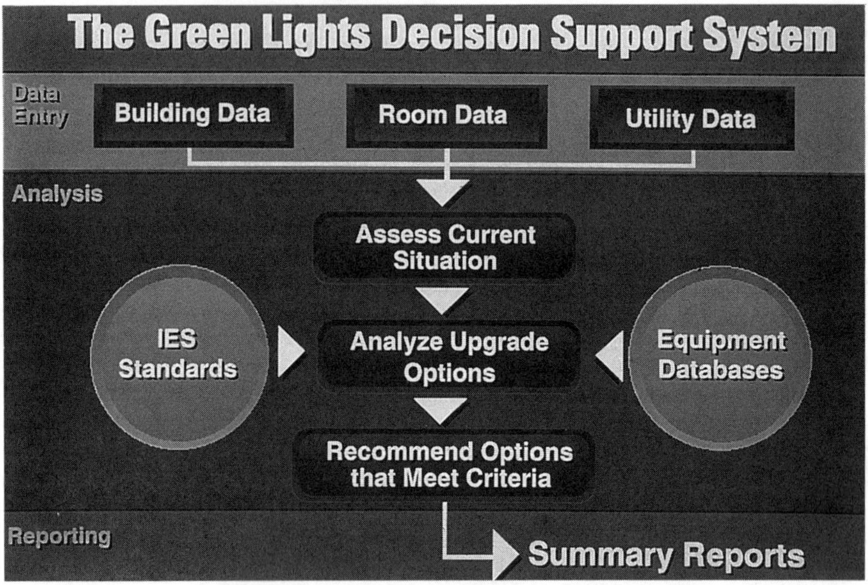

Figure 7-4. The Green Lights Decision Support System is a computerized survey and analysis tool designed to help Green Lights Partners make proper lighting upgrade choices. The Decision Support System typically advises choice of the option which presents the highest energy savings and thereby prevents the most pollution. *Source: U.S. Environmental Protection Agency, Green Lights Program.*

Figure 7-5. A public recognition program advertises the success of Green Lights and Green Lights Partners. *Source: Source: U.S. Environmental Protection Agency, Green Lights Program.*

More information is available by contacting:

GREEN LIGHTS
Global Change Division
Office of Air and Radiation
U.S. Environmental Protection Agency
401 M Street, SW (6202J)
Washington, DC 20460
(202) 775-6650
FAX: (202) 775-6680

Green Lights offers a brochure and introductory video which discusses the merits of the program and how corporations can become Partners.

MEASURING DEFERRED EMISSIONS OF AIR POLLUTION

According to the U.S. Environmental Protection Agency, generating one kWh of electricity results in the byproduct production of an average quantity of carbon dioxide, sulphur dioxide and nitrogen oxides, depending on the fuel source:

Table 7-1. Air pollution emissions per kWh.

	AIR POLLUTION PRODUCED PER kWh*		
Fuel Source	Carbon Dioxide	Sulphur Dioxide	Nitrogen Oxides
Coal	2.4 pounds	9.0 grams	4.4 grams
Oil	1.9 pounds	3.7 grams	1.5 grams
Natural gas	1.4 pounds	0.0 grams	2.4 grams
Average all sources	**1.6 pounds**	**5.3 grams**	**2.8 grams**

*Environmental multipliers which are the EPA-rated average for each state are published Appendix I.

To determine average prevented air pollution emissions, complete Environmental Assessment Worksheet #1.

Environmental Assessment Worksheet #1
Calculating Deferred Emissions

Area description: WORD PROCESSING DEPT. Area No.: 4
Fixture before upgrade: 4 F40/CW LAMPS + 2 MAGNETIC BALLASTS
Fixture after upgrade: 4 F32T8 LAMPS + 1 ELECTRONIC BALLAST

DEFERRED EMISSIONS

	Carbon Dioxide	Sulphur Dioxide	Nitrogen Oxides
1. Reduction in energy consumption (kWh) per fixture	312	312	312
2. Appropriate environmental multiplier	x 1.6 lbs.	x 5.3 g.	x 2.8 g.
3. Resulting emissions	= 499.2 lbs.	= 1,653.6 g.	= 873.6 g.
4. Conversion rate to obtain pounds		÷ 454	÷ 454
5. Resulting emissions		= 3.64 lbs.	= 1.92 lbs.
6. Number of fixtures receiving this upgrade	x 90	x 90	x 90
7. Total emissions	= 44.928 lbs.	= 327.6 lbs.	= 172.8 lbs.
8. Conversion rate to obtain tons	÷ 2,000	÷ 2,000	÷ 2,000
9. Resulting emissions	= 23 tons	= 0.16 tons	= 0.09 tons

To complete Environmental Assessment Worksheet #1, first identify the type of fixture which received retrofit and what type of retrofit occurred.

1. Calculate annual energy savings in kWh per fixture resulting from the retrofit.

2 and 3. Multiply this by the appropriate environmental multipliers (see Table 7-1). Environmental multipliers which are the EPA-rated average for each state are listed in Appendix I. The resulting calculation provides figures representing deferred emissions (pounds of carbon dioxide and grams of sulphur dioxide and nitrogen oxides) per fixture.

4 and 5. If the resulting figure is in grams, divide by 454 to determine pounds.

6 and 7. Multiply by total number of fixtures in the area receiving this retrofit. The resulting calculation provides figures representing total deferred emissions in pounds.

8 and 9. Divide pounds by 2,000 to determine tons.

OPTIONAL. Once you have calculated deferred carbon dioxide emissions in tons, divide by 5.5 to determine how many "cars were taken off the road." (One automobile produces an average 5.5 tons of carbon dioxide per year. This image will help people such as marketing professionals visualize the numbers.)

CALCULATING CONSERVED RESOURCES

Fossil fuels arc in finite supply. According to some estimates, the United States could experience serious shortages or oil and natural gas early in the next century. A lighting upgrade not only reduces air pollution, but also conserves the resources that were to produce the electric energy.

The heat energy content of fossil fuels is measured in British Thermal Units (BTU). Table 7-2 shows the units of measure and heat content for popular fuels.

Table 7-2. Heat Content of Typical Fuels.

FUEL	TYPE/GRADE	UNIT OF MEASURE	HEAT CONTENT IN BTU
Oil	No. 1	Gallon	135,000
Oil	No. 2	Gallon	139,000
Oil	No. 4	Gallon	145,000
Oil	No. 6	Gallon	153,000
Natural Gas		Cubic Foot	1,000
Natural Gas		Therms	100,000
Propane		Pound	21,560
Butane		Pound	21,180
Coal	Anthracite	Short Ton	28-30,000,000
Coal	Bituminous	Short Ton	25-28,000,000
Electricity		Kilowatt-hour	3,414

Sources: U.S. Bureau of Mines, U.S. Department of Energy, 1993 Building Efficiency Standards, *Federal Register, Vol. 54, No. 18, January 1, 1989; the American Society of Heating, Refrigeration and Air Conditioning Engineers (ASHRAE)* Handbook of Fundamentals.

To determine the approximate amount of fuel saved, complete Environmental Assessment Worksheet #2 on the next page.

Environmental Assessment Worksheet #2:
Calculating Conserved Resources, Page 1

Area description: WORD PROCESSING DEPT. Area No.: 4
Fixture before upgrade: 4 F40/CW LAMPS + 2 MAGNETIC BALLASTS
Fixture after upgrade: 4 F32T8 LAMPS + 1 ELECTRONIC BALLAST

	Coal	Oil	Natural Gas
1. Type of fuel	ANTHRACITE	NO. 2	
2. Heat content (BTU) per unit	8,000,000 BTU/Ton	139,000 BTU/Gal.	1,000 BTU/cub.ft.
3. Conversion	÷ 2,000 lbs./Ton		
4. BTU per unit	= 4,000 BTU/lb.		
5. Conversion	÷ 3,414 BTU/kWh	÷ 3,414 BTU/kWh	÷ 3,414 BTU/kWh
6. kWh per unit	= 4.1 kWh/lb.	= 40.72 kWh/Gal.	= 0.29 kWh/cub.ft.
7. Efficiency losses	x 0.33	x 0.33	x 0.33
8. Adjusted kWh per unit	= 1.35 kWh/lb.	= 13.44 kWh/Gal.	= 0.1 kWh/cub.ft.
9. Units per kWh	1 ÷ 1.35 kWh/lb.	1 ÷ 13.44 kWh/Gal.	1 ÷ 0.1 kWh/cub.ft.
10. Units per kWh	= 0.74 lbs./kWh	= 0.074 Gals./kWh	= 10 cub.ft./kWh
11. Reduction in energy consumption per fixture	x 312 kWh	x 312 kWh	x 312 kWh
12. Amount of fuel saved	= 230.88 lbs.	= 23.09 Gals.	x 3.120 cubic feet
13. Number of fixtures receiving this upgrade	x 90	x 90	x 90

Environmental Assessment Worksheet #2:
Calculating Conserved Resources, Page 2

Area description: WORD PROCESSING DEPT. Area No.: 4

	Coal	Oil	Natural Gas
14. Total fuel saved	=20,779.2 lbs.	=2,078.1 Gals.	=280,800 cub.ft.
15. Percentage of electric generation at the utility for this fuel source	x 0.6 %	x 0.2 %	x 0.1 %
16. Actual fuel saved	= 12,468 lbs.	= 416 Gals.	= 28,080 cub.ft.
17. Conversion determine larger unit of measure	÷ 2,000 lbs./Ton	÷ 42 Gals./Barrel	
18. Fuel saved in large unit	= 6 Tons	= 10 Barrels	

To complete Environmental Assessment Worksheet #2, again identify the type of fixture which received retrofit and what type of retrofit occurred.

1. For each fuel (coal, oil or natural gas), write in the type of fuel used (whether the coal is anthracite or bituminous, etc.). Obtain from the local utility.

2. Determine the appropriate heat content in BTU per unit of fuel (tons, gallons or cubic feet). Obtain from Table 7-2.

3 and 4. Divide tons of coal by 2,000 to determine heat content of the fuel per pound.

5 and 6. Divide heat content per unit of fuel by 3,414, the heat content in BTU per kWh of electricity. The result of this calculation is the kWh per unit of fuel.

7 and 8. Multiply by 0.33 to determine losses due to inefficiencies in production, transmission and distribution of power.

9 and 10. 1 divided by the result determines units of fuel per kWh.

11 and 12. Multiply by energy savings (kWh) per fixture to determine amount of fuel prevented from consumption per fixture.

13 and 14. Multiply conserved fuel per fixture by the total number of fixtures receiving this retrofit to determine total fuel saved.

15. Few utilities use exactly the same percentage of fuels in their gen-

eration mix. For example, a midwestern utility might rely on coal for as much as 75% or more of electric generation; an eastern utility might rely on an even split of coal, oil and natural gas. Fuel sources in the "other" category include nuclear, hydroelectric and solar power, which don't produce or produce a negligible quantity of air pollution. Write the percentage of the local utility's reliance on each fuel source for electric generation. Obtain from the utility or the Energy Information Administration, U.S. Department of Energy.

16. The resulting figure represents the actual fuel saved in pounds, gallons or cubic feet.

17 and 18. If in pounds, divide by 2,000 to determine tons. If in gallons, divide by 42 to determine barrels.

AN INNOVATIVE IDEA: USING AN AVOIDED ENVIRONMENTAL COST (AEC) FACTOR

In Chapter 6, we discussed environmental externalities as a motivating factor for utilities to embrace demand side management. In review, these are environmental costs required by a growing number of public utility commissions to be internalized by utilities during resource bidding and least-cost planning.

New York State, for example, assigned an environmental cost of 1.4 cents per kWh to coal-fired power plants (as of November 1991). The state levied $1,959 per ton for nitrogen oxides and $0.98 per ton for carbon dioxide. These figures represent costs assigned to fish killed from acid rain, lost coastal area due to global warming, increased health care costs due to pollution, etc.

Doanh Van, Ph.D., PE, Energy Utilization and Technical Information Engineer for Warner-Lambert in Morris Plains, New Jersey, determined an environmental cost per kWh to be incorporated into the simple payback analysis, a demand side version of the environmental externality. This figure, called by Van the Avoided Environmental Cost (AEC) factor, recognizes a temperature rise in accordance with the global warming theory. It represents the future cost to society of cooling spaces in a warmer climate and of lost agricultural yield due to drought.

Here's an example of a scenario in which an investment can avoid a future cost. Suppose an oil tanker's hull is damaged in an accident and the cleanup costs $500 million to the company's stockholders. If this accident could have been avoided through a certain ship design, $500 million in savings could be annualized over the life of that design and create a payback period. This is the rationale behind using the AEC factor; making an investment in energy efficiency saves the company not only money on the electric bill, but defers future payment of the costs of environmental damage caused by global warming.

COSTS INCORPORATED INTO AEC

Even in conservative scenarios, the impact of global warming is expected to be expensive. The damage caused by a warmer climate includes loss of coastal areas due to flood, loss in tourism, recreation, health effect and the overall effect of dislocation of large populations. Van's AEC deals solely with two costs that can be reasonably estimated: 1) increase in energy demand and 2) loss in agriculture.

Cost of increasing energy demand

In many global warming scenarios, the earth is projected to warm 0.9 to 5.6 degrees Fahrenheit by the year 2025. As average temperature increases, humans are expected to compensate through technology. This means more artificial cooling.

According to *Britannica World Data* (1988), there are one billion dwelling units. Van starts with the assumption that each will have at least a one-kilowatt cooling machine. Due to warmer and more humid weather, people are likely to run these machines harder and longer, we assume for an hour per day.

Van therefore estimates cost for additional energy demand:

1 BILLION UNITS X 1 HOUR/DAY X 365 DAYS/YEAR X 1 kW/UNIT = 365 BILLION kWh

If an economic value of one U.S. penny is assigned to each kWh, the total cost of this increase in energy use would be:

365 BILLION kWh X $0.01/kWh = $3.65 BILLION/YEAR

Cost of decrease in agriculture

As the agricultural industry depends on climate for production, it is expected that worldwide production will suffer in a Greenhouse future.

Britannica World Data (1988) reports that total world agricultural value in U.S. dollars is about $2.62 trillion each year. During the 1980s, U.S. agriculture suffered from frequent drought. In 1988, during which the hottest summer on record was recorded, agricultural loss amounted to about $14 billion — 15% of total U.S. agricultural value.

Based on these facts, Van estimates that the world's farmers will lose 10% of their yield in the face of a 5.9-degree Fahrenheit rise in global temperature. This amounts to $262 billion each year.

Calculating AEC

Together, both costs add to roughly $265 billion per year. In the 13 post-oil-embargo years, worldwide oil consumption increased about 10% annually. If $265 billion were spent annually on energy efficiency upgrades to prevent a 10% increase in energy consumption, or 7.65 trillion kWh, the AEC factor is:

$$\text{Avoided Environmental Cost Factor} = \frac{\$265 \text{ Billion/Year}}{7.65 \text{ Trillion kWh}} = \$0.035/\text{kWh}$$

Van calculates the AEC per deferred kWh to be 3.5 cents. This figure is comparable with the results of ongoing research at the Pace University Center for Environmental Legal Studies.

IMPACT ON THE PAYBACK FORMULA

As discussed in Chapter 4, the simple payback formula is:

$$\text{Payback (in years)} = \frac{\text{Capital Investment (\$)}}{\text{Resulting Savings (\$/Year)}}$$

The AEC affects the formula:

$$\text{Payback (in years)} = \frac{\text{Capital Investment (\$)}}{\text{(Resulting Savings + AEC) (\$/Year)}}$$

where AEC = AEC factor x kWh savings

Van developed this cost and its application because, in his words: "A problem that we as engineers face today is that too many good energy efficiency and conservation projects are rejected on the grounds that they do not warrant a fast-enough payback on investment. This study proposes that the AEC be calculated into payback measures as a valid — albeit indirect — economic return. With new support coming from this modified payback formula, which effectively shortens the payback period of lighting and any other energy use improvement projects, energy professionals may be able to receive commitment from senior management to projects that would otherwise be denied."

Sources:

Blake, Norm, "Conserving Our Nation's Resources: Measuring the Environmental Benefit," *MainLighter: The Lighting Management & Maintenance Magazine*, November 1991.

Blake, Norm, "How Saving Energy Saves the Environment," *MainLighter: The Lighting Management & Maintenance Magazine*, August 1991.

Blake, Norm, "Measuring the Environmental Benefit," *MainLighter: The Lighting Management & Maintenance Magazine*, September 1991.

Bobrow, Edwin E. and John Fritz, "Corporate Strategies for Energy and the Environment," *Strategic Planning for Energy and the Environment*, Fall 1991.

Christensen, Nancy, "Lighting Usage and the Environment," *MainLighter: The Lighting Management & Maintenance Magazine*, December 1991.

Flynn, Jack, "Green Marketing: How to Put an Environmental Spin on Your Energy Product, Service, Program or Project," *Strategic Planning for Energy and the Environment*, Fall 1992.

Harvey, Hal and Bill Keepin, "Energy, Acid Rain and Air Pollution," MainLighter: *The Lighting Management & Maintenance Magazine*, October 1991.

Harvey, Hal and Bill Keepin, "Energy's Relationship to the Environment: Global Climate Change," *MainLighter: The Lighting Management & Maintenance Magazine*, November 1991.

Garelik, Glenn, "It's Not Easy Being Green," *Time*, 18 December 1989.

Kirkpatrick, David, "Environmentalism: The New Crusade," *Fortune*, 12 February 1990.

"Last Decade Warmest on Record," *Science News*, 10 February 1990.

Lemonick, Michael D., "Scrub That Smokestack," *Time*, 18 December 1989.

Lighting Upgrade Manual, First Edition. Washington, DC: U.S. Environmental Protection Agency, Green Lights Program, November 1991.

Lovins, Amory, "Abating Global Warming — At a Profit," *Strategic Planning for Energy and the Environment*, Spring 1990.

Lyman, Francesca. *The Greenhouse Trap*. Boston: The Beacon Press, 1990.

Priest, Joseph, *Energy: Principles, Problems, Alternatives*, Third Edition. Menlo Park, California: Addison-Wesley Publishing Company, 1984.

Raloff, J., "Acid Assessment: The State of the Science," *Science News*, 24

February 1990.

Roberts, Leslie, "How Fast Can Trees Migrate?", *Science News*, 10 February 1989.

Schneider, Stephen H. *Are We Entering the Greenhouse Century?* San Francisco: Sierra Club Books, 1989.

Schneider, Stephen H., "Greenhouse Effect: Science and Policy," *Science News*, 10 February 1989.

Sun, Marjorie, "Tighter Ozone Control Urged by Scientists," *Science*, 24 June 1988.

Van, Doanh, Ph.D., PE, Energy Utilization and Technical Information Engineer, Warner-Lambert, Morris Plains, New Jersey.

White, Robert M., "The Great Climate Debate," *Scientific American*, July 1990.

Further reading:

Kerr, R.A., "New Greenhouse Report Puts Down Dissenters," *Science*, 3 August 1990.

Moller, Detlev, "Estimation of the Global Man-Made Sulphur Emission," *Atmospheric Environment*, 1984.

National Air Pollutant Emission Estimates, 1940-1985. Research Triangle Park, North Carolina: U.S. Environmental Protection Agency, 1987.

Nierenberg, W.A., R. Jastrow and F. Seitz. *Scientific Perspectives on the Greenhouse Problem*. Washington, DC: George C. Marshall Institute, 1989.

Ottinger, Richard L., David R. Wooley, Nicholas A. Robinson, David R. Hodas and Susan E. Babb, *Environmental Costs of Electricity*. New York: Pace University Center for Environmental Legal Studies, Oceana Publications, Inc., 1990.

Raloff, J., "New Acid Rain Threat Identified," *Science News*, 30 April 1989.

Shea, C.P. "Protecting the Ozone Layer," in *State of the World 1989*. New York: W.W. Norton, 1989.

Waggoner, P.E. *Climate Change and U.S. Water Resources*. New York: Wiley Interscience, 1990.

World Resources 1988-89. New York: Basic Books, Inc., 1988.

U.S. Environmental Protection Agency, Green Lights Program.

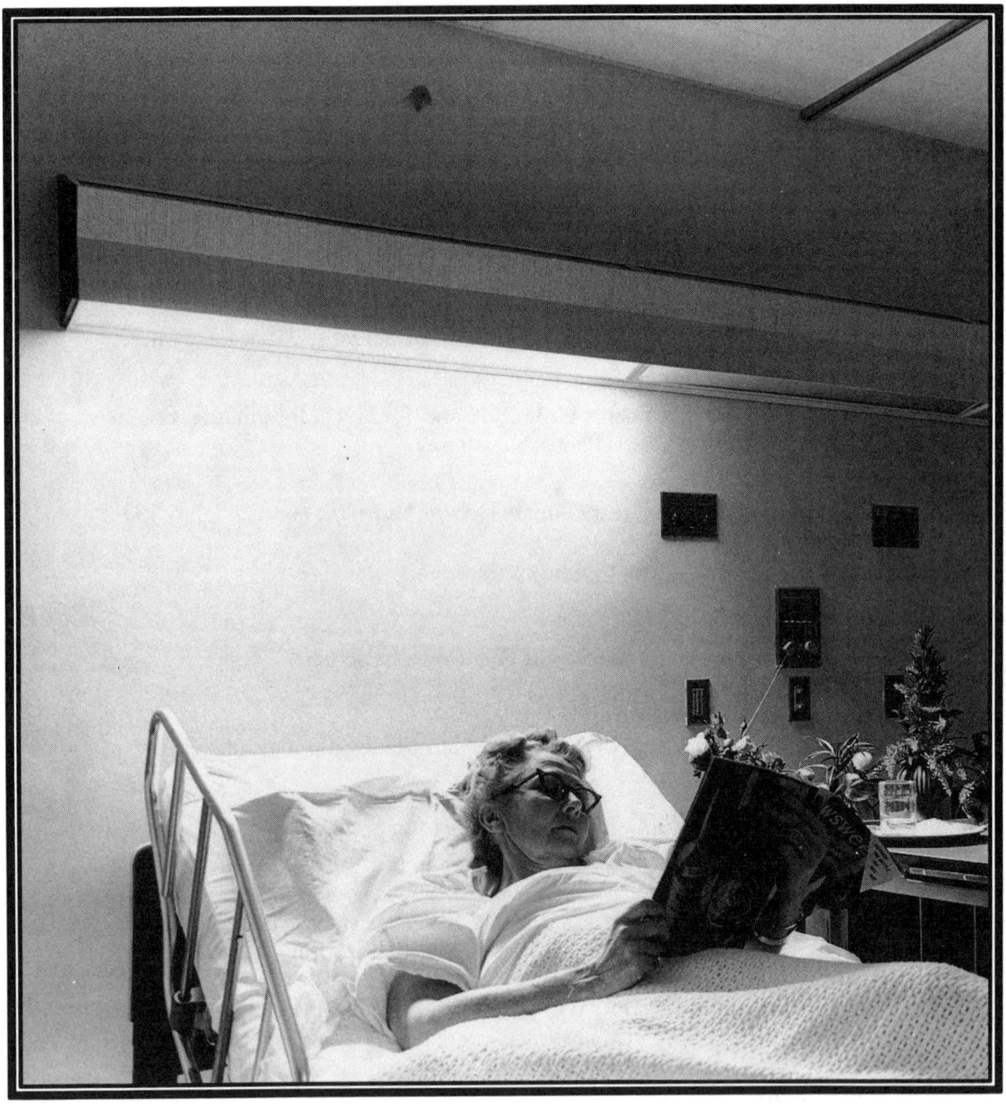

Source: National Lighting Bureau

CHAPTER 8

Getting Help: Associations, Trade Groups, Professionals, Government, Publications

GETTING HELP

Numerous organizations offer information about professional services and the benefits of lighting management. Listed in this chapter are professional and trade associations, Federal government resources, and energy and lighting magazines. A comprehensive listing of these organizations and other energy/environmental groups, demonstration centers, and education centers are available from at least two sources. *Lighting Listings: A Guide to Lighting Publications, Research Organizations, Educational Opportunities, and Associations* (order code LRC-TB) is available from:

Lighting Research Center (LRC)
Attention: Publications
115 Greene Building
Rensselaer Polytechnic Institute
Troy, NY 12180-3590
(518) 276-8716
Fax: (518) 276-2999

Another listing, *Commercial Lighting Efficiency Resource Book* (CU-7427), is available from:

Electric Power Research Institute (EPRI)
Distribution Center
207 Coggins Drive
Pleasant Hill, CA 94523
(510) 934-4212

PROFESSIONAL AND TRADE ORGANIZATIONS

American Lighting Association (ALA)
435 North Michigan Avenue, Suite 1717
Chicago, IL 60611
(312) 644-0828

Association of Demand Side Management Professionals (ADSMP)
7040 West Palmetto Park Road
Suite 2315
Boca Raton, FL 33433
(407) 361-0023

Association of Energy Engineers (AEE)
4025 Pleasantdale Road, Suite 420
Atlanta, GA 30340
(404) 447-5083

Illuminating Engineering Society of North America (IESNA)
345 East 47th Street
New York, NY 10017
(212) 705-7926

International Association of Lighting Designers (IALD)
18 East 16th Street, Suite 208
New York, NY 10003
(212) 206-1281

interNational Association of Lighting Management Companies (NALMCO)
34-C Washington Road
Princeton Junction, NJ 08550
(609) 799-5501

National Association of Energy Service Companies (NAESCO)
1440 New York Avenue, NW
Washington, DC 20005
(202) 371-7000

National Association of Independent Lighting Distributors (NAILD)
320 Grote Street
Buffalo, NY 14207
(716) 875-3670

National Electrical Contractors Association (NECA)
7315 Wisconsin Avenue
Bethesda, MD 20814
(301) 657-3110

National Electrical Manufacturers Association (NEMA)
2101 L Street, Northwest
Washington, DC 20037
(202) 457-8400

National Lighting Bureau (NLB)
2101 L Street, NW, Suite 300
Washington, DC 20037
(202) 457-8437

FEDERAL GOVERNMENT RESOURCES

Energy Information Administration (EIA)
U.S. Department of Energy (DOE)
EI-231, Forrestal Building
Washington, DC 20585
(202) 586-8800

National Appropriate Technology Assistance Service (NATAS)
U.S. Department of Energy
P.O. Box 2525
Butte, Montana 59702-2525
(800) 428-2525
Montana: (800) 428-1718

U.S. Department of Energy (DOE)
Office of Conservation and Renewable Energy Inquiry and Referral Service
Office of Building Technologies
1000 Independence Avenue, SW
Washington, DC 20585
(202) 586-9220

U.S. Environmental Protection Agency (EPA)
Global Change Division
Office of Air and Radiation
Green Lights Program
401 M Street, SW (6202J)
Washington, DC 20460
(202) 775-6650

PUBLICATIONS

Architectural Lighting
Gralla Publications
1515 Broadway
New York, NY 10036
(212) 869-1300

Architectural Record Lighting
McGraw-Hill, Inc.
1221 Avenue of the Americas
New York, NY 10020
(212) 512-2841

Energy Focus
651 W. Washington Street, Suite 300
Chicago, IL 60661
(312) 993-0929

Energy User News
One Chilton Way
Radnor, PA 19089
(215) 964-4000

Journal of the Illuminating Engineering Society
Illuminating Engineering Society of North America
345 East 47th Street
New York, NY 10017
(212) 705-7926

Lighting Design and Application (LD&A)
345 East 47th Street
New York, NY 10017
(212) 705-7926

Lighting Dimensions
135 Fifth Avenue
New York, NY 10010
(212) 677-5997

Lighting Management & Maintenance (LM&M)
interNational Association of Lighting Management Companies
34-C Washington Road
Princeton Junction, NJ 08550
(609) 799-5501

Today's Lighting Distributor
National Association of Independent Lighting Distributors
c/o CMA Publications
34-C Washington Road
Princeton Junction, NJ 08550
(609) 799-4900

Source: GE *Lighting*

CHAPTER 9

Success Stories: Quality Lighting Upgrades

CUTTING THROUGH RED TAPE AT THE UNIVERSITY OF CENTRAL OKLAHOMA

Even public institutions have been able to circumvent red tape to implement effective lighting strategies. The University of Central Oklahoma in Edmond, Oklahoma has long been involved in environmental issues. Now the University has gotten high marks for a retrofit program that conserves energy — and saves enough money to help pay for other campus projects.

"I think it's the responsibility of the University to set the pace for public institutions," says Joyce Mounce, vice president of administration. "We thought we had a real opportunity here, and through a lot of hard work by a lot of people, we were right."

Mounce began by contracting with Johnson Controls for a $1.8 million upgrade of campus heating, air conditioning, wiring and lighting systems. Several projects paid for themselves and yielded enough additional savings to pay for improvements which did not have an immediate payback — like rewiring several buildings and adding a UPS system in the math and science building.

The lighting upgrade was handled by IllumElex Corporation, a lighting management company, and involved 27 buildings encompassing 950,000 square feet. A total of 11,520 fluorescent fixtures got either new lamps, new ballasts, new reflectors or a combination of those items.

"The lighting retrofit," says Mounce, "generated some of the most immediate savings: a total of $110,000 per year. And overall, light levels are better than they were before." Mounce adds that it "takes a combination of factors to get a project of this magnitude through the approvals required for public institutions: having sufficient facts to justify the project, tenacity to push it through the process, and a lot of teamwork at the University level.

"It's not always easy," she says, "but it's worth it."

Source: U.S. Environmental Protection Agency's Green Lights program.

USING A MOVE TO RE-THINK LIGHTING STRATEGY AT KNORR BRAKE

It is important to recognize milestones that provide an opportunity to revamp lighting management strategy. When Knorr Brake Corporation, the manufacturer of braking systems for train and subway cars, outgrew its facility, an extensive search located a complex in Westminster, Maryland. Unfortunately, the major renovation budget allocated little money to the existing lighting system. When approached by Primo Lighting Management, however, Knorr recognized the opportunity to initiate an effective lighting management strategy.

Primo, a lighting management company, conducted a major renovation of the lighting system, resulting in a lighting load reduction of over 96 kilowatts and over $26,000 annual savings in energy costs. As an added benefit to Knorr, no initial expenditures were required and 100% of the project costs are funded by monthly energy savings over a pre-established period of time.

Source: U.S. Environmental Protection Agency's Green Lights program.

TEAMWORK AT AUGUSTA COLLEGE

It is often difficult for one individual to shoulder the responsibility for all aspects of lighting management, from building surveys and analysis to implementation and financing. Teamwork can ensure that every facet of the strategy receives professional attention. August College in Georgia involved experts from a variety of disciplines when designing its lighting strategy.

Joseph Mele, vice president for business and finance, recognized the potential for considerable energy savings at the August campus, which covers 70 acres and includes 50 buildings. He obtained a federal grant to assist with a major lighting renovation project, then turned the implementation over to Mimmo Guerrieri, director of operations.

Guerrieri worked with consulting engineers Jones Nall & Davis (JN&D) to develop a plan that would improve the level of light in 19 classroom and lecture buildings. The plan called for relamping, reballasting and reflectorizing of 6,000 fluorescent fixtures.

To implement the plan, the college chose IllumElex Corporation, a lighting management company. The recommendations resulted in higher color rendition and more lumens per lamp. Without increasing energy consumption, light levels in classrooms were increased from 40 to 60 footcandles. Levels in other areas were maintained without lowering energy consumption.

JN&D engineer Gary Gabriel noted that Augusta realized even more than higher lighting levels — substantial cost savings. "We're predicting $70,000 annual savings on a job that cost $260,000. Payback will take place in less than four years."

The team approach helped the college use the expertise of a number of individuals to implement a successful lighting management plan.

Source: U.S. Environmental Protection Agency's Green Lights program.

CUSTOMIZED UPGRADES OF MULTIPLE FACILITIES AT BANK OF AMERICA

The task of designing customized lighting plans for hundreds of facilities can seem arduous. Bank of America, however, recently undertook a lighting upgrade of over 400 locations throughout the state of California. Energy savings well in excess of $1 million per year has led to an average payback period of about two years per branch.

Originating as a delamping and reflector project by Parrish Lighting and Engineering, the scope of the installation was widened by Amtech Lighting Services, a lighting management company, to include other energy-efficient lighting equipment. Amtech Lighting Services surveyed the buildings and recommended electronic ballasts, compact fluorescent units and custom-made reflectors on a branch-by-branch basis. Light level measurements were taken, and care was used in engineering the project to ensure that adequate light levels were maintained after the installation.

The cost and savings information for each location was reconciled with rebate incentives available from local utility companies, and Bank of America received a complete cost, rebate and simple payback calculation for each location.

Based on the combined effect of utility rebates in excess of $750,000, reduced energy costs and reduced maintenance expenses, Bank of America plans to upgrade an additional 300 branches in 1992.

Source: U.S. Environmental Protection Agency's Green Lights program.

ACTON SCHOOLS SCORE HIGH IN EFFICIENCY

The Acton Public School System in Acton, Massachusetts contains seven schools. Wishing to reduce costs without sacrificing light levels important for students to read and write, the school system decided to start with its "overhead" costs — its lighting system. Aetna Corp., a lighting management company, was contracted to upgrade the lighting system.

Aetna Corp. first thoroughly cleaned the system, then installed electronic ballasts and energy-saving lamps in the two-lamp fluorescent fixtures. Wattage per fixture was reduced by about 40%, from 94 to 58. In fixtures containing four traditional fluorescent lamps, a reflector was installed and two lamps removed after they were cleaned. Energy use was reduced by 50% while permissible lumen output was produced. The remaining two lamps were replaced with energy-saving lamps and the magnetic ballast was replaced with an electronic ballast. Wattage per fixture was reduced more than 60%, from 188 to 71 watts. In addition, Aetna Corp. replaced 60-watt and 150-watt incandescent lamps with 15-watt compact fluorescent adapter units.

Energy savings exceeded 1.8 million kWh, or $126,000 per year. The payback period was immediate, thanks to a Boston Edison Utility Company energy rebate program called ENCORE, which presented a check to Acton public schools to pay for the entire program.

Source: interNational Association of Lighting Management Companies.

ELMBROOK SCHOOL DISTRICT MAKES THE GRADE: STARTS WITH A PILOT PROJECT

The School District of Elmbrook, Wisconsin decided to explore whether the good news about lighting management was true. Elmbrook contacted Stay-Lite Lighting, a lighting management company, to audit the lighting system in one elementary school as a pilot project. If the project went well, all nine schools in the district would be upgraded. To make the upgrade even more cost-effective, Stay-Lite took advantage of a rebate program from Wisconsin Electric Power Company.

The first step in this project was to evaluate the current lighting system and isolate any problems or inconsistencies. The school was lighted by a mixture of standard and energy-saving lamps powered by standard magnetic ballasts, and incandescents in exit signs and other fixtures.

Stay-Lite recommended installing cathode cut-out ballasts powering energy-saving F40T12/ES lamps and fluorescent retrofit kits for the exit signs. The company also prescribed replacing twenty-six 300-watt incandescent lamps in the gymnasium with only 12 metal halide lamps, which alone increased light level from 13 to 55 footcandles. Twenty-five 200-watt incandescent fixtures used to light the outside of the building were replaced with a system of just twelve 35-watt high pressure sodium lamps.

When the project was finished, the schools showed an estimated annual energy savings of $6,000. With the annual utility rebate from Wisconsin Electric Power Company of $9,500, the school realized a payback period of only 26 months. With these results, Stay-Lite received permission to upgrade the other schools in the district.

This presented new challenges. The time element was one major factor as was the focus of this phase of the project — lamp and ballast conversion. A total of 12,000 cathode cut-out ballasts were installed to replace traditional ballasts, and 24,000 lamps were replaced with energy-saving lamps. Exit signs and other incandescent applications were upgraded to compact fluorescent lighting. The most success, however, came from upgrading the high school gymnasiums.

In one gym, one hundred and twenty-eight 400-watt mercury high-bay fixtures lighted the area from a recessed ceiling. Existing light levels were measured at 35 footcandles at mid-court. The lamps were replaced with lay-in metal halide fixtures utilizing a 400-watt, 40,000-lumen, high-output clear lamp. Only 36 fixtures were needed to cover the 12,000-square-foot gym. With the help of a new coat of light-colored paint, light levels jumped to 120 footcandles. The estimated annual electric bill savings amounted to $7,800. Once again, a utility rebate helped offset the cost of the project by 50%.

The total upgrade of the Elmbrook schools took just 11 weeks. Total kilowatt-hour reduction totaled 1,444,000 kWh, which resulted in an estimated annual energy savings of $100,000. With utility rebates reaching $200,000, the school district realized a payback period for all schools of just 1.9 years.

Source: interNational Association of Lighting Management Companies.

OAK PARK HOSPITAL CURES LIGHTING ILLS

Oak Park Hospital is located in Oak Park, Illinois. The hospital's administration determined that increasing security through increased light level was the primary goal of the upgrade project.

Low light levels in the hospital's multilevel parking facility and parking lots were to be upgraded. Existing mercury and incandescent lamps were replaced more-efficient high pressure sodium lamps. This resulted in the required light levels with the bonus of a significant reduction in energy consumption.

A secondary focus of the upgrade was to replace the numerous incandescent lamps in the hospital's stairwells. All incandescents were replaced with compact fluorescent lamps for additional energy savings.

As a result of the upgrade, the hospital reduced energy use by 155,100 kilowatt-hours. This reduced electric load created an annual energy savings of $12,300 against a project cost of $36,000, allowing the hospital to realize a three-year payback.

Source: interNational Association of Lighting Management Companies.

LIGHTING CONSULTANT WORTH HIS WEIGHT IN GOLD, CLIENT SAYS

A Wisconsin lighting consultant has been worth his weight in gold to Oak Creek-based Superior Die Set Corporation, says the company's president, Casey S. Janiszewski. Suggestions made by the 2,880-ounce lighting consultant, Witold Zalewski of Intex, will save Superior Die Set about $1 million over the 20-year life of the equipment he recommended.

According to Janiszewski, a number of energy conservation improvements were made by his company in the early 1980s, many inspired by articles he read in trade journals. One of the stories he came upon centered on a winner of the National Lighting Awards Program, sponsored by the National Lighting Bureau. A new lighting system installed in a drafting room had cut energy and other operation and maintenance (O&M) costs by about 70% annually while helping employees achieve a 7.5% productivity improvement, due to the better quality of the new lighting. The O&M cost savings were worth about $10,000 annually, but the productivity enhancement had an annual value of more than $200,000 per year.

According to Janiszewski, "The basic lesson I learned from the article was that there's a major difference between lighting energy conservation and lighting management. Saving energy is all well and good, but making energy savings the goal can waste money because it ignores lighting's purpose. It's installed so people can see what they're doing. When you use better lighting — lighting that's more suited to the work people are performing — they can see better. The end result is more productivity and more productivity translates into more dollars, far many more dollars than can be obtained from energy savings alone."

He added, "Not that lighting management ignores the energy issue. Energy savings still are obtained ... but the dollars to be saved by reducing energy waste are small compared to those to be had as a result of more productivity, fewer errors, less product waste, and so on. By concentrating on better lighting — through lighting management — a shrewd manager can obtain the best of both worlds."

Based on a recommendation, Janiszewski called Intex of Mequon and later met with Zalewski. The two toured Superior Die Set's facilities: about 12,000 square feet of production area, with an additional 30,000 square feet for administration in offices, drafting rooms, computer rooms and other rooms. Over the next two weeks, Zalewski monitored the activities of employees and reported that, as a general rules, they were not as energy-conscious as desirable. In addition, he noted that several areas were receiving too much light and the excess was interfering with work. The modifications recommended by Zalewski relied on conventional technology but nonetheless produced startling results. He suggested installation of occupancy sensors in three lavatories to reduce energy consumption.

Zalewski also recommended installation of new ballasts and current limiters in drafting areas to reduce excessive lumen output and energy consumption. The new ballasts were needed because the existing ones were 25 years old and beginning to fail at a rapid rate. The lower lighting levels were recommended because existing levels were too high and were causing serious glare problems. Said Janiszewski, "There was so much light bouncing off the [drafting] paper that contrast was being eroded. The details on the surface on the paper were being washed out. By lowering the light level, people would have just the right amount of light, and the reflected glare would be cut back. Basically, what we did was provide better seeing conditions so drafters could reduce their errors."

According to the National Lighting Bureau, glare typically is caused by poorly designed lighting systems or improperly shielded fixtures, not too much light. As such, reducing lighting levels does not usually correct glare problems. In this instance, however, far too much light was being provided for the space, tasks and workers involved, as one year's monitoring revealed. Prior to the conversion, it was taking approximately 7.1 hours for a drafter to turn out one drawing. After the lighting levels were lowered, the time to produce a drawing was cut by 11.3% to just 6.3 hours.

The company's computer room comprised another area of concern. There, a disk drive area was separated from the operations area by a large window. Light from the disk drive room turned the window into a glare source, causing light to scatter into the operations area. "Whenever the lights were on in the computer operations area, they were on in the disk drive area, and vice versa," said Janiszewski. "But observations showed that, with few exceptions, when the operators went into the disk drive area, few people were in the operations area."

Zalewski recommended installation of occupancy sensors in the disk drive area so that it would be illuminated only when people were using it. In that way, the window separating the disk drive area from the computer operations area would be a glare source for only a relatively few minutes each day.

Said Janiszewski, "This was another area where we began keeping very close records of productivity, and Witold's observations were right on the money, figuratively and literally. With the glare greatly reduced, computer operations downtime fell by 4 hours per month. At $112 per hour, that comes to $5,736 per year saved. In addition, computer errors were reduced by 50%. This is mea-

sured in 'incident rate' rather than time, so we have no way of attaching an accurate dollar amount to it. 'Thousands of dollars' is a reasonable way of expressing it, however. And as an added plus, our security personnel tell us that the use of motion detectors makes their work that much more effective. 'When the lights are on, someone's in there so I know to check it out,' one guard told me."

Table 10-1 comprises consolidated data for the three areas. As can be seen, the total cost of the retrofits amounted to just under $3,000, while the value of the annual savings — including $1,748 of annual O&M savings — came in at more than $44,584 per year. As such, the simple payback was less than 24 days, with a simple return on investment of almost 1500%.

Table 10-1: Drafting area, computer rooms and bathrooms.

	Before	After	Diff.	% Diff.
GENERAL				
Connected load	9.996 kW	8.232 kW	-1.764 kW	-18%
Hours of use (averaged)	4,752 h/y	3,095 h/y	-1,657 h/y	-46%
Energy consumption	47,499 h/y	25,480 kWh/y	-22,019 kWh/y	-46%
O&M COSTS				
Utility costs	$2,993/y	$1,605/y	$-1,388/y	-46%
Replacement lamps	$100/y	$49/y	-$51/y	-51%
Lamp replacement labor	$544/y	$265/y	-$280/y	-51%
Luminaire cleaning	$110/y	$125/y	+$15/y	+14%
Ballast replacement	$149/y	$105/yr	-$44/y	-30%
Total	$3,896/y	$2,148/y	$-1,748/y	-45%
COST BENEFIT				
Modification cost		$2,980		
O&M savings		$1,748/y		
Productivity improvements		$37,460+/y		
Error reduction savings		Not calculated		
Downtime savings		$5,376/y		
Reduced absenteeism		Not calculated		
Enhanced security		Not calculated		
Total		$44,584+/y		
SIMPLE PAYBACK		Less than 24 days		
SIMPLE RETURN ON INVESTMENT		More than 1496%/y		

Source: National Lighting Bureau.

APPENDIX I

GENERAL TECHNICAL INFORMATION

STATE	AVERAGE COMMERCIAL COST PER kWh ($)	AVERAGE INDUSTRIAL COST PER kWh ($)	ENVIRONMENTAL EMISSIONS PER kWh CO_2 POUNDS	SO_2 GRAMS	NOx GRAMS	CITY	COOLING SEASON IN WEEKS
ALABAMA	$0.068	$0.043	1.6	7.0	2.5	Birmingham	32.9
						Huntsville	30.9
						Mobile	38.4
						Montgomery	35.3
ALASKA	0.088	0.077	0.2	0.2	0.2		
ARIZONA	0.081	0.054	1.2	2.0	1.9	Flagstaff	18.6
						Phoenix	41.3
						Tucson	40.1
ARKANSAS	0.067	0.049	1.4	1.8	2.1	Blytheville	29.7
						Fort Smith	30.5
						Little Rock	29.7
CALIFORNIA	0.091	0.071	0.6	0.1	1.0	Barstow	32.2
						Bishop	30.4
						Los Angeles	32.6
						Sacramento	28.4
						San Diego	29.8
						San Francisco	22.2
						Santa Barbara	12.2
COLORADO	0.058	0.045	2.3	3.7	3.4	Colorado Springs	21.6
						Denver	22.6
						Grand Junction	23.7
						Trinidad	25.4
CONNECTICUT	0.087	0.072	0.8	1.9	0.7		

STATE	AVERAGE COMMERCIAL COST PER kWh ($)	AVERAGE INDUSTRIAL COST PER kWh ($)	ENVIRONMENTAL EMISSIONS PER kWh CO_2 POUNDS	SO_2 GRAMS	NOx GRAMS	CITY	COOLING SEASON IN WEEKS
DELAWARE	$0.067	$0.044	2.5	11.1	3.6	Dover	23.6
						Wilmington	23.7
DISTRICT OF COLUMBIA	0.063	0.051	2.6	8.2	2.7	Washington	26.0
FLORIDA	0.066	0.051	1.7	5.4	2.5	Jacksonville	41.6
						Miami	50.1
						Orlando	46.2
						Pensacola	38.4
						Tampa	46.0
GEORGIA	0.073	0.047	1.5	8.9	2.3	Atlanta	30.0
						Augusta	35.1
						Macon	34.8
						Savannah	38.0
						Valdosta	38.9
HAWAII	0.092	0.066	1.9	3.3	1.7		
IDAHO	0.043	0.026	0.0	0.0	0.0	Boise	19.7
						Lewiston	18.9
						Pocatello	18.8
ILLINOIS	0.078	0.054	1.1	7.2	2.6	Champaign	23.6
						Chicago	20.9
						Peoria	24.0
						Rockford	21.0

STATE	AVERAGE COMMERCIAL COST PER kWh ($)	AVERAGE INDUSTRIAL COST PER kWh ($)	ENVIRONMENTAL EMISSIONS PER kWh CO_2 POUNDS	SO_2 GRAMS	NOx GRAMS	CITY	COOLING SEASON IN WEEKS
INDIANA	$0.061	$0.041	2.4	15.5	4.8	Fort Wayne	22.5
						Indianapolis	23.9
						South Bend	21.4
						Terre Haute	24.8
IOWA	0.064	0.040	2.2	7.0	3.7	Council Bluffs	23.0
						Mason City	19.7
						Sioux City	22.2
KANSAS	0.065	0.048	2.0	5.2	3.3	Dodge City	25.6
						Goodland	23.6
						Kansas City	25.7
						Wichita	27.0
KENTUCKY	0.053	0.043	2.2	10.3	4.3	Covington	24.4
						Hopkinsville	28.4
						Louisville	26.6
LOUISIANA	0.070	0.043	1.4	1.4	2.0	Alexandria	37.2
						Lake Charles	39.2
						New Orleans	39.6
						Shreveport	35.2
MAINE	0.074	0.054	0.5	1.3	0.3	Portland	15.5
MARYLAND	0.066	0.047	2.0	8.9	3.0		
MASSACHUSETTS	0.081	0.073	1.6	6.3	2.3	Boston	19.8
						Springfield	20.1

STATE	AVERAGE COMMERCIAL COST PER kWh ($)	AVERAGE INDUSTRIAL COST PER kWh ($)	ENVIRONMENTAL EMISSIONS PER kWh CO_2 POUNDS	SO_2 GRAMS	NOx GRAMS	CITY	COOLING SEASON IN WEEKS
MICHIGAN	$0.079	$0.056	1.7	4.4	3.1	Detroit	19.2
						Grand Rapids	19.0
						Lansing	19.5
						Sault St Marie	12.8
						Travarse City	17.0
MINNESOTA	0.060	0.041	1.8	3.6	2.7	Duluth	12.7
						Internat'l Falls	14.1
						Minneapolis	18.8
MISSISSIPPI	0.072	0.048	1.3	5.5	2.2	Biloxi	37.6
						Columbus	33.8
						Jackson	35.3
MISSOURI	0.065	0.049	2.0	13.4	4.4	Columbia	25.7
						Kansas City	25.7
						Springfield	26.9
						St. Louis	26.3
MONTANA	0.046	0.031	1.5	1.6	2.2	Billings	18.4
						Glasgow	17.5
						Great Falls	16.9
						Helena	15.5
NEBRASKA	0.057	0.042	1.5	2.3	3.4	Grand Island	22.7
						North Platt	22.0
						Omaha	23.0

STATE	AVERAGE COMMERCIAL COST PER kWh ($)	AVERAGE INDUSTRIAL COST PER kWh ($)	ENVIRONMENTAL EMISSIONS PER kWh CO_2 POUNDS	ENVIRONMENTAL EMISSIONS PER kWh SO_2 GRAMS	ENVIRONMENTAL EMISSIONS PER kWh NOx GRAMS	CITY	COOLING SEASON IN WEEKS
NEVADA	$0.060	$0.044	2.2	3.1	3.2	Ely	20.2
						Las Vegas	35.4
						Reno	21.0
						Winnemucca	22.3
NEW HAMPSHIRE	0.088	0.067	1.2	6.3	2.2	Manchester	19.0
NEW JERSEY	0.088	0.072	0.7	2.1	1.5	Trenton	22.9
NEW MEXICO	0.082	0.053	2.4	2.3	3.0	Alburquerque	27.3
						Alamogordo	32.5
						Clovis	29.4
NEW YORK	0.099	0.053	1.2	3.8	1.3	Albany	19.5
						Buffalo	18.8
						New York City	20.0
						Syracuse	19.4
NORTH CAROLINA	0.063	0.047	1.3	4.2	2.0	Greensboro	28.1
						Raleigh	30.0
						Wilmington	33.6
NORTH DAKOTA	0.065	0.049	2.5	6.5	4.0	Bismarck	18.3
						Fargo	17.0
						Grand Forks	16.9
						Minot	16.2

STATE	AVERAGE COMMERCIAL COST PER kWh ($)	AVERAGE INDUSTRIAL COST PER kWh ($)	ENVIRONMENTAL EMISSIONS PER kWh CO_2 POUNDS	SO_2 GRAMS	NOx GRAMS	CITY	COOLING SEASON IN WEEKS
OHIO	$0.072	$0.039	2.2	17.6	4.0	Cincinnati	24.2
						Cleveland	21.0
						Columbus	23.8
						Dayton	24.3
						Toledo	21.3
OKLAHOMA	0.059	0.037	1.8	2.3	2.8	Altus	31.2
						Enid	28.4
						Oklahoma City	29.5
						Tulsa	29.7
OREGON	0.048	0.033	0.1	0.1	0.1	Burns	17.3
						Eugene	15.0
						Medford	21.2
						Pendleton	20.0
						Portland	15.8
PENNSYLVANIA	0.078	0.058	1.4	7.4	2.2	Philadelphia	23.7
						Pittsburgh	21.9
						Scranton	20.1
						Williamsport	21.0
RHODE ISLAND	0.080	0.075	1.9	1.7	3.3	Providence	18.7
SOUTH CAROLINA	0.062	0.043	0.8	2.5	1.2	Charleston	36.0
						Columbia	33.4
						Myrtle Beach	32.3

STATE	AVERAGE COMMERCIAL COST PER kWh ($)	AVERAGE INDUSTRIAL COST PER kWh ($)	ENVIRONMENTAL EMISSIONS PER kWh			CITY	COOLING SEASON IN WEEKS
			CO_2 POUNDS	SO_2 GRAMS	NOx GRAMS		
SOUTH DAKOTA	$0.067	$0.047	1.1	4.8	2.9	Huron	20.4
						Rapid City	19.6
						Sioux Falls	20.5
TENNESSEE	0.061	0.050	1.5	10.6	2.6	Knoxville	29.0
						Memphis	30.4
						Nashville	28.4
TEXAS	0.061	0.041	1.8	2.5	2.5	Amarillo	28.4
						Corpus Christi	43.0
						Dallas	34.6
						Houston	42.0
						Lubbock	30.8
						San Antonio	41.3
UTAH	0.068	0.040	2.2	2.8	2.8	Salt Lake City	19.9
						Wendover	21.6
VERMONT	0.081	0.065	0.1	0.8	0.2	Burlington	16.7
VIRGINIA	0.060	0.042	1.0	3.3	1.5	Richmond	26.8
						Roanoke	26.3
WASHINGTON	0.041	0.027	0.2	0.7	0.4	Seattle	9.4
						Spokane	15.6
WEST VIRGINIA	0.054	0.036	2.2	12.4	3.9	Charleston	26.1
						Clarksburg	22.5

STATE	AVERAGE COMMERCIAL COST PER kWh ($)	AVERAGE INDUSTRIAL COST PER kWh ($)	ENVIRONMENTAL EMISSIONS PER kWh CO_2 POUNDS	SO_2 GRAMS	NOx GRAMS	CITY	COOLING SEASON IN WEEKS
WISCONSIN	$0.059	$0.041	1.8	6.2	3.1	Green Bay	17.5
						Madison	20.2
						Milwaukee	20.9
WYOMING	0.052	0.036	2.5	2.2	3.7	Casper	19.2
						Cheyenne	18.3
						Rock Springs	16.7

Sources:

Commercial and industrial kilowatt-hour costs ... Electric Sales and Revenue 1989, *Department of Energy, Energy Information Administration,* DOE/EIA-05140(89). *Note that each figure represents a blend of demand charges, energy use charges and other charges.*

Environmental multipliers ... *U.S. Environmental Protection Agency's Green Lights Program, 1992.*

Cooling seasons ... Controlling Energy Consumption in Single Buildings, *U.S. Department of the Navy, Naval Civil Engineering Lab*, CR82.028, 1982.

APPENDIX II

WORKSHEETS

Survey Worksheet: Area-by-Area Survey, Page 1

Area description: ________________________ Area No.: __________
Room Length ________' x Room Width ________' = Room Area ____________ft²
Daylight _____ is ______ is not present.
Occupancy hours _______ a.m./p.m. to ______ a.m./p.m. ____ days per week

room ceiling reflectance: _________%

room walls reflectance: _________%

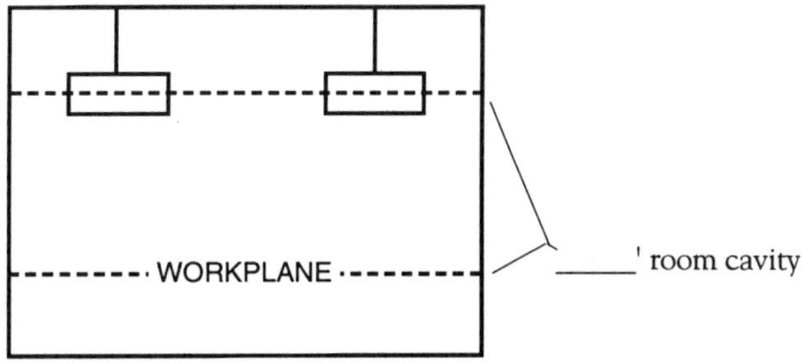

	Fixture Type #1	Fixture Type #2	Fixture Type #3
1. Type of fixture	________	________	________
	________	________	________
	________	________	________
2. Nominal size (2' x 4', etc.)	________	________	________
3. Number of lamps per fixture	________	________	________
4. Type of lamps	________	________	________
5. Type of ballasts	________	________	________
6. Type of shielding material	________	________	________
7. Physical condition (excellent, good, fair, poor)	________	________	________

Survey Worksheet: Area-by-Area Survey, Page 2 **Area No. ________**

	Fixture Type #1	Fixture Type #2	Fixture Type #3	TOTAL
8. Degree of dirt accumulation (very clean, clean, medium, dirty, very dirty)	________	________	________	
9. Total number of burnouts	________	________	________	
10. Number of fixtures	________	________	________	
11. Input watts per fixture	x ________	x ________	x ________	
12. Total watts	= ________	= ________	= ________	= ________
13. Square footage				÷ ________
14. Watts per square foot				= ________
	÷ 1,000	÷ 1,000	÷ 1,000	
15. Demand for power in kilowatts	= ________	= ________	= ________	= ________
16. Annual operating hours	x ________	x ________	x ________	
17. Energy consumption in kWh	= ________	= ________	= ________	= ________

VISUAL TASKS PERFORMED	LIGHT LEVEL ON TASK	RANGE OF AGE OF OCCUPANTS	LIGHTING QUALITY PROBLEMS
________			________
________	________ Fc	________	________
________			________
________	________ Fc	________	________
________			________
________	________ Fc	________	________
________			________
________	________ Fc	________	________

Light Level Worksheet #1: Specifying Minimum Light Level

Area description: ______________________________ Area No.: __________

1. Task ______________________________

2. Generic task ______________________________

3. Illuminance Category ______________________________

4. Illuminance range ______________________________

Weighting Factors: If Illuminance Category is A, B or C

5. Range of occupant ages and weighting factor __________/__________

6. Room surface reflectances + __________

7. Total weighting Factor = __________

a. For Illuminance Categories A through C			
Room and Occupant Characteristics	Weighting Factor		
	-1	0	+1
Occupants ages	Under 40	40-55	Over 55
Room surface reflectances*	Greater than 70 percent	30 to 70 percent	Less than 30 percent

Weighting Factors: If Illuminance Category is D, E, F, G, H or I

8. Range of workers' ages and weighting factor __________/__________

9. Speed + __________

10. Accuracy + __________

11. Reflectance on task background + __________

12. Total weighting factor = __________

13. Recommended light level ______________________ Fc

b. For Illuminance Categories D through I			
Task and Worker Characteristics	Weighting Factor		
	-1	0	+1
Workers ages	Under 40	40-55	Over 55
Speed and/or accuracy**	Not important	Important	Critical
Reflectance of task background***	Greater than 70 percent	30 to 70 percent	Less than 30 percent

* Average weighted surface reflectances, including wall, floor and ceiling reflectances, if they encompass a large portion of the task area or visual surround. For instance, in an elevator lobby, where the ceiling height is 7.6 meters (25 feet), neither the task nor the visual surround encompass the ceiling, so only the floor and wall reflectances would be considered.

** In determining whether speed and/or accuracy is not important or critical, the following questions need to be answered: What are the time limitations? How important is it to perform the task rapidly? Will errors produce an unsafe condition or product? Will errors reduce productivity and be costly? For example, in reading for leisure there are no time limitations and it is not important to read rapidly. Errors will not be costly and will not be related to safety. Thus, speed and/or accuracy is not important. If however, prescription notes are to be read by a pharmacist, accuracy is critical because errors could produce an unsafe condition and time is important for customer relations.

***The task background is that portion of the task upon which the meaningful visual display is exhibited. For example, on this page the meaningful visual display includes each letter which combines with other letters to form words and phrases. The display medium, or task background, is the paper, which has a reflectance of appoximately 85 percent.

Source: IES Lighting Handbook, *1987 Application Volume*.

Light Level Worksheet #2, Page 1:
Calculating Required Lumen Output Per Fixture

Area description: ______________________ Area No.: __________

		Reflector retrofit
1. Average light level to be maintained	______________ Fc	
2. Area	x ______________ ft^2	
3. **Variable A**	= ______________	
4. Number of fixtures	______________	
5. Coefficient of utilization	x ______________	x ______________
NON-RECOVERABLE LIGHT LOSS FACTORS		
6. Fixture ambient temperature factor	x ______________	
7. Supply voltage variation factor	x ______________	
8. Lamp position factor	x ______________	
RECOVERABLE LIGHT LOSS FACTORS		
9. Lamp burnouts factor	x ______________	
10. Lamp lumen depreciation	x ______________	
11. Fixture dirt depreciation	x ______________	x______________
12. Room surface dirt depreciation	x ______________	
13. **Variable B**	= ______________	= ______________
14. Required lumens per fixture	= ______________	= ______________

Variable A ÷ Variable B

Light Level Worksheet #2, Page 2:
Calculating Required Lumen Output Per Lamp

Area description: ______________________ Area No.: __________

14. Required lumens per fixture	__________	__________
	÷ 2 lamps	÷ 2 lamps
15. Required lumens per lamp	= __________	= __________
16. Required lumens per fixture	__________	__________
	÷ 3 lamps	÷ 3 lamps
17. Required lumens per lamp	= __________	= __________
18. Required lumens per fixture	__________	
	÷ 4 lamps	
19. Required lumens per lamp	= __________	

	Option #1	Option #2	Option #3	Option #4
	______	______	______	______
	______	______	______	______
	______	______	______	______
	______	______	______	______
20. Lumens/Lamp	______	______	______	______
21. Ballast factor	x ______	x ______	x ______	x ______
22. Optical factor 1	x ______	x ______	x ______	x ______
23. Optical factor 2	x ______	x ______	x ______	x ______
24. Lumen maintenance adjustment factor	x ______	x ______	x ______	x ______
25. End lumens/lamp for upgrade option	= ______	= ______	= ______	= ______
26. Does it comply?	______	______	______	______

Light Level Worksheet #3: Determining Maintained Light Levels

Area description: ______________________ Area No.: ________

	Option #1	Option #2	Option #3	Option #4
	________	________	________	________
	________	________	________	________
	________	________	________	________
	________	________	________	________
1. Number of lamps per fixture	________	________	________	________
2. Lumen output per lamp	x ________	x ________	x ________	x ________
3. Lamp lumen output per fixture	= ________	= ________	= ________	= ________
4. Total number of fixtures	x ________	x ________	x ________	x ________
5. Total lamp lumen output	= ________	= ________	= ________	= ________
6. Coefficient of utilization (CU)	x ________	x ________	x ________	x ________
7. Ballast factor	x ________	x ________	x ________	x ________
8. Fixture ambient temperature factor	x ________	x ________	x ________	x ________
9. Supply voltage variation factor	x ________	x ________	x ________	x ________
10. Lamp operating position factor	x ________	x ________	x ________	x ________
11.Optical factor1	x ________	x ________	x ________	x ________
12.Optical factor2	x ________	x ________	x ________	x ________
13. Burnout factor	x ________	x ________	x ________	x ________
14. Lamp lumen depreciation	x ________	x ________	x ________	x ________
15. Fixture dirt depreciation	x ________	x ________	x ________	x ________
16. Room surface dirt depreciation	x ________	x ________	x ________	x ________
17. End lumens	= ________	= ________	= ________	= ________
18. Area in square feet	÷ ________	÷ ________	÷ ________	÷ ________
MAINTAINED LIGHT LEVEL (FOOTCANDLES)	= ________	= ________	= ________	= ________

Lighting Quality Worksheet

Area description: ______________________________ Area No.: ________

Problem: ______________________________

Corrective action: ______________________________

Economics Worksheet:
Determining Simple Payback Per Fixture/Per Facility, Page 1

Area description: ______________________________ Area No.: __________

Fixture type: ____

LIGHTING ENERGY SAVINGS	Existing	Option #1	Option #2	Option #3	Option #4
1. Number of lamps per fixture and type	________	________	________	________	________
	________	________	________	________	________
	________	________	________	________	________
2. Number of ballasts per fixture and type	________	________	________	________	________
	________	________	________	________	________
3. Other equipment (occupancy sensors, reflectors, etc.)	________	________	________	________	________
	________	________	________	________	________
4. System input watts per fixture	________	________	________	________	________
5. Operating hours per year	x ________ ÷ 1,000	x ________ ÷ 1,000	x ________ ÷ 1,000	x ________ ÷ 1,000	x ________ ÷ 1,000
6. Lighting energy consumption (kWh)	= ________	= ________	= ________	= ________	= ________
7. Lighting energy consumption (kWh) for existing system		– ________	– ________	– ________	– ________
8. Lighting energy savings (kWh) per fixture (make positive)		= ______(+)	= ______(+)	= ______(+)	= ______(+)
9. Number of fixtures		x ________	x ________	x ________	x ________
10. Total lighting energy savings (kWh)		= ________	= ________	= ________	= ________
11. Operating hours from **Step 5**		÷ ________	÷ ________	÷ ________	÷ ________
12. Total demand savings (kW)		= ________	= ________	= ________	= ________
13. Total lighting energy savings (kWh) from **Step 10**		________	________	________	________
14. Cost per kWh		x $______	x $______	x $______	x $______
15. TOTAL LIGHTING ENERGY SAVINGS PER YEAR ($)		= $______	= $______	= $______	= $______

Retrofit Economics Worksheet: Determining Simple Payback, Page 2

Area description: ______________________ Area No.: __________

	Existing	Option #1	Option #2	Option #3	Option #4
A/C ENERGY SAVINGS					
16. Total lighting energy savings (kWh) from **Step 10**		________	________	________	________
17. Cooling season in weeks ÷ 52		x ________	x ________	x ________	x ________
		x 0.9	x 0.9	x 0.9	x 0.9
		÷ 2.7	÷ 2.7	÷ 2.7	÷ 2.7
18. Total A/C energy savings (kWh)		= ________	= ________	= ________	= ________
19. Cost per kWh from **Step 14**		x ________	x ________	x ________	x ________
20. Total A/C energy savings ($)		= $______	= $______	= $______	= $______
21. Heating season (1–**Step 17**)		________	________	________	________
		x 0.2	x 0.2	x 0.2	x 0.2
22. Variable A		= ________	= ________	= ________	= ________
23. Total lighting energy savings (kWh) from **Step 10**		________	________	________	________
		x 3,414	x 3,414	x 3,414	x 3,414
		x **Var. A**	x **Var. A**	x **Var. A**	x **Var. A**
		÷ 0.8	÷ 0.8	÷ 0.8	÷ 0.8
24. Increase in needed heat load in BTU		= ________	= ________	= ________	= ________
		÷ 100,000	÷ 100,000	÷ 100,000	÷ 100,000
25. Increase in needed heat load in therms		= ________	= ________	= ________	= ________
26. Utility cost per therm		x $______	x $______	x $______	x $______
27. Additional heating cost		= $______	= $______	= $______	= $______
28. A/C energy savings from **Step 20**		$ ________	$ ________	$ ________	$ ________
29. Additional heating cost from **Step 27**		– $______	– $______	– $______	– $______
30. TOTAL NET H.V.A.C. SAVINGS		= $______	= $______	= $______	= $______

Retrofit Economics Worksheet:
Determining Simple Payback Per Fixture/Per Facility, Page 3

Area description: ______________________________ Area No.: ________

	Existing	Option #1	Option #2	Option #3	Option #4
INSTALLATION COST					
31. Cost per lamp		$_______	$_______	$_______	$_______
32. Number of lamps per fixture		x _______	x _______	x _______	x _______
33. Lamp cost per fixture		= $_______	= $_______	= $_______	= $_______
34. Cost per ballast		$_______	$_______	$_______	$_______
35. Number of ballasts per fixture		x _______	x _______	x _______	x _______
36. Ballast cost per fixture		= $_______	= $_______	= $_______	= $_______
37. Reflector cost		$_______	$_______	$_______	$_______
38. Equipment cost per fixture (33 + 36 + 37)		= $_______	= $_______	= $_______	= $_______
39. Electrical labor rate per hour		$_______	$_______	$_______	$_______
40. Hours to install components in one fixture		x _______	x _______	x _______	x _______
41. Labor cost per fixture		= $_______	= $_______	= $_______	= $_______
42. Total cost per fixture (38 + 41)		= $_______	= $_______	= $_______	= $_______
43. Number of fixtures from **Step 9**		x _______	x _______	x _______	x _______
44. Total installation cost for fixtures		= $_______	= $_______	= $_______	= $_______
45. Cost of other:____________		$_______	$_______	$_______	$_______
46. Number of other		x _______	x _______	x _______	x _______
47. Total equipment cost of other		= $_______	= $_______	= $_______	= $_______
48. Electrical labor rate per hour from **Step 39**		$_______	$_______	$_______	$_______
49. Time to install all other		x _______	x _______	x _______	x _______
50. Labor cost of other		= $_______	= $_______	= $_______	= $_______
51. Total cost of other (47 + 50)		= $_______	= $_______	= $_______	= $_______
52. TOTAL INSTALLATION COST (44 + 51)		= $_______	= $_______	= $_______	= $_______

Retrofit Economics Worksheet:
Determining Simple Payback Per Fixture/Per Facility, Page 4

Area description: ____________________ Area No.: __________

	Existing	Option #1	Option #2	Option #3	Option #4
UTILITY REBATES					
53. Total demand savings (kW) from **Step 12**		= ______	= ______	= ______	= ______
54. Rebate per kW		x $______	x $______	x $______	x $______
55. kW rebate		= $______	= $______	= $______	= $______
56. Total lighting energy savings (kWh) from **Step 10**		______	______	______	______
57. Rebate per kWh		x $______	x $______	x $______	x $______
58. Total kWh rebate (assumes no rebate for A/C savings)		= $______	= $______	= $______	= $______
59. Rebate per lamp		$______	$______	$______	$______
60. Total number of lamps (lamps per fixture x fixtures from **Step 9**)		x ______	x ______	x ______	x ______
61. Total lamp rebate		= $______	= $______	= $______	= $______
62. Rebate per ballast		$______	$______	$______	$______
63. Total number of ballasts (ballasts per fixture x no. of fixtures from **Step 9**)		x ______	x ______	x ______	x ______
64. Total ballast rebate		= $______	= $______	= $______	= $______
65. Rebate per reflector		$______	$______	$______	$______
66. Number of reflectors		x ______	x ______	x ______	x ______
67. Total reflector rebate		= $______	= $______	= $______	= $______
68. Rebate per other:________		$______	$______	$______	$______
69. Number of other		______	______	______	______
70. Total other rebate		= $______	= $______	= $______	= $______
71. TOTAL INSTALLATION COST from **Step 52**		$______	$______	$______	$______
72. TOTAL REBATES **(55 + 58 + 61 + 64 + 67 + 70)**		– $______	– $______	– $______	– $______
73. NET TOTAL INSTALLATION COST		= $______	= $______	= $______	= $______

Retrofit Economics Worksheet:
Determining Simple Payback Per Fixture/Per Facility, Page 5

Area description: ______________________________ Area No.: __________

	Existing	Option #1	Option #2	Option #3	Option #4
SIMPLE PAYBACK					
74. TOTAL LIGHTING ENERGY SAVINGS from **Step 15**		$______	$______	$______	$______
75. NET HVAC SAVINGS from **Step 30**		+ $_____	+ $_____	+ $_____	+ $_____
76. TOTAL ENERGY SAVINGS		= $_____	= $_____	= $_____	= $_____
77. NET TOTAL INSTALLATION COST from **Step 73**		$______	$______	$______	$______
78. TOTAL ENERGY SAVINGS from **Step 76**		÷ $_____	÷ $_____	÷ $_____	÷ $_____
79. PAYBACK (YEARS)		= ______	= ______	= ______	= ______
SIMPLE RETURN ON INVESTMENT					
80. TOTAL ANNUAL ENERGY SAVINGS from **Step 76**		$______	$______	$______	$______
81. NET TOTAL INSTALLATION COST from **Step 73**		÷ $_____	÷ $_____	÷ $_____	÷ $_____
		x 100	x 100	x 100	x 100
82. RETURN ON INVESTMENT (%)		= _____%	= _____%	= _____%	= _____%
83. Does the option comply with company financial policy?		________	________	________	________

Maintenance Worksheet #1:
Determining Maintenance Interval, Page 1

Area description: ______________________ Area No.: __________

Fixture type: ______________________________________

IES maintenance category: ________ Dirt conditions: ______________

1. Rated life of lamps
@ ______ hours per start ____________

2. Annual operating hours ÷ __________

3. Years before rated life = __________

x 0.6

4. Maintenance interval at
60% of lamp life = __________ years

5. Round off to closest
six-month interval ____________ years x 12 = ____________ months

6. Repeat maintenance
interval at 60% of lamp
life from **Step 4** ÷ __________ years

7. Adjustment factor = __________

x 60

8. New maintenance interval as
percentage of rated life = __________%

9. Repeat years before rated
life from **Step 3** ____________

x 0.7

10. Maintenance interval at
70% of lamp life = __________ years

11. Round off to closest
six-month interval ____________ years x 12 = ____________ months

12. Repeat maintenance
interval at 70% of lamp
life from **Step 10** ÷ __________ years

13. Adjustment factor = __________

x 70

14. New maintenance interval as
percentage of rated life = __________%

CONTINUED

Maintenance Worksheet #1:
Determining Maintenance Interval, Page 2

Area description: ______________________ Area No.: __________

15. Repeat years before rated life from **Step 3** __________

x 0.8

16. Maintenance interval at 80% of lamp life = __________ years

17. Round off to closest six-month interval __________ years x 12 = __________ months

18. Repeat maintenance interval at 80% of lamp life from **Step 16** ÷ __________ years

19. Adjustment factor = __________

x 80

20. New maintenance interval as percentage of rated life = __________%

OPTION #1, "60%" of life: Group relamp and clean @ __________% of life at

__________ years (__________ months); spot relamp early failures.

OPTION #2, "70%" of life: Group relamp and clean @ __________% of life at

__________ years (__________ months); spot relamp early failures.

OPTION #3, "80%" of life: Group relamp and clean @ __________% of life at

__________ years (__________ months); spot relamp early failures.

OPTION #4: Spot relamp all failures; no cleaning (assume 10-year fixture dirt depreciation factor).

	OPTION #1	**OPTION #2**	**OPTION #3**	**OPTION #4**
Lamp lumen depreciation factor	__________	__________	__________	__________
Fixture dirt depreciation factor	__________	__________	__________	__________

Maintenance Worksheet #2: Maintenance Cost, Page 1

Area description: ______________________ Area No.: __________

	OPTION #1	OPTION #2	OPTION #3	OPTION #4
SUPPORTING DATA				
1. Operating hours per year	________	________	________	________
2. Rated lamp life @ ______ per start				÷ ________
3. Variable A				= ________
4. Operating hours between group relamping intervals	÷ ________	÷ ________	÷ ________	
5. Variable B	= ________	= ________	= ________	
GROUP RELAMPING				
6. Labor cost to group relamp one lamp	$________	$________	$________	
7. Lamp cost to group relamp one lamp	+ $________	+ $________	+ $________	
8. Group relamping cost for one lamp	= $________	= $________	= $________	
9. Number of lamps	x ________	x ________	x ________	
10. Group relamping cost	= $________	= $________	= $________	
SPOT RELAMPING				
11. Percentage of lamps failing before group relamping	________	________	________	
12. Number of lamps	x ________	x ________	x ________	________
13. Number of lamps to be spot relamped	= ________	= ________	= ________	
14. Labor cost to spot relamp one lamp	$________	$________	$________	$________
15. Lamp cost to spot relamp one lamp	+ $________	+ $________	+ $________	+ $________
16. Spot relamping cost for one lamp	= $________	= $________	= $________	= $________
17. Number of lamps to be spot relamped from **Step 13**	x ________	x ________	x ________	x ________
18. Spot relamping cost				= $________
19. **Variable A** from **Step 3**				x ________
20. Final spot relamping cost	= $________	= $________	= $________	= $________

Maintenance Worksheet #2: Maintenance Cost, Page 2

Area description: ______________________________ Area No.: __________

	OPTION #1	OPTION #2	OPTION #3	OPTION #4
CLEANING				
21. Labor cost to clean one fixture	$________	$________	$________	
22. Number of fixtures	x ________	x ________	x ________	
23. Total cleaning cost	= $________	= $________	= $________	
TOTAL COSTS				
24. Group relamping cost from **Step 10**	+ $________	+ $________	+ $________	
25. Final spot relamping cost from **Step 20**	+ $________	+ $________	+ $________	$________
26. Total cost	= $________	= $________	= $________	
27. **Variable B** from **Step 5**	x ________	x ________	x ________	
28. TOTAL ANNUALIZED COST	= $________	= $________	= $________	= $________

Environmental Assessment Worksheet #1
Calculating Deferred Emissions

Area description: ______________________________ Area No.: ______________
Fixture before upgrade: ____
____
Fixture after upgrade: ____
____

	DEFERRED EMISSIONS		
	Carbon Dioxide	Sulphur Dioxide	Nitrogen Oxides
1. Reduction in energy consumption (kWh) per fixture	______________	______________	______________
2. Appropriate environmental multiplier	x ________ lbs.	x __________ g.	x ___________ g.
3. Resulting emissions	= ________ lbs.	= __________ g.	= ___________ g.
4. Conversion rate to obtain pounds		÷ 454	÷ 454
5. Resulting emissions		= ________ lbs.	= _________ lbs.
6. Number of fixtures receiving this upgrade	x _____________	x _____________	x ______________
7. Total emissions	= _________ lbs.	= ________ lbs.	= _________ lbs.
8. Conversion rate to obtain tons	÷ 2,000	÷ 2,000	÷ 2,000
9. Resulting emissions	= _________ tons	= ________ tons	= _________ tons

Environmental Assessment Worksheet #2:
Calculating Conserved Resources, Page 1

Area description: ______________________________ Area No.: ______________
Fixture before upgrade: ____
____
Fixture after upgrade: ____
____

	Coal	Oil	Natural Gas
1. Type of fuel	______________	______________	
2. Heat content (BTU) per unit	______ BTU/Ton	______BTU/Gal.	______BTU/cub.ft.
3. Conversion	÷ 2,000 lbs./Ton		
4. BTU per unit	= ________BTU/lb.		
5. Conversion	÷ 3,414 BTU/kWh	÷ 3,414 BTU/kWh	÷ 3,414 BTU/kWh
6. kWh per unit	= ______kWh/lb.	= ______kWh/Gal.	= ______kWh/cub.ft.
7. Efficiency losses	x 0.33	x 0.33	x 0.33
8. Adjusted kWh per unit	= ______kWh/lb.	= ______kWh/Gal.	= ______kWh/cub.ft.
9. Units per kWh	1 ÷ ______kWh/lb.	1 ÷ _____kWh/Gal.	1 ÷ ______kWh/cub.ft.
10. Units per kWh	=______lbs./kWh	=______Gals./kWh	= ______cub.ft./kWh
11. Reduction in energy consumption per fixture	x_________kWh	x_________kWh	x _________kWh
12. Amount of fuel saved	= __________lbs.	= __________Gals.	x _________cubic feet
13. Number of fixtures receiving this upgrade	x ___________	x ___________	x ___________

Environmental Assessment Worksheet #2:
Calculating Conserved Resources, Page 2

Area description: ______________________________ Area No.: ______________

	Coal	Oil	Natural Gas
14. Total fuel saved	= __________lbs.	=__________Gals.	= _________cub.ft.
15. Percentage of electric generation at the utility for this fuel source	x ___________%	x___________%	x ___________%
16. Actual fuel saved	= _________lbs.	= _________ Gals.	=________cub.ft.
17. Conversion determine larger unit of measure	÷ 2,000 lbs./Ton	÷ 42 Gals./Barrel	
18. Fuel saved in large unit	= __________Tons	= ________Barrels	

APPENDIX III

PRODUCT GUIDES

- Efficient fluorescent lamps and fixtures (September 1992)
- Efficient exit sign fixtures (September 1992)
- Fluorescent reflectors (September 1992)
- Electronic ballasts (September 1992)
- Lighting controls (May 1992)
- Occupancy sensors (May 1992)

Source: *Energy User News*, May 1992, September 1992.

EFFICIENT FLUORESCENT LAMPS AND FIXTURES

COMPANY	MODEL	PRICE & PAYBACK	FUNCTIONS & FEATURES	LEAD TIME & INSTALLATION	WARRANTY & SERVICE
Berghelli Inc. 11221-2 St. Johns Industrial Pkwy South Jacksonville, FL. 32246 (904) 646-4416 FAX: (904) 646-1949	Regina 86236-HT/US	The Regina is a NEMA 4x energy-saving lighting fixture in 120v or 277v with electronic ballast. Degree of protection: IP-65 (water, corrosion, dust, and weather resistant). Utilizes T-8 or T-12 lamps. Substantial energy savings (up to 30%). Consult factory representatives for prices. Payback is within 3 years.	Regina is an energy-saving fixture constructed of polycarbonate plastic to assure vandal-resistance and durability. Regina has a NEMA 4x rating and is adaptable for indoor or outdoor applications. Regina utilizes fluorescent lamps which are powered through reduce energy consumption up to 30%. Liquid silicon gasketing is injected during the molding process to insure resistance. Through a patented "plug-in" connection system, Regina can be effortlessly installed and maintained.	Lead time 2-4 weeks. Installation by electrical contractor.	3 year factory warranty.
The Bodine Company P.O. Box 460 Collierville, TN 38027-0460 (901) 853-7221 FAX: (901) 853-5009	B30/B35, B50, B60, B70A, B90, B100, B94, B113, B413, B426, B226, HID1600	Products operate most 20W-215W fluorescent lamps, including U-shaped circline, energy saving and compact fluorescents, depending on model selected. They are compatible with standard, energy saving, dimming, and electronic AC ballasts. HID1600 operates high intensity discharge lamps in emergency. Prices quoted by local manufacturer's representative, payback depends on model selected.	Hi-temp no-maintenance ni-cad battery charger and electronic circuitry in compact red case, charging indicator light, test switch & installation hardware. Products meet NEC and UL 90-min. code requirements, and do not affect normal fixture operations. Initial lumen output depends on lamp type and model selected. HID1600 consists of a central battery cabinet with no-maintenance 12V batteries, charger, rectifier/filter, fusing and disconnect means and remote invertor ballasts installed at the fixture, replacing the standard AC ballast. Emergency lumen output is approximately 50% of normal light output.	Normally no lead time on most products, installation by contractor of lighting fixture manufacturer.	Products warranted fully (not pro-rata) from 5 yrs to one, depending on model selected. HID1600 warranted for 5 years.
Brownlee Lighting 3071-K North Orange Blossom Trail Orlando, FL 32804 FAX: (407) 297-3705 (407) 297-3677	Large selection of energy saving fixtures and retrofit products for commercial and residential commercial applications.	Contact Brownlee for pricing info.	Utilizing compact fluorescent and HID light sources. Decorative fixtures-some with brass and glass-some with emergency back-up. UL listed ropelight and PL converters for exit signs. Some custom capability.	Normally 2 weeks, 3-4 weeks for very large orders.	1 year warranty from date of shipment.
Edison Price Lighting 409 East 60th St. New York, NY 1002[illegible] (212) 838-5212 FAX: (212) 868-798[illegible]	Super Baflux Series	Two efficient 26 watt compact fluorescents that consume only 60 watts but deliver the same light output as a 200 watt incandescent with 13 times the life. Payback in less than 15 months.	Only 4 3/4" deep, very shallow low brightness recessed compact fluorescent fixture designs with parabolic louvers to provide excellent cut off. Fits well into today's architecture. Anti-iridescent reflector finish and HPF ballasts are standard.	The products will be ready or shipment in early September. Orders will require about 3 days lead time for less than 100 fixtures and about 2 weeks lead time for larger quantities. These products are to be installed by an electrical contractor.	2-year warranty against manufacturing defects. We will, at our discretion, either replace or repair.

GE Lighting Nela Park Cleveland, Oh 44112 1-800-GELAMPS	GE Lighting carries a complete selection of compact Fluorescent lamps, both electronically ballasted and screw-in incandescent replacements and plug-in types for commercial application.	Varies is quantity and type purchased; before you relamp always check with local utility for rebate types.	GE compact Fluorescents use up to 75% less electricity than incandescent they replace, last 10-13 times longer and provide excellent color rendition.	Screw-in compact Fluorescents require no lead time and are retrofits to incandescents; plug-in types require a compatible fixture.	Product services available through local GEL sales office.
Lights of America 611 Reyes Drive Walnut, CA 91789 (800) 321-8100	C2013, C2004, C2027, C2020 TP, C2022 TP, C2030 TP	Modular construction of quad and circline lamp format. Lamp rated life 9-12K hours. Ballast life approx. 65K hours. Up to $108/yr energy savings per compact.	Electronic ballast up to 30% more efficient than conventional magnetic. Only compacts offering 115 watts & 150 watts of light. Power factor of .99 and total harmonic distortion of less than 10% available.	Stock. Installation by contractor or end user.	1 year limited.
Luxo Lamp Corp. 36 Midland Avenue Port Chester, New York, 10573 (914) 937-7016	FL-13, FL-18, FL-24 Sonnet, JAC	Energy-efficient compact fluorescent task lamps available in the following wattages: 13-watt two-tube compact fluorescent, 13-watt four-tube compact fluorescent, and the 18 and 26-watt blax compact fluorescent. All lamps can be mounted on edge, table, walor furniture panel system with an infinite number of spring-balanced arms and brackets. These products can save 70 to 80% on energy load versus incandescent lamps. Can also be accepted for Utility Rebate programs. List prices: $150-$350. Payback- 1-3 years.	These task lamps are used in conjunction with other manufacturer's ambient systems to help reduce electrical loads. By using task lamps with energy efficient fluorescent light source, the total number of ceiling fixtures or light bulbs can be reduced to meet new energy codes and enhance worker productivity. These task lamps are designed to reduce eye strain and fatigue, eliminate glare (direct or indirect) reduce shadows and insure contrast on the work surface without computer screen washout.	Lead time 2-4 weeks. Can be individually installed by contractor or your own maintenance personnel.	Lifetime mechanical warranty. Serviced by factory.
Osram Corp. 110 Bracken Rd. Montgomery, NY 12549 (914) 457-4040 FAX: (914) 457-4142	Dunlux, EL, 7W, 11W, 15W, 20W, 23W Dunlux EL Globe, Dunlux EL Reflector	$19.95-$24.00 depending upon model	Low power consumption; high output; long life make them the logical choice for energy conscious application.		

EFFICIENT FLUORESCENT LAMPS AND FIXTURES

COMPANY	MODEL	PRICE & PAYBACK	FUNCTIONS & FEATURES	LEAD TIME & INSTALLATION	WARRANTY & SERVICE
Phillips Lighting Co. 200 Franklin Sq. Dr. Somerset, NJ 08875 (980) 563-3000 FAX: (980) 563-3941	Earth Light Collection of Compact Fluorescent Lamps (SL*17, SL*18 and SLS)	Varies with quantity purchased. Check local utility for rebates.	Energy-efficient alternatives to incandescent lamps. These lamps, available in a variety of wattages, replace 60 to 90-watt incandescents at equivalent light output. While using 75% less energy and lasting up to 13 times longer. Not for use on dimmable circuits.	Earth Lights are coated with trichromatic phosphors which produce a light quality and appearance similar to incandescent (CRI-82). They are direct retrofits for incandescent sockets with their integrated electronic ballast.	UL Listed.
	TL 70/TL 80 T-8 Energy Efficient Fluorescent Lamps	Price varies with quantity, check local utility for rebates.	Energy-efficient alternatives to standard fluorescent systems. When these lamps are used with an electronic ballast, energy-savings of up to 43% is achievable for equivalent light output performance.	TL 70/80 lamps are coated with trichromatic phosphors, enabling high color rendering cability with CRI of 75 for TL 70 and CRI of 85 for TL 85 series. Lamps are available in 2', 3', 4', and 5' sized and color appearance of 3000k, 3500k, and 4100k.	
	PL Compact Fluor. lamps. PL-S 5 watt-13 watt; PL-C (cluster) 13W-26W; PL-L (long) 18W-40W	Varies with quantity purchased. Check local utility for rebates. Payback perios varies between 6 months and 2 years.	Energy-efficient alternative to incandescent and/or standard fluorescent lamps. PL lamps replace 15 to 100 W incandescent while using up to 75% less energy and lasting up to 13 times longer.	All of these lamps are coated with trichromatic phosphors that produce light (CRI-82) similar in appearance to incandescent and high color rendering fluorescents. They require a ballast and dedicated lamp holder or incandescent socket adapter for their operation.	
ProLight, Inc. 581 Ottawa Ave. Holland, MI 49423 (616) 396-2556 FAX: (616) 396-0686	Compact Fluorescent Floodlamp	Average $25	ProLight's UL-listed Compact Fluorescent Floodlamp is the right choice for many ceiling downlights found in offices, lobbies, elevators and track/display lighting. The Floodlamp unit contains a ballast adapter, a fluorescent lamp and a reflector. Available pre-assembled, these units screw into any standard incandescent socket. The simple "turn and lock" feature provides for easy separation of the reflector from the ballast adapter.	Each Floodlamp adapter is designed for a specific wattage, from 5 to 28 watts. They accept either twin tube or double-twin tube fluorescent lamps. Replacing a 60-watt incandescent with ProLight's 13-watt flooding saves almost $20 per year.	Different series adapters have warranties ranging from 3-5 years.

ProLight, Inc. (continued)	Light Fantastic Table Lamp Converter Kit.	$16-$27.	UL-listed Light Fantastic table Lamp Convertor Kits can replace lamps in almost any lighting fixture that utilized a cord and a wall plug. These fixtures can be table lamps; floor lamps; desk lamps; wall lamps; portable lights or worklights. The complete convertor unit is composed of three parts: a screw socket adapter, a fluorescent lamp and a wall plug convertor. All three simply plug in or screw in with no other parts or tools required.	The convertor kits are available in 5 to 22 watts to replace of 25-75 watt incandescent lamps. Replacing a 75-watt incandescent lamp with a 22-watt convertor kit saves over $18 per year.	Light Fantastic Convertor Kits have a 3 year warranty.
Quadrotech Ltg. Co. 47 St. James Place Piedmont, CA 94611 (510) 482-1727 (510) 482-0843	KEN-LITE® KL and SKL Series for down lights. KT Series for tracks.	9 Watt Models replace 40 to 60 Watt incandescent. 13 Watts advancement Models replace 75 to 100 Watt incandescent. Based on 10 cents for KWH rate. Payback is approx. 6 months for 24 hrs. per day operation.	KEN-LITE® represents the latest design in compact fluorescent lamp (Quad) application. The overall lengths of KEN-LITES® are equal to PAR 38 or R40 incandescent bulbs fin downlight application.	2 days. Products are in stock.	10 years ballast warranty for KL Models. 3 Years ballast warranty for SXL Models.
SIMKAR 601 East Cayuga St. Philadelphia, PA 19120-4691 (215) 831-7700 FAX: (215) 831-7703	Para-Spec TEP 2x2 Series Twin Tube Parabolic and Standard 2x2 Troffers, Suffix B-10 TT	TEP models for 2x2 Lay-in Ceiling/ Depth 5 5/8" TE or Standard 2x2 Lay-in Model/Depth 3.75". Choice of 2 or 3 40 watt Twin-Tube lamps. Payback less than 2 years. Consult local factory rep. for market pricing.	SIMKAR offers customers flexibility in lamp configurations and options (like emergency lighting) on Twin Tube Troffers in the popular 2x2 lay-in size. TEP Deep Cell Parabolic and TE Cell offer Aluminum Louvers, or select a standard 2x2 troffer with your choice of "interchangeable" shielding. VCP comfort and Photometrics are impressive.	Stock to 4 weeks, installation by contractor.	2 year warranty. Field service available.
	Vandgard V5007 & 5"x8" PL7.	V5007/V5014 Indoor or Outdoor backplates or optional Aluminum backplates. SIMKAR ships Vangard models with both a Tamper-proof & Standard screws. V501A Backbox hardware can be used for conduit connections, photocell, or outlet (indoor units only). Payback year compared to incandescent models. Consult Local Factory Rep. for Market Pricing.	Vangard Selections feature Vandal Polycarbonate Lens. Compact depth of 4", and multipatterned diffuser blends in with any structure. Gasketing seals out moisture, dust & insects.	Stock to 4 weeks, installation	Field Service Available
Osram Sylvania, Inc. 10 Endicott St. Danvers, MA 01923 (508) 777-1900	Sylvania FE-18 and FE-22 Electronic compact fluorescent lamps.	Suggested User Price: $20-$25. At $0.10 Kwh, you save up to $57 over lamp life with the 18 watt compact fluorescent, and $68 with the 22 watt lamp.	The compact fluorescent lamps are an energy saving alternative to incandescent lamps, last thirteen times than that of incandescent lamps and no adaptor is needed.	UL Listed CSA Certified FCC-47 FR18 consumer limits	10,000 hours* FE-18-1100 lumens, FEE-22-1400 lumens *Average rated hours based on 3 hours/start.

EXIT SIGNS

COMPANY	MODEL	PRICE & PAYBACK	FUNCTIONS & FEATURES	LEAD TIME & INSTALLATION	WARRANTY & SERVICE
Lithonia Lighting	Signature Electroluminescent Exit Sign	N/A	Uses less than one watt of power. 8-year lamp allows emergency battery and electronics to fit in housing. Perfectly uniform lighting. Five times brighter in emergency mode.		3-year total customer satisfaction warranty
ProLight, Inc. 581 Ottawa Ave. Holland, MI 49423 (616) 396-0686	Exit Retrofit Kit	$15-$26	The Exit Retrofit Kit is available in six styles, allowing conversion of virtually any exit fixture. A double-face mounting tape eliminates the need for any special tools for installation. One ProLight Exit Retrofit Kit contains two compact fluorescent lamps. When one lamp burns out after 10,000 hours, the other becomes lit, doubling the operating time before relamping to 20,000 hours. Applications for the Exit Retrofit Kit include any public or private building where exit signs are required. These buildings include industrial/commercial office buildings; apartment complexes; hotels/motels; government; school and retail.	The Exit Retrofit Kit contains a ballast and a single nine-watt lamp. It replaces incandescent exit sign systems which contain two 20- to 25-watt lamps for savings of 30 to 40 watts per sign.	Exit Retrofit Kits have a 3 year warranty
Scientific Component System, Inc. Peter C. Kreft VP Sales & Marketing 2651 Dow Avenue Tustin, CA 92680 (714) 730-3555 (800) 654-3498 FAX: (714) 730-3817	X18-CRS9Q 4*CAN, X-18-CR7 6*CAN, X18-CR13 8*CAN, Recessed Downlight Retrofit Conversion Fixtures. New construction models also now available (all sizes).	Anodized Aluminum Fixtures with 2 dedicated electromagnetic ballasts operating 2 standard PL lamps in a "Cross-Tube" lamp configuration. Consumes 76% less energy. Delivers maximum Lumens per watt. 96% efficient. Price $30.00 to $40.00 range. Payback: 6 months to 1 year.	Installs in seconds, fits flush to ceilings. Improved photometrics, trim ring as a heat sink to constantly draw damaging heat away from lamps and ballasts, 5 year warranty, one-way zip-locking mechanism prevents theft or easy removal, and most utilities have approved the X18 as permanently hardwired, qualifying for maximum rebate/incentives.	Lead time 2 to 4 weeks. Installation do it yourself or contractor.	5 years all components.
Shield Source, Inc Rte. 2 Box 19C VFW Road Grasonville, MD 21633 (410) 827-6022 FAX: (410) 827-9657	101PRW-1, 151WR-1, 210PRW-1, Self-luminous EXIT signs	10, 15, or 20 year life, Tritum Powered Self-luminous: no electrical usage installed cost $150-$250 . Payback within one year.	Self-luminous Tritum powered EXIT signs requiring no electrical energy. Offered in Plastic, Aluminum, Steel, and Stainless Steel Frames. Red, Green, or Black legends, with Clear Acrylic (Optional Lexan) .13 thick faceplate, sealed unitized construction. Meet NFPA 101 Light Safety Code, NEC: Article 500, Class 1 & 2 UL Listed, OSHA and Uniform Basic & Standard Building Codes.	Lead time from stock to 4 weeks. Installed by user or contractor.	Warranted to replace free of charge any product in which the luminosity is found to be defective during its specified luminous life.
Siltron Illumination Inc. 7915 Center Ave. Rancho Cucamonga, California 91730	AUX, APS, AWX ADX & LX L.E.D. exits	Light Emitting Diodes replace incandescent or fluorescent lamps as the light source in these rugged maintenance-free electronic exit signs. With	• Light Emitting Diodes • Electronic Source • Never change another Light Bulb • (80-100 Year Life)	Lead time: Units are in stock. Installation: By contractor or user.	20 year warranty on the exit sign. 10 year warranty on the bat

(800) 874-3392 FAX: (800) 284-1789 (714) 911-3500 FAX: (714) 980-4477		100 year life, these LEDs greatly reduce the man hours required to relamp and service conventional signs. Energy & maintenance savings will provide pay-back in 1-2 years in most areas. Fully UL Listed. NYC Approved 35707.	• NiCad Battery backup, Lead Optional • Superior Visibility • Attractive Rugged Enclosure, Heavy Extrusions • Polycarbonate Face Plate • Energy-Saving (Only 4.5 watts) • Slim profile-Design		tery. Expected lamp life 100 years. Service by authorized Siltron Service contractor.
SRB Technology Inc. 2580 Landmark Drive Winston-Salem, North Carolina 27103 (800) 552-0098 FAX: (919) 768-7720	Betalux, Lumin Exit, Self Luminous Exit Signs, Explorer LED Exit Signs	Depending upon energy and labor rates. Company payback in 1-3 years.	Self Luminous Exit Signs use no electricity, no wiring, no lamps, no energy consumption. Maintenance for up to 20 yrs. They are explosion proof and weather proof and comply with U.L. NFPA 101, State of California codes. LED consumes only 4.5 watts of energy and LEDs almost never need to be replaced. Comes with a NicCad battery for emergency power (as needed) and can be ordered either prismatic in direct view series.	Self luminous on average under 5 working days. Installs with just 4 screws. LED 1-2 weeks. Installed by Electrician.	Self Luminous for the rated life of the sign 10-15-20 yrs. LED 25 yrs.
Standard Enterprises, Inc. 14849 Build America Drive Woodbridge, VA 22191 (703) 490-3300 (800) 296-EXIT (703) 497-2316	ExternaLamp Exit bulb	New technology 15 year lamp designed for exit lighting. Burns 9 watts. Available in bayonet, intermediate, edison and cande-labra bases. Also FAX available in 120 or 277 volts. Installed cost, approx $15.00. Payback under 1 year.	Incandescent Exit Lamp with rated life of 130,000 hours (15 years) and burns 9 watts. Manufactured in USA. Proven results. ETL tested. Snaps or screws into existing fixture-no rewiring involved. Unconditional 3 year warranty. This product offers tremendous sav-ings on lamp replacement, energy, and labor time. Qualifies for most utility rebates. Protected against power surges. Non-breakable.	Available stock. No hard wiring required. Installation by user.	Unconditional 3 year warranty.
Sylvania Lighting Services Danvers, MA 01923 (508) 750-2169	Twin Tube Exit Sign Conversion Kit with F9TT/D827 lamp.	Suggested user price $18-$20. Installed price $25-$30. At $0.10 Kwh your annual power savings alone will range from $12.50 to almost $35 per sign. Total operat-ing costs will be reduced from between $24 to $10 per sign.	The Twin Tube Exit Sign Conversion Kit is a sub-assemble designed to field-convert 120V incandescent exit sign fixtures to compact fluo-rescent. The twin tube lamp is highly efficient, delivering 50 lumens per watt. The kit comes with a choice of incandescent base adaptors; either DC Bayonet, intermediate or medium screw base.	UL Listed.	10,000 hours/lamp* F9TT-600 lumens. *Average rated hours based on 3 hours/start.

FLUORESCENT REFLECTORS

COMPANY	MODEL	PRICE & PAYBACK	FUNCTIONS & FEATURES	LEAD TIME & INSTALLATION	WARRANTY & SERVICE
Alcoa, Aerospace/ Commercial, Rolled Products, P.O. Box 8001 Battendorf, IA 52722 (800) 225-3678	Everbrite lighting sheet. Available in flat sheet and rolled coils.	Contact company for distributor, fabricator or installer in your area. Paybacks typically occur in less than two years.	Exceptional durability, resistance to static electricity, cleanability. Represents nearly 60 years of applied technology in lighting sheet products. Will not crack, peel, delaminate or discolor.	Reflectors made of EverBrite lighting sheet are available in custom design or off the shelf..	25 year warranty, installer offers additional warrants.
American Illuminetics Carisbad, CA (619) 438-3338	Custom designed Polished Aluminum, Silver and enhanced aluminum reflectors. New reflectorized fixtures for tenant improvement and new construction.	Approx. $22-$35. Typical payback 8-24 months.	Custom designed reflectors for all sized and types of fluorescent fixtures. Reflectivity 88% aluminum to 95% silver. For ease of maintenance and installation we use a tablock system which requires no screws. 3 piece reflector allow ballast change without removing lamps.	10 to 20 days from date of order.	25 years aluminum, 5 years silver. Service by company or distributor.
Badger USA Baraboo, WI (608) 356-7744	Megalux. Three types-anodized aluminum, aluminum film, silver film.	$14-$30 typical 2 by 4 fixture (silver or aluminum). Typical payback between 8 to 18 months.	Reflectivity: aluminum 86-91%, silver film 95%. All models are designed for maximum performance.	Stock to 30 days. Installation by authorized Badger USA contractors.	10-25 years, limited parts and labor. Service by authorized Badger USA contractor.
Brayer Lighting San Francisco, CA (415) 255-1793	Silver (95% Reflectivity) Enhanced Aluminum (92%). Polished Aluminum (86%-88%). Can provide any material specified.	Contact Brayer for price info. Typical paybacks from 3 to 24 months.	Total Project Engineering/Retrofit component integration. Tool free access to ballasts. UL Classified. Anti-static, dust repellent composition. Silver 95% reflective. Enhanced aluminum 92%., Polished Aluminum (86%-88%).	1 month. Installed by distributor network. Lead time 1-2 weeks.	Warranty 5-25 years. Materials, Labor & Transportation. Complete satisfaction guaranteed.
Bright Side Lighting Merced, CA (209) 723-4142	Aluminum and silver reflectors, also fixtures.	2 by 4, $20 to $30 paybacks within 2 years.	Roll-form and custom designed, troffer or strip, also Gull-Wing kits (4 or 8 ft. strip). Reflectivity 87% aluminum, 95% silver. All type fluorescent fixtures available.	Installed by distributor network. Lead time 1-2 weeks.	10 years silver, 25 years aluminum. Service by distributor network/contractor.
C.E.W. Lighting 14040 Welch Road Dallas, TX 75244	CAD-STAR	$55.06 up to 400 watt. $93.67 for 100 watt.	CAD reflectors offer both glare reduction and lighting improvement. EYE CAD-STAR reflectors designed for use on high intensity discharge indoor fixtures, are faceted shades which collect light without waste. The facet design eliminates absorption of reflected light by the light source, providing an effective utilization of light.		1 year.
Electronic Advanced Systems Co. West Conshoshocken, PA (215) 828-5060; FAX: (215) 834-8588	Aluminum and silver reflectors and socket bars, specializing in 2 by 4 and Channel fixtures.	$15-20 for 2 by 4 aluminum, $20-25 for 2 by 4 silver, typically $2/sq.ft. aluminum, $2.75/sq.ft. silver. Payback 6-18 months.	Custom-made, any size. Independently tested by Ben Franklin Technology Center using IES procedure.	Same day to one week. Installation by dealer, contractor or end-user.	10 years silver, 25 years aluminum. Service by company and dealers.

Kilowatt Saver Corapolis, PA (800) 899-2988	Mirror Tech aluminum and Silver Specular.	Contact company for price information. Payback within 2 years.	Custom-made, any size. Independently tested byBen Franklin Technology Center using IES procedures.	3-5 days. Installed by user, contractor or distributor.	10-25 year warranty. Service by Kilowatt Saver or subcontractor.
KW/2 Lighting Division of Solar Kinetics, Inc. 10635 King William Drive Dallas, TX 75220 (214) 556-2376 FAX: (214) 869-4158	KW/2 Reflector Retrofit Kits & Magnum series custom designed fluorescent fixtures utilizing specular silver or specular aluminum or enhance aluminum material.	Competitive pricing results in paybacks typically 2 yrs. or less. Contact KW/2 for area rep. and information.	KW/2 silver kits utilize Limelight film; aluminum kits are Alcoa EverBrite and require no parts or tools for removal. Custom kit design is recommended; stock kits are available. KW/2 Products are represented by a nationwide network. Special services offered are a follow: Luminaire Pre/Post Retrofit Testing via IES LM-41 Format, Lumen-Micro Lighting Performance Predictions, Technical Product Specification Assistance, Detailed Economic Payback Analysis, Lighting Systems Package Testing and Recommendation.	Stock items shipped next day, all other products 2-5 wks. Installed by end user, network or contractor.	5 yr. full labor and material for KW/2 silver kits direct from film manufacturer. Serviced by KW/2 or subvendor.
Megalite Somerville, NJ (908) 874-0511	Megalite aluminum reflectors, fixtures, kits.	\$9-\$15 on 2 by 4 retrofit reflector. Payback immediate to 2 years.	All sized. Retrofit reflectors (two and three piece), fixtures (with T-8s, reflectors, electronic ballasts), reflectorized fixture kits (with aforementioned components). Regular aluminum 89% reflectivity, Megabrite aluminum 91% reflectivity.	Installation by user, contractor or manufacturer.	25 year warranty. Service by sub-vendors or manufacturer.
MetalOptics, Inc. 8222 N. Lamar Suite E44 Austin, TX 78753 (512) 832-0025	Silver and aluminum reflectors for all types of applications featuring Optikit (TM) fixture retrofit kits and Optilux (TM) reflectorized fixtures.	Contact MetalOptics for pricing information. Typical paybacks from 0-24 months.	Custom designed and independently tested reflectors and fixtures for any size an application. Reflectivity: Silver- 95%, Aluminum 85%-87%. Tab-lock retrofit system for easy installation. All lamp and ballast configurations available. In house audit, design and manufacturing services.	1-3 weeks. Materials only or turn-key services available.	5-year reflector warranty, service by MetalOptics. Material warranty: 10-25 years.
Mirrorlight Inc. New York, NY (212) 245-1140	Lumenen (clear anodized aluminum) reflectors. Also Silver (Specular Plus).	Approx. \$24-\$35. Paybacks usually within 1 year.	Two grades of aluminum, 87%-89% and 89%-91% reflectivity. Silver 95%. Custom- all configurations, also reflectorized fixtures. Design, manufacture, full service.	2-6 weeks. Installed by Mirrorlight contractors.	25 year warranty (10 years silver). Service by Mirrorlight.
ML Systems Valhalla, NY (914) 741-0400	Aluminum Everbrite reflectors	Contact company for price info. Payback within 2 years.	Custom and CAD designed. 86% reflectivity. For any application, \ in NY area and Boston. Company is subsidiary of Alcoa.	3 day lead time. Installed by subcontractor, user or ML Systems.	25 years Service by ML Systems.
The NOVA Group Smithfield, RI (800) 553-8468	3M Silverlux™ Reflectors.	Contact company for price information. Payback 6 months to 2 years.	95% Reflective, Custom Designed for all types of fixtures. 3M Direct Lease or Financed. Established since 1985. serviced by National Network.	1 to 3 weeks delivery. Installation by Authorized Network.	5 years 3M Direct Labor and Material (Fortune 100 Company).

FLUORESCENT REFLECTORS

COMPANY	MODEL	PRICE & PAYBACK	FUNCTIONS & FEATURES	LEAD TIME & INSTALLATION	WARRANTY & SERVICE
Orion Industries Hayward, Ca. (415) 487-5980	Orion Mirror, also specular aluminum.	Orion Mirror $40-$50, specular aluminum $25-$35. 6 months to 2 years payback, $6-$10 installation fee per fixture.	Custom designed for all fluorescent and HPS fixtures. Available to fit 1 by 4, 2 by 4, 1 by 8, 2 by 2, and 3 by 3 fixtures. Orion Mirror is a mirror surface with a multi-layer coating, vacuum-deposited on high-grade, anodized aluminum substrate. Omega Mirror 95% reflective, specular aluminum 86% total reflectivity.	Delivery in 2-6 weeks. Installation by contractor.	10 year warranty on Orion Mirror and aluminum.
Parke Industries 2246 Lindsey Way Glendora, CA 91740 (714) 599-1204 FAX: (714) 599-1208	Custom silver specular reflectors, Aluminum and aluminum film all retrofits. Also, Powerlite series of reflectorized fixtures for new construction, commercial and industrial.	Reflectors installed, $30-$50. Payback typically 8 months-2 years. Reflectorized fixtures $20-$30 more, payable 6-18 months.	Available in all configurations, reflectivity 85-88% on aluminum, 94% on silver, custom-designed. Offices nationwide. Project finance available. All designs independently tested for maximum efficiency and proper light distribution. Complete systems design.	2-3 weeks, installed by Parke or project managers supplied.	10 yr. warranty on silver. 25 yr. warranty on aluminum service by Parke.
Reflect-A-Light Palacka, Fla. 1-800-537-1629	Reflect-a-Light aluminum reflectors.	Competitive pricing-contact company. Payback 6 months-2 years.	13 different styles and custom designed, any application. Two types of aluminum-minimum 87% reflective.	7-14 days. Installation by Universal Lighting Maintenance.	25 years, service by Universal Lighting Maintenance.
Reflective Light Technologies Redondo Beach, Ca. (213) 538-5256	Max-Flux, silver, domestic aluminum, and enhanced aluminum or imported aluminum.	Price based upon material and quantity.	Aluminum is Alcoa Everbrite. Alanod, or O.C.L.I. Configurations include 2 x 4 , 2 x 2, strip, others, also reflectorized fixtures. Reflectivity 95% on silver, 86% on domestic aluminum, 89% on imported, and 94% on enhance aluminum.	4 weeks. Installation by RLT nationally.	7 years silver, 25 years aluminum. Service by RLT.
Silverlight Corp. Burr Ridge, Ill. (708) 986-1651	Sterling Silver reflectors, also Alcoa Everbrite reflectors.	Silver $35-$50 suggested, aluminum $28-$40 suggested (2 by 4 prices). Paybacks 1-2 years.	Sterling Silver made by "sputtering process", sputtered onto DuPont Type D Mylar polyester, also has anti-static inhibitor to prevent dust accumulation. 95% reflective. Custom-designed reflectors.	2-3 weeks. Installated by contractor.	Full 5 year warranty on silver, aluminum 25 years. Service by contractor.
Tamarack Corporation 282 Shafer Rd. Corapolis, PA 15108 (800) 837-9464	Alcoa EverBrite aluminum and Specular + silver reflectors.	Depending on reflector model and material price may vary from $6.25 to $36.00. Payback: 6 months to 2 years.	Custom-designed and standard configurations available for retrofitting troffers, strips, and HIDs. New fixtures with reflectors also available. Aluminum reflectivity: 86% Silver reflectivity: 95% Full service company: design, manufacturer, and installation with special lamp	Lead time: 5-10 days. Installed by customer, distributor, or Tamarack.	25 years for aluminum, 5 years for silver. Service by distributor or Tamarack.

			and ballast options. Tamarack is an EPA Green Lights Ally.		
3M Construction Markets Department St. Paul, Minn. (612) 736-2388	Silverlux silver reflectors.	Varies with individual manufacturers (see features column).	Custom-designed, uses 3M silver film. 95% reflective. Individual manufacturers of Silverlux reflectors include Roth Bros., Youngstown; Mor-Lite, Largo, Fla; Jedco, Salt Lake City; Nova Conservation, Smithfield, R.I.; Dynamic Energy Products, Chicago; X-Tra Light, Houston and Costa Mesa, Ca.; and Gelsenberger Mfg., Woodridge, NJ.	Varies with individual manufacturers.	5 years. Service by dealers or contractor.
Warner Technologies Los Angeles, CA (310) 444-0488 New York, N.Y. (212) 541-3711	Silver and aluminum reflectors, also Speculume reflectorized fixture.	Aluminum $30-$35, silver $35-$45 (retrofit). Payback within 1-2 years.	Reflectors and reflectorized fixtures in sized 1 by 4, 1 by 8, 2 by 2, 2 by 4. Reflectors are 3-piece — user can take ballast cover off without using screws. Reflectivity: 88% aluminum, 95% silver. Alanod aluminum used.	For reflectors, 2 weeks. For fixtures, 4 weeks shipping. Installation by Warner.	Warranty 5 years silver, 25 years aluminum. Service by Warner.

ELECTRONIC BALLASTS

COMPANY	MODEL	PRICE & PAYBACK	RATED LIFE & FEATURES	WARRANTY & SERVICE
Advance Transformer Co. Rosemont, Ill. (708) 390-5000	Advance Discrete Electronic, Advance Reduced Harmonic Discrete Electronic, Advance Mark V Integreated Circuit, and Advance mark VII Controllable Integrated Circuit electronic ballasts. Ballasts are available for 1, 2, 3 and 4 lamps. Lamps operated include T8, T10, T12, Compact, and Slimline. Both Parallel and Series designs available.	Prices range from $25-$80, depending on model. Energy savings up to 43% over a standard system; up to 75% with the controllable Mark VII. Payback can be from 6 months to 3 years.	Average rated design life of 20-25 years. Light output of all models meets proposed ANSI standards. All ballasts are high power factor (greater than .9), most greater than .95. All models 75% quieter than A rated Magnetic ballasts. THD of reduced harmonic Discrete Electronic ballasts less than 20%. THD of Mark V and Mark VII ballasts less than 15% for 277V, less than 10% for 120V. Mark V and Mark VII have constant light output at plus/minus 25% voltage fluctuations, "soft start" rapid start design, crest factor of 1.4 or less. Mark VII can be controlled from 100A% to 20% light output by a wide range of compatible devices: wall dimmer, occupancy detector, ambient light sensor, building management system, etc.	3-year replacement warranty. Service through Advance and Advance Authorized Stocking Distributors.
Duralux Industries Austin, TX (512) 443-3292	EconaLux electronic ballast. 120V and 277 V two-lamp rapid start models for T-12 lamps. Instant start and multi-lamps ballast introduced in near future.	Typical price is $20 to $30, depending on volume. Paybacks will usually range from six months to two years.	Average rated life 25 years, T-12 and T-8 models operate in true rapid start mode. T-8 unit is full light ballast. Harmonics less than 13%, reduced flicker.	3-year warranty. Full replacement service by distributors.
Elba USA City of Industry, CA. (800) 626-3522	Elba electronic ballasts, main models dedicated T-8, dedicated T-10, T-12, 8 ft. slimline. Each is 2-lamp ballast.	Prices start around $30 (to end user). Paybacks less than two years.	Rated life 25 years. All ballasts are parallel circuit. Most have low harmonic distortion (low-THD), below 10%. Low EML	Warranty three years and labor allowance. Service by contractor.
Electronic Ballast Technology Inc. Torrance, CA (310) 784-2000	1-, 2-, 3- and 4-lamp ballasts for F40, T-12 and T-8 lamps in both 120 and 277V. Also Slimline and high-output ballasts. Ballast for 39, 40, and 50 watt Biax. High power factor ballast for Quad 4-13, 18 and 26 watt compact fluorescent lamp. Dimming ballast and dimming control system for 2 lamp 32 W and 40 W applications.	Contact company for price information. Paybacks range from 6 months to 2 years, depending on hours of operation and electricity costs.	All EBT ballast feature a minimum of less than 20% Total Harmonic Distortion. (THD). 100,000 hours rated life. No flicker, very low audible noise, meets FCC requirements for electromagnetic interference and radio frequency interference. 1-,2- and 3-lamp ballasts are for both rapid start (preheated) and instant-start operation. Quad ballast feature THD 20% and 98% power factor. Dimming ballast control features no additional wiring requirement. The unit wires directly into existing wall switch junction box.	3 years, full replacement, service factory- directed through distributor.
Etta Industries Boulder, CO (303) 444-2244	Sinusoidal Electronic Dimming Ballasts. Models include 2-lamp 120 and 277 V ballasts for 32 W, 34 W, 36W, 39 W, 40 W T-12 lamps 4-foot, T-10 and T-8 lamps (3 lamp to be available). Also available 40W Biax (dimming), 50- and 100-W HPS. Continuous dimming capabilities.	Typical price for ballasts; Non-dimming $25-$40. Dimming $40-$75/ballast, dimming controls and photocells available through Etta or approved control manufacturers. Quantity discounts available. Paybacks 1-2 years, saves 25-85% over standard core and coil.	Average rated life more than 20 years. 1.41 crest factor, flicker-free operation. Sinusoidal lamp drive voltage, soft-starting, soft heating of filaments. True Rapid Start-protects lamp life. Meets FCC guidelines on EMI, surge-protected in accordance with IEE Standard587.Powerline harmonics under 5% voltage regulation and brownout protection. Continuous dimming capabilities from 100 to 5% of full light output. Power factor 0.99. Digital dimming control system.	3-year labor allowance $10. Nationwide service. Through distributors and technical sales representatives.

King Technology, Inc. Carlstadt, NJ (201)356-8711	F-40 T-12 2-lamp ballast, 120V and 277V versions: F-32 T-8 2-lamp ballast, 120V and 277V versions. 3-lamp, 4-lamp, F-96. Integrated Circuit Ballast with 3-5% THD to be introduced June 1992.	Prices range from $17-25 for 2-lamp, 4-ft rapid stars ballast. Quantity discount available. Contractors' financing available.	Rated life 25 years. THD 3-12%. BEF 1.52. Crest factor 1.5.	3 year warranty and labor allowances.
MagneTek, Lighting and Electronic Products 200 Robu Road Paranus, NJ 07652 (800) BALLAST Huntington, IN (219) 356-7100	Triad electronic, 1-4 lamps T-12, 1-3 lamp T-10, 1-4 lamp T-8, also 1-3 lamp compact, 2-lamp slimline, 2-lamp high-output, 1-3-lamp light-level switching. Reduced Harmonics "RH" full light output (20% THD, High Performance "HP" full light output (10% THD and Low Wattage "L" reduced light output (20% THD versions of above also available.	Prices from $22 to $50. Energy savings potential up to 60%. Premium payback over magnetic typically less than two years.	Rated life 25 years. Meets FCC guidelines for RFI and EMI meets IEEE Standard 587 (for surge protection). Depending on type of lamp being used, ballast-lamp combination has a lumen-to-watt ratio between 75 and 105. Lowest wattage T-8 systems in the industry. Light level switching systems available. Most models parallel lamp configured allowing companion lamps to remain fully lit when 1 or more lamps fail.	Exclusive 5 (five) year replacement warranty. Nationwide service, call 1-800-BALLAST.
Motorola Lighting Inc. 887 Deerfield Parkway Buffalo Grove, Il 60089 1(800) MLI-0089	High performance electronic ballasts Rapid Start-1,2,3 and 4 lamp for T8 and T12 lamps. Instant Start-1, and 2 lamp T12. All models available in 120 or 277 voltages.	Simple payback can be less than one year depending on burning hours, Kwh rate and rebate programs. Two year range is typical. Please contact a Motorola Lighting Inc. Authorized Distributor for pricing information. For a distributor listing or other information, please call 1(800) MLI-0089.	Power factor greater than .99; total harmonic distortion less than 10%; crest factor less than 1.5; no visible flicker; no audible noise; meets all applicable regulatory requirements including UL and FCC. Rated life 20 years plus; features poke-in connector for simplified installation.	Three year warranty includes on-site service through our national Authorized Service Companies. Call 1(800) MLI-0089 for more information.
Royce Electronic Products Inc. Ontario, Ca. (714) 467-1400	Royce ballasts, 2-, 3-, or 4-lamp ballasts, each in 120 and 277V.	$22-$35. Payback less than two years.	Meets FCC noise, EMI, RFI requirements. Comes in variety of output (low-, medium and high) for each lamp.	Service by contractor. 3-year warranty.
Smallwood Technologies Littleton, Co. (303) 973-7401 (800) 458-1635	Sever Plus, sinusoidal electronic ballast. Models include SEB24120, SEB24277 (recommended for T-10, T-12), SEBT8120 (for T-8x), SEB-18120, SEB-18127.	Saver Plus in low $20s. Payback within one year. Savings up to 40%. SEB models $15.	Rated life 20 years. Flicker free, weighs approx. 1 pound, sound rated A, high power factor. No EMI/RFI interference. Saver Plus has less than 20% harmonic distortion. SEB models; 98% power factor, harmonic distortion under 10%.	3-year warranty. Service by contractor. $10 labor replacement-each ballast.
Tektron Enterprises Inc. 3122 West Alpine Ave. Santa Ana, CA 92704 (714) 641-1988 FAX: (714) 641-2005	Solid state full range dimmable ballast model 23A0000. 2 Lamp ballast for T8, T10 and T12 4ft. lamps.	Price from $40-$70/ballast depending on options and quantities. Payback 1-2 years. Saves 25-85% of energy over standard core and coil.	Average rated life 25 years. Patented design, custom IC. Dimmable 0-100% (with very low flicker) by a standard incandescent dimmer. One dimmer can control up to 600 ballasts (1200 lamps). Manual, automatic, remote, or multi-function control. Typical power factor 97%. 3rd harmonic distortion less than 5%. Less than 18% thd. Can interface with occupancy sensors, daylighting etc. Quiet operation class a sound rated.	3 year replacement warranty. Service by contractor.

ELECTRONIC BALLASTS

COMPANY	MODEL	PRICE & PAYBACK	RATED LIFE & FEATURES	WARRANTY & SERVICE
Terralux, Inc. P.O. Box 1093 Oshkosh, Wi 54902 (800) 637-6909	Terralux Electronic Ballasts, Rapid start for 1 or 2 T-12 of T-10 lamps. Instant start for 1 or 2 T-8 lamps. All models available in 120 and 277 volts.	Contact company for price information. Payback will range from 6 months to 2 years depending upon application and hours of operation.	Average life rated 20 years. THD less than 10%. Crest factor 1.5. Meets all FCC requirements for EMI and RFI. High Light output available for T-8 lamps.	3 year replacement warranty. $10 labor allowance. Service by contractor.
Thomas Industries Tupelo, Miss. (601) 842-7212	REB microchip electronic ballast. 1-, 2-, and 3-lamp, 4-foot rapid start, T-12 and T-8, standard and dimming, 120 or 277V.	Approx. $30 each, payback approx. 1 year compared to standard magnetic ballast. Dimming adder approx. $30 each.	Rated life 60,000 hours. Adjusts input voltage to allow constant wattage and light levels to lamps. Compensates for power fluctuation of plus or minus 10% of normal input voltage. Constant light output over life of lamp, weighs only 14 ounces. Dimming accessories available, can be dimmed to 25%. Meets IEE criteria for RFI and EMI, THD less than 10%.	3-year warranty, service by Thomas.
Toshiba America Consumer Product, Inc. Buffalo Grove, Ill. (800) 453-4242	Toshiba Electronic Ballasts, 1, 2, 3, lamp F40, T12 in both 120 and 277V. Also, Electronic F96, T12 Slimline and high output in 120 and 277V.	Contact company for price information. Typical payback is 6 months to 2 years. Energy savings 20-40% over standard.	Average rated life over 20 years. Flicker free operation. Meets FCC standards for RFI and EMI. Power factor 98%. Meets federal ballast efficiency factor. Operates both rapid start and instant start F40 lamps. Operates 30° cooler which can lower air conditioning costs.	Warranty 3 years and $10 labor allowance.
Valmont Electric Co. Danville, Ill (217) 446-4600	Ultra-Miser line of 1-, 2-, 3- and 4-lamp ballasts for F40T12 and T-8 lamps. 1- and 2-lamp slimline and high-output ballasts available for 60-96-inch lamps. All models available in 120- and 277V ratings.	Typical value prices for F-40 models range from mid-$20s to mid-$30s. Input watts for 3-lamp F40T12watt rapid start is 90 saving 58 watts compared to standard lamps and ballasts. Input watts for 4-lamp F32T8 model are 104. Paybacks generally less than 2 years.	25 years average life expectancy. Meets FCC requirement for EMI and RFI. All ratings provide full light output. Lamp Current Crest factors range from 1.5 to 1.7 Ratings with total harmonic distortion of less than 20% available in all models.	Wrap-around warranty service program. 3-year replacement warranty. Up to $10 labor allowance.

LIGHTING CONTROLS

COMPANY	MODEL	SYSTEM HARDWARE & PRICE & PAYBACK	SYSTEM FUNCTIONS & FEATURES	LEAD TIME & INSTALLATION	WARRANTY & SERVICE
Andover Controls Corporation 300 Brickstone Square Andover, MA 01810 (508) 470-0555	Infinity LCX 890 Lighting Controller.	Stand-along microprocessor-based controller for up 24 zones with 20 amp, 277VAC mechanically-held relays. Networks and shares information with complete line of Infinity HVAC, security, and access controllers. Installed cost $150-$400 per zone. Payback within 3 years.	Programs zones with Plain English Language. TOD, holiday scheduling: daylighting, cleaning crew override; runtime analysis; tenant billing. 24 direct control inputs for motion senors, wall switches; 24 programmable inputs for photocells, dimming switches, wall switchies, motion sensors. Extensive data logging and reporting for each zone and input point. Optional color graphic workstations provide global alarm management, dynamic graphics, data archiving, year-in-advance scheduling.	Lead time 4-8 weeks, installation by user, contractor, or Andover Facility Automation Specialist.	18 months parts warranty. Service through Andover Facility Automation Specialists.
Advanced Control Technologies, Inc. 1936-B S. Lynhurst Dr. Indianapolis, IN 46241 (317_248-2640 (800) 866-2281 FAX: (317) 248-6898	Lighting control Panel (LCP)	8, 16, or 24 relay (20 amp) panels. Controllable in groups or individually. Accepts 5 signal types plus RS-485 port directly from BAS.One to three year payback.	8, 16, and 24 relay enclosures, fully assembled and tested. LCD-display allows programming of relays into groups via non-volatile memory. 20 amp relay circuits are controllable by groups or individually from analog, pulse width modulation, digital, or binary signals. RS485 port available for direct control of up to 150 panels from Building Automation system. Flash and local override available.	Lead time 2-6 weeks. Installation by contractor.	1 year factory warranty.
Davis Controls Corp. 5420 Newport Dr. Rolling Meadows, Ill. 60068 (708) 253-5314	Model LCM 20/40. Fluorescent Lighting Circuit Module.	Solid-state controllers for 120V or 277Y circuits-can control 16 amps of a 20-amp circuit. Operates with standard and energy-saving lamps, ballasts. Compatible with standard wall switches, interfaces directly with any BAS or EMS. Price $600 list, payback within 1 1/2 years.	Reduces energy consumption to 30% without noticably affecting light levels. Works by interrupting voltage 4 times each cycle. Maintains RMS lamp and ballast voltages within ANSI standards. Uses existing wiring and wall switches. Banks of lights can be switched on and off without light level change elsewhere. Reduces ballast temperatures.	Lead time stock to 4-6 weeks. Installation by authorized distributor.	1 year warranty, service by contractor.
Day-Brite Lighting Division Thomas Industries 1015 South Green St. Tupelo, MS 38801 (601) 842-7212	Energy Savings control system.	Day-Brite/Benjamin HI/LO HID Luminaires, Metal Halide or High Pressure Sodium, HI/LO Motion Detectors. PLC Controlled. 120 Volt thru 480 Volt Lighting Systems. Payback 2 yrs or less. Consult factory reps.	Day-Brite/Benjamin HID Lighting fixtures are automatically controlled by HI/LO motion detectors in warehouse aisles, storage rooms, etc. Automatically switches to 100% light when someone enters aisle, etc. When no one is present system switches to low, a savings of up to 70% can be achieved. No control wiring is required for installation of the HI/LO system.	4 to 6 weeks lead time, installation by electrical contractor.	5 year warranty on HI/LO fixtures and control system, service by electrical contractor or Day-Brite/Benjamin field service personnel.

LIGHTING CONTROLS

COMPANY	MODEL	SYSTEM HARDWARE & PRICE & PAYBACK	SYSTEM FUNCTIONS & FEATURES	LEAD TIME & INSTALLATION	WARRANTY & SERVICE
General Electric Co. Wiring Devices Div. 225 Service Avenue Warwick, RI 01886 (401) 886-6200	TLC (Total Lighting Control)	Up to 24,000 Os and Is. Uses 24-V, 20-amp panels. Each panel up to 48 20-amp relays. $150 per point, payback within 1-3 years.	Fully distributed. Can integrate occupancy sensors. Consistent buildup from standard low-voltage through automated lighting BAS. Supports integration with other BAS devices. User can access information anywhere on network. Blink alert of scheduled light level reduction. Computer cards supply expert scenarios. Telephone override.	Panels from stock. Installed by contractor or user.	1 year warranty. Installed by contractor.
Holophane Company 214 Oakwood Avenue Newark, Ohio 43055 (614) 345-9631	Automatic Energy Reduction System. For HID Lighting.	Composed of Holophane luminaire, occupancy sensor and control module.	Holophane luminaire is designed with a modified ballast and relay. The ballast is designed to operate and HID lampat Two output wattages, normal output like a standard luminaire and a reduced output. The system is wired to a contact closure through a control module allowing the luminaire to switch between output wattages.	Lead time 4 weeks. Installation by contractor.	Warranty 2 years, 5 on electrical components. Service by Holophone field service personnel.
Honeywell, Inc. Home & Building Control 1985 Douglas Drive N. Golden Valley MN 55422-3992 (800) 345-6670 ext. 7148	Honeywell Low voltage Lighting Control Systems.	Major components: Two-wire 24V communication bus-transmission unit, relays, realy controllers, dry contract interface units, switches. Price is function of system configuration. Payback as short as 1 1/2 years.	Provide on/off control of all types of lighting and other electrical loads, up to 256 individually addressed control points, 2048 branch circuits; feature EEPROM memory protection (no battery back-up). Individual, group, and pattern switches offer different control strategies. Add components by a simple network connection at any point. Reduced installation cost over other low voltage systems plus energy savings.	Lead time: Stock to 4 weeks. Installed by contractor.	1 year warranty: Service by factory.
Illumination Control Systems P.O. Box 034084 Indialantic, FL 32903 (407) 768-6238	FS Series Dimming controller. Series 25 and 100 thru 500 Automatic Lighting Controls. LPS: Loop Powered Sensor.	Over 40 solid state 120 and 277 VAC designs with manual, programmable, TOD and sensor controlled light levels for one or more 20 amp circuits. Prices start at $450 and paybacks are 1-1.5 years. Over 20 solid state daylighting controls. Continuous light level selection either manual, programmed or TOD. One or two sensor input with multiple logic options. Controls any reasonable number of 20 amp circuits. Prices start at $550. Typical paybacks 3-	Savings range between 40% to 95% Compatible with all BAS and EMS. Unaffected by individual wall switch operation regardless of circuit dimming level. Iron core ballasts become 15 to 35 times more reliable than electronic ballasts. Savings range 80% to 94% based upon NASA and NIST testing and 10 years of documented field use. Reliability demonstrated over 10 years of documented field use. Reliability demonstrated over 10 years of actual field use is in excess of 3 million hours Mean Time Between Failure.	Stock to 6 weeks. Installed turn key, by contractor, or by in-house staff.	3 year warranty parts and labor. Service by dealer. LPS, 5 year warranty. Service by dealer. Series 25 and 100 thru 500 Automatic Lighting controls.

Illumination Control Systems (continued)		months-1.5 years. Self-powered (2 wire) sensor for use on standard 4-20 ma signal loop on BAS/EMS. Provides current or voltage output in proportion to light level. Models for indoor and outside use. Prices start under $200.	Compatible with all BAS and EMS. Meets 0.1% linearity and 0.1% repeatability. Specs.		
Johnson Controls Corp. 507 E. Michigan Strett Milwaukee, WI 53201-0423 (414) 274-4000	Intelligent Lighting Controller (stand-alone component of Metasys FMS).	Price $200 per output, installed. Controller has up to 40 outputs per panel. Payback within 3 years.	Control of local or remote on/off outputs; weekly and holiday time scheduling; telephone and switch overrides; controls HID, fluorescent, and incandescent; stores normal and overtime use for billing purposes; 32 zones; switch interlock and cleaning crew interlinking; password protection; available for all common line voltages; can be networked together and with other FMS functions.	Lead time 4 weeks.Installed by local branch office or in-house staff.	1-year warranty. Service by 170 branch offices throughout North America.
Paragon Electric Co. 606 Parkway Blvd. P.O. Box 28 Two Rivers, WI 54241 (414) 793-1161 FAX: (414) 793-3736	EC128, Sun Tracker.	EC128 controller has 12 Os, 8 A or DIs, can expand to 196 points. Sun Tracker elec. lighting control handles 1 or 2 channels, 15 amp switch capacity. EC 128, $1,159. Sun Tracker $177-$266. Paybacks within 2 years.	EC128 user-programmable. 365-day calendar, automatic daylight savings time and leap year correction. Contains "Astro Dial" which compensates for changes in sunrise & sunset when input with latitude, date. Also can accept photocell and light-level sensor, connects with PC, security codes. Sun Tracker has Astro-Dial, battery backup, user-prommable.	8-10 weeks. Can be performed by electrician.	1-year warranty, service by contractor.
Peschel Energy Inc. P.O. Box 71 Salisbury, CT 06068 (203) 435-0808 FAX: (203) 435-8186	Peschel PL/5000 Series.	Enables user to vary output voltage to save energy. Average cost between $90 and $200/wk. Typical 480 277-Y, 80-amp system would cost about $7,000. Payback typically 1.5 years.	Autotransformer, installed at building's central switch. Allows lamps to warm up before voltage is reduced. User can program TOD with built-in timer option, or can be used with controller. Photocell option available. Has manual override switch, bypass switch, input circuit-breaker. Can be used with fluorescents, incandescents, HIDs.	8-10 weeks. Can be performed by electrician.	2 years, parts and service; units are field-serviceable.
PLC-Multipoint 11001 31st Place West Everett, WA 98204 (206) 353-3353	Model LC8U, HID and Fluorescent, Control.	Photosensor and controller. Switches 120V or 277V 20A circuits. Price $310.00 List. Payback 1.5 years.	Indoor/Outdoor Controller, with dual set-points. Includes Hold-On timer for HID lighting. Hand/Off/Auto Switch and Nema 3R Enclosure. Expandable up to 2 20A circuits.	Stock products installed by contractor.	2-year warranty factory service.

LIGHTING CONTROLS

COMPANY	MODEL	SYSTEM HARDWARE & PRICE & PAYBACK	SYSTEM FUNCTIONS & FEATURES	LEAD TIME & INSTALLATION	WARRANTY & SERVICE
Powerline Communications, Inc. 7 Green Tree Dr. South Burlington, VT 05403 (800) 767-3674	WatchKeeper Control System.	WatchKeeper has 8,16, 24, or 32 relay outputs, 32 digital inputs and 1-8 analog inputs. May have 255 panels on one network to provide over 8000 20 amp @ 277 VAC rated relays in one system. Software supports multiple networks. Price range is $40,00 to $110.00 per point of control. Payback is less than 2 years.	80 schedules per day, 365 day calendar, automatic daylight savings adjustment. Switch inputs are two or three wire. Voice guided telephone overrides, remote location touch screen control. All inputs may be shared with all panels on the network. Data and trend logging provide tenant billing based on time or kw use per circuit. Override control may be blocked for certain periods or lights may warn off to indicate an off command pending. Photocell control directed through the network and any panel may use the actual ambient light level for control. Seven day battery/capacitor back-up.	3-4 weeks installed by electrician.	1 year warranty & 10 year warranty on relays. Service by contractor with factory support.
Staefa Control System Inc. 8515 Mirlani Drive San Diego, CA 92126 (619) 530-1000	Smart II controller.	4 Is for standard Applications, 4 universal Is for unique applications and host I/O, 4 Is and 8 DIs. Price $347-$657. Payback within 1 year.	Five controller types-VAV, fan coil, heat pump, DDC and MUX, all with lighting capability. Localized control of small lighting zones, single point of operation for lights and HVAC. Night override, occupancy sensing, light dimming interface. Programmable for handheld control.	Lead time six weeks, installation by Staefa or authorized contractor.	12-month warranty; service by Staefa or authorized contractor.
Triad Technologies 6080-F McDonough Dr. Norcross, GA 30096 (404) 242-1922 FAX: (404) 242-1944	LP-2000 and LP-2500 series, lighting interface panel.	Panel designed to be connected to EMS or BAS. Assigns relays (no limit), telephone overrides (8,191 zones). Price $50-$125 per point (the more relays, the lower the cost). paybacks 6 months-3 years.	User bar-chart LED, 3 pushbuttons on actual processing unit (for setting up relays) contributes to ease of setting up relays. Option-flickering lights (warning) will keep lamps on for programmable period, with telephone override. Barrier between high and low-voltage components. Scheduling available through BAS. UL listed, FCC approved.	Stock-4 weeks. Installation by electrician.	1-year warranty. Service by factory or EMS distributor.
Wide-Lite P.O. Box 606 San Marcos, TX 78667 (512) 392-5821 (800) 235-2314 FAX: (512) 353-5822	ZoneMate HID Dimming with Automatic Energy Control (AEC) daylight control. Also Bi-Level HID hi-low level HID control.	Systems typically contain controller, ballasts, light sensors, manual switches. Timeclocks, occupancy sensors are option. Most controllers can handle one to hundreds of fixtures. Cost of a simple dimming control system starts at $500.00 for the controller, plus ballast cost.	ZoneMate and Bi-Level systems control lamp input at lamp side of ballast. ZoneMate with AEC controls, user-set levels to match daylight. Bi-Level switches HIDs to low standby levels when full output not needed. Can be used with Wide-Lite fixtures or utilizing remote ballasts for other r fixtures. ZoneMate controls mercury vapor, metal halide, HPS from 150-1,000W; Bi-Level controls, 150-1,500W.	Lead time 2-6 weeks. Installed by contractor.	5-year limited warranty. Service by contractor with factory support.

OCCUPANCY SENSORS

COMPANY	MODEL	OPERATION & INSTALLATION	CAPACITY	PRICE & PAYBACK	FEATURES, WARRANTY & SERVICE
Bryant Electric Co. Division of Westinghouse Electric Co. 185 Plains Rd. Milford, CT 06460 (203) 876-3600 FAX: (203) 876-3675	Motion Switch sensors. MSWV0-1600 wideview; MSWS-1277 wall switch; MSH-130 hallway model; MSH-WS wall switch, hallway; MSCM-600 ceiling mounted.	All models infrared. Wide-view, hallway models are wall-mounted. Most operate with separate power pack transformer (included in price); wall switch models have internal power pack transformer.	MSWV-1600, 1,600 sq. ft; MSWS-1277, 1,200 sq. ft; both hallway models; 130 ft; MSCM, 600 sq. ft. Power packs can take up to 20 amps; wall switch models up to 10 amps.	MSWV-1600, $92: both wall switch models. $60; MSH-130, $82; MSCM0-600, $60. Paybacks within 18 months.	MSWS-1277 will work with all Electronic Ballasts with 1700 Field of View. On MSH-130 and MSWV-1600, photocell option at extra cost. All sensors have LED, adjustable time delay 30 seconds- 20 minutes. Two-year warranty, service by Bryant.
Earlwood Technologies 1055 W. Germantown Norristown, PA 19403 (215) 539-4210 (800) 233-5193 FAX: (215) 539-5921	Lights Out. Includes LO-300WS (wall switch), LO-310SW (dual wall switch), Lo-550 (recessed ceiling), models 1,000, 2,000. All ultrasonic.	Except for wall switch and LO-550 (recessed ceiling), all can be installed on ceiling or wall. Ceiling models work with interface package (includes transformer).	Wall switch 300s can sense up to 350-800 sq. ft; 550, in excess of 550 sq. ft; Model 1,000, 1,000 sq. ft; Model 2,000, 2,000 sq. ft.	End user price typically $65-$113, control unit for ceiling models, $25-$43. Paybacks two years or less.	Adjustable time delay, sensitivity. On 55, four delay times. 310WS controls two lighting circuits. Wall switch models accept both 120v and 277v; ceiling models have 24vDC relay. Three-year warranty, service by local dealer network.
Hubbell Inc. Wiring Device Div. P.O. Box 3999 Bridgeport, CT 06605 (203) 337-3100 FAX: (203) 368-6245	H-Moss family of infrared sensors: models 100, 200, 750, 700, 1500 and 1200. Motion alert OS-120 outside sensors.	H-Moss 1500 and H-Moss 750 are wall switch replacement models. The remainder are wall or ceiling mount sensors.	750 wall switch covers 1450 sq. ft. 1500 wall switch covers 2,7 sq. ft. Overhead units cover from 200 to 2,500 sq. ft. Switching capacity differs among units (range: 640-2,500W). See brochure for details.	End user prices range from $64-$112 depending on model. Payback 1-3 years.	New H-Moss 1200 covers 380-2,500 sq ft., depending on ceiling height. Capacity can be added with Add-a-Relay. 135 local sales representatives off free energy audits, free layout design. Defective units replaced by company, no time limit.
Leviton Manufacturing 5925 Little Neck Pkwy. Little Nech, NY 11362 (718) 229-4040 FAX: (718) 631-6598	Model 6773 and 6774, ceiling mounted infrared sensors; 6775, wall switch infrared. 6777-Wide View.	6775: replaces regular wall switch. Ceiling models: sensor head can be aimed to whatever area is needed. These are used with switching relay (included in price).	All sensors cover up to 2,700 sq. ft., 180-degree angle. 6775 can cover 600W at 120v, 1,200W at 277v. Ceiling models can run to 15 amps at 120v, 10 amps at 177v, 660W incandescent. 6779- power control box rated at 20 amp 120/277v.	Suggested user prices: Ceiling models slightly under $100, wall switch under $100, wall switch under $65. Paybacks within 1 year at full load.	Adjustable delays 15 seconds-15 minutes. Wall switch very small (2 3/8 x 1 1/4 inches). All units conform to Calif. Title 24 standards (indicator light). Power Control Box can take more than one sensor. Service by authorized distributor, two-year limited warranty.
Lithonia Control Systems One Lithonia Way Decatur, GA 30035 (404) 987-4400 FAX: (404) 987-1002	Litronic ultrasonic occupancy sensors. Wall switch: LUS WS 120, LUS WS 277. Ceiling mount: LUS 250, LUS 500, LUS 700, LUS 1000, and power control station only, LUS PCS.	Wall switch models replace toggle switch. Ceiling mount models operate with a power control station, which powers up to 4 sensing heads, or sensing head operates with up to four power control stations.	LUS WS 120, 250 sq ft, 800W max; LUS WS 277 250 sq ft, 1,500W max; LUS 250, 250 sq ft, 20 amps max, ballast 13 amps tungsten; LUS 500, 500 sq ft, 20 amps max ballast, 13 amps tungsten; LUS 750, 750 sq ft, 20 amps max ballast, 13 amps tungsten; LUS 1000, 1000 sq ft, 20 amps max ballast, 13 amps tungsten.	Price: contact company. Payback 1-2 years.	Wall switch: automatic/off switch, adjustable sensitivity, adjustable “lights off” timer, no exposed screws or mounting hardware, low profile, five standard colors. Ceiling mount: on/automatic/off switch, adjustable sensitivity, adjustable “lights off” timer, no exposed screws or mounting hardware, low profile. 3 year warranty.

OCCUPANCY SENSORS

COMPANY	MODEL	OPERATION & INSTALLATION	CAPACITY	PRICE & PAYBACK	FEATURES, WARRANTY & SERVICE
MyTech Corporation 10125 Metropolitan Austin, TX 78758 (512) 836-9400 FAX: (512) 836-9548	America's most comprehensive line of Ultra-sonic and PIR sensors and Energ-y Analyzers. Wall switch replacement sensors in-clude: six application-specific Double Density ultrasonic SupreSensor models; and 7/16: profile passive infrared LowPro 4. Ceiling sensors: eleven Fail Safe OmniSensor models for Medium and Large Areas, Corridors and Stairwells.	SupraSensor and LowPro 4 wall sensors control from 120 VAC thru 277 VAC have adjustable Timer and Sensitivity, are tamper proof and do not require Grnd. Class II OmniSensors control multiple 20 amp circuits, network to cover areas of any size, and are Fail Safe. Energy Analyzers measure lighting burn hours, unoccupied lighting hours, and run time of electrical equipment.	Wall sensors cover up to 800 SF, control up to 1500 Watts @ 277 VAC, and can control two circuits. OmniSensor models cover from 350 SF to 3300 SF with six models, and four models controlling from 65 to 150 lineal feet in corridors. Pulse/latching low voltage lighting system models available.	Pricing varies by coverage and technology. Payback ranges from 6 to 18 month. ROI ranges from 125% to 250%.	Ultrasonic SupraSensors have flexible Timer adjustment from 8 sec. to 32 min. Restroom SupraSensor does not require line of sight. If OmniSensor system is damaged or fails, lighting remains On. OmniSensor options include: Dual Output (HVAC); adjustable DayLight Control; and Gain Setback for zoning.
Novitas Inc. 1657 Euclid St. Santa Monica, CA. 90404 (213) 452-7890 FAX: (213) 452-6168	Novitas ultrasonic and infrared sensors. Wall switch models: 01-122, 01-133, 01-041. Ceiling models: 01-071/072 (one way); 01-082/082 (two way); 01-091 /092 (corridor & warehouse); 01-074/084/094 (EMS control). Switch-packs: 13-012 (low power lighting controller); 13022 (high power EMS controller).	Wall switches replace ordinary switch, 2-wire installation. Ceiling sensors operate with Novitas switchpack or GE relay. Suggested mounting heights 8-25 ft. depending on model. Switch-pack can power two sensors; one sensor can power 10 switchpacks, EMS versions can also control HVAC.	01-122/01-133, 500 sq ft; 01-122-120V, 500W max; 01-141, 500 Sq. ft.; 01-141-120V, 800W max; 01-041-277V, 1,200W max; 01-133-277V; 1,000W max; 01-071/072/074, 900 sq. ft.; 01-082/083/084, 2,100 sq. ft.; 01-091/092/094, 100 linear ft. Fluorescent or incandescent.	Suggested user prices: wall switches, $52.50-$65, depending on quantity. Paybacks within 1.5 years. Ceiling sensors, $94-$150, depending on quantity. Paybacks 1-1.9 years.	Wall switches: Automatic or manual on. Pushbutton on avoids activation when adequate daylight is needed. Pushbutton off for periods when room is occupied. Manual on/off override switch. Adjustable time delay 90 seconds-30 minutes. Ceiling sensors: Automatic on. Multiple frequencies available for independent control of adjacent areas. Adjustable time delay 30 seconds-12 minutes, manual override switch. All have sensitivity adjustment, 5 year warranty, service by company.
Pace Technologies P.O. Box 30394 Walnut Creek, CA 94598 (415) 934-6274 FAX: (415) 934-6778	Micromatic infrared wall switch sensor model MS 180-A10.	Replaced existing toggle switch. Can be applied at either 120V or 277V.	Covers 1,000 sq. ft. Switching capacity of up to 1,500W at 277V or 800W at 120v.	Micromatic: $35-$45 depending on quantity. Payback under 1 year.	Low profile device using a standard switchplate. 180-degree field of view. Adjustments provided for timing, sensitivity, and photocell. Included test LED. 3 year warranty, service by manufacturer.
RAB Electric Manufacturing, Inc. 70 Ludlow Avenue Northvale, NJ 07647 (201) 784-8600 FAX: (201) 784-0077	LightAlert!, PIR Sensors; LOS2400, Ceiling Sensor; LOS2400, Hallway/ Aisle Sensor; LOS300, Wall Switch Replacement; LOS900, Wall Switch Replacement; £OS1000, Wall Switch Replacement; Complete Line of Outdoor PIR sensors.	All units are passive-infrared and are compatible with electronic ballasts and dimming systems. The LOS2400 ceiling sensor has an integral relay and interface with line voltage. Wall sensors have 180 degree field of view.	LOS300- 300 sq. ft, 300 watts. LOS900- 1,500 sq ft, 900 watts. LOS1000- 1,000 sq. ft., 800 watts. LOS 2400- 2500 sq ft, 2,400 watts.	For price information contact factory. Payback under 18 months.	Adjustable Time/Sensitivity. Tamper-proof controls. LED includes capture of movement. Wall Switches for 120 VAC, 277 VAC. Standard one-year warranty and low cost. Service available. Protection against RF interference.

Sensor Switch, Inc. P.O. Box 1088 Branford, CT 06405 (203) 265-2842 FAX: (203) 269-9621	Complete line of Sensor Switch Occupancy Sensors designed to control all interior areas, models include line voltage, 24 volt and 15 volt systems.	All models infrared. Line voltage replace existing toggle; 15 volt operates with control unit dual top; 24 volt interfaces with GE RR7 relays.	Wide Views, 1,600 sq. ft.; Ceiling Mouts 300 or 600 sq. ft.; Hallways 130 linear ft; Wall Switches 2,400 sq. ft.: 1200 watts: Control Units 20 amps.	Consult factory for price. Payback within 18 months.	24 volts sensors may combine daylight control. Sensors have LED. Time delay is adjustable from 30 seconds to 20 minutes. One year warranty, service by Sensor Switch. All made in U.S.A.
TORK One Grove Street Mt. Vernon, NY 10550,(914) 664-3542 FAX: (914) 664-5052	Model PC10 Ceiling Mount. Model PS51A In line wall switch.	Both models are passive infrared, and control fluorescent or incandescent lighting as well as motor loads.	PC10 covers 1000 sq. ft. PS51A covers 800 sq. ft.	For price contact factory. Payback is 1-2 years.	Use adjustable turn-off delay. Shortened time delay for test purposes. Two year limited warranty. Service provided by TORK.
Unenco, Inc. 2555 Nicholson St. San Leandro, CA 94577 (800) 227-0452	Complete line of Passive infrared and ultrasonic occupancy sensor both wall and ceiling mounted. Wall switches: sOM -500-A, SOM-1000-A, SOM-1000-A-2 (2 circuit switching, complies with Calif. Title 24), SOM-1200-2-HD. Infrared ceiling mounted: PIR-1000, PIR-2000. Models PIR-1000-P and PIR-2000-P are self-contained with power pack. Ultrasonic ceiling mounted: C-600-R, C-1000 and C-500-2000.	SOM models are all passive infrared wall switches, 2 wire installation and replace ordinary wall switches. Passive infrared ceiling sensors are semi-flusch, low profile for HVAC or lighting and cover from 1000 to 2000 sq. ft. All ceiling mounted sensors work with power pack 120/277 VAC. Unenco sensors work in conjunction with electronic ballast and can interface with EMS.	Wall switches, SOM series coverage range is from 800 sq. ft. up to 4000 sq. ft. dependent upon model. Load capacity range is from 1000 watts up to 4000 watts per circuit dependent upon model. Passive infrared ceiling mounted sensors cover anywhere from 1000 to 2000 sq. ft.; load capacity range is up to 4800 watts at 277 VAC. Ultrasonic ceiling mounted sensors range from 400 to 2000 sq. ft. load capacity is up to 4800 watts at 277 VAC.	Suggested user prices for wall switches range from $52 to $88 dependent upon model. Passive infrared ceiling mounted sensors range from $52 to $85 dependent upon model. Ultrasonic ceiling mounted sensors range from $61 to $88 dependent upon model. Power switch packs, all models $18. All prices for all models $18. All prices for all models dependent on quantity. Payback is 1-2 years.	SOM wall switches, automatic or off, neon lighted switch, LED indicator, by-passkey, 3-15 minute time delay. 30 second time delay for testing, 180 degree view coverage. Ceiling sensors, by-pass key, sensitivity adjustment, LED indicator, PIR series 360 degree viewing coverage. All products offer 3-5 year warranty, free layout and design, toll free 800# for technical support and a 90 day money back guarantee. Conforms to OSHA, U.L. listed, all units are metal enclosed.
Visonic LTD. 20 Northwood Drive Bloomfield, CT 06002 (203) 243-0833	Super sensitive occupancy sensors. No more lights-off while you work!SRN 2000E Series Passive Infrared Sensors; up to 168 beams; 500 foot corridors. Power packs for lighting and HVAC control. PIR guards are also available for areas in which the detector could be damaged physically.	SRN 2000E Series IR's are wall mounted and are generally used with power pack.	Sensors cover areas up to 13,500 square ft. (100 degree field-of-view, 125 x125 ft); 500 ft. hallways. 40 coverage patterns available with one unit. Timer, range, and light sensitivity controls on-board. 30 degree vertical and horizontal calibration. Power packs supply 250mA VDC and switch 30 amps. Secured in junction lighting box.	SRN 2000E Series from $50,00-$68.00. Payback varies on application size.	Toll-free technical support and design service standard. 6 year repair warranty.
The Watt Stopper 296 Brokaw Road Santa Clara, CA 95050 (408) 988-5331 (800) 879-8585 FAX: (408) 988-5373	DT-100L Dual Technology has PIR, Ultrasonic and Lightlevel in one sensor; PIRD Automatic Wall Switch WI-120-277 for single or bi-level lighting, WI-R7P for low voltage, WS-120/277 Decorator Wall switch (includes light level); WPIR and CI-100 ceiling sensors, Ultrasonic Ceiling sensors W-500-A, W-1000A, W-2000A, W-2000H, Ultrasonic wall switch W-120C, and lightlevel sensors.	Wall switch sensors have dry contact relay for switches, 4 wire model available to hand two loads of lighting; Ceiling sensors are 24VDC and control lighting or HVAC, work with power switchpacks (5-20amps) DT-100L and CI-100 have isolated relay to communicate with EMS or HVAC.	DT-100L and CI-100 have choice of 3 lens patterns of up to 1500 sq. ft. wall switches are 180 degree and up to 1000 sq. ft., ceiling sensors range from under 500 to more than 2000 sq ft; Complete line of PIR, ultrasonic, and dual technology to cover all building spaces; All wall switch versions 50-1200 watts.	Wall switches $50-65, ceiling sensors $70-100, DT-100L $140, paybacks within 1.5 years.	Adjustable time delay from 15 seconds to 30 minutes, sensitivity adjustable on all sensors; customs models available; daylighting feature on wall switches, DT-100L and CI-100 will hold one level of lighting off when adequate natural light is present; 5 year warranties, service by manufacturer, toll-free customer service/technical support.

Index

This Index is provided for general access capability. Page numbers indicate major mentions or significant passages concerning the topic. It is suggested that the reader also consult the Table of Contents and locate sections that can provide the most complete information regarding a given topic.

Associations, 234-235

Avoided Environmental Cost
Factor, 227-229

Ballasts
Traditional fluorescent ballast composition, 10
Service life, 17

Ballast circuits, fluorescent
Preheat, 12-13
Instant start, 13
Rapid start, 13-14

Ballast retrofit, 21, 24
Energy-efficient ("premium") magnetic ballasts, 24
Cathode cut-out ("hybrid") ballasts, 24-25
Electronic ballasts, 25-26

Brightness (luminance), 6, 82

Capacitor in ballasts, 10-11

Class P ballasts, 10

Color, 87-92
As part of visible light spectrum, 87-89
Color rendering, 6-7, 20, 21, 23, 90-92
Color Rendering Index, 6-7, 23, 90-92
Color temperature, 89-90, 91

Color rendering, 6-7,

Coefficient of utilization (CU), 62-65

Controls, 20

Design, 21

Demand for power (kilowatts), 99, 103

Efficiency, 7-8
U.S. energy efficiency, 97, 98

Energy service companies, 196-198

Energy use (kilowatt-hours), 98-99, 102, 103-104

Environmental benefits of lighting upgrades, 201-229
Calculating conserved resources, 223-227
Calculating deferred emissions, 221-223
Carbon dioxide and global warming, 202-203
Environmental Assessment Worksheet #1: Calculating Conserved Resources, 225-227
Environmental Assessment Worksheet #2: Calculating Deferred Emissions, 222-223
Green marketing, 206-207
Nitrogen oxides and smog, 204-205
Sulphur dioxide and acid rain, 204

F32T8 fluorescent lamp, 23-24

F40T10 fluorescent lamp, 22-23

F40T12/ES fluorescent lamp, 19-20

Federal assistance, 235-236

Federal Ballast Energy Law, 18

Financing, 163-198

Fluorescent ballast disposal, 116, 154, 156-160

Fluorescent lamp disposal, 116, 154-155

Fluorescent lamps
Identification, 14-16
Service life, 16-17
Traditional lamp composition, 9-10

Fluorescent lighting, 8-17

Fluorescent lighting system
Composition, 9
Theory of operation, 11-12

Footcandles (illuminance or light level), 4-5

Glare, 6, 82-87
Direct glare, 6, 82
Disability glare, 82, 84, 85
Discomfort glare, 82
Reflected glare, 6, 83

Green Lights Program, 208-221
List of Partners, Allies and Endorsers, 211-219

Heating, ventilation and air conditioning (HVAC), 98, 119-125

High Output (HO) fluorescent lamps, 14

IESNA *Lighting Handbook*, 38

Illuminating Engineering Society of North America (IESNA), 38

Lamp retrofit, 19, 21

Light
composition, 87-89
definition, 1

Light loss factors
Non-recoverable, 65
Ballast factor, 66
Ambient fixture temperature, 66, 67
Supply voltage variation, 66, 67
Optical factor, 88
Fixture surface deprecia-

tion, 68
Lamp position (HID systems), 68
Recoverable, 68
Lamp burnouts, 68-69, 132, 134-137
Lamp lumen depreciation, 69-70, 132, 133
Fixture dirt depreciation, 70-73, 139
Room surface dirt depreciation, 73-74, 75, 139-140

Light meters, 42-44

Lighting management, xiii

Lighting system
composition, 2-3
mission, 3

Lighting upgrades/retrofit, xiv-xv, 21, 24, 26, 101, 104-105
Retrofit economic analysis and worksheet, 109-119

Lighting upgrade software, 106

Lumen method, 61-62

Lumens (luminous flux or light output), 4

Maintained illumination method, 61, 81
Light Level Worksheet #3, 81

Manufacturer rebate, 193

Measuring energy savings, 107

National Energy Policy Act of 1992, 17

Occupancy sensors, 20

Occupant survey, 40-42

Parabolic louvers, 83, 85, 86, 87

Payback formula, 107

PCB ballasts, 156-160

Performance contracting, 196

Planned maintenance, 22, 127-154
Benefits, 128-131, 140
Cleaning, 128, 129, 130, 138
Group relamping, 128, 130, 131-138, 139
Maintenance Worksheet #1: Determining Maintenance Interval, 145-147
Maintenance Worksheet #2: Maintenance Cost, 147-153
Managing a maintenance program, 143-154
Mission, 127
Operations and methods, 141-143

Planning an upgrade, 37-39, 105, 131

Power factor charge, 99-100

Publications, 236-237

Quality lighting, xiii, 51-52, 82
Lighting Quality Worksheet, 92-93
Value to bottom line, 52
Visual comfort probability (VCP), 83, 85

Ratchet clause, 99

Recycling of lamps and ballasts, 155, 160

Reflectors, 20, 26, 76, 77, 78, 79, 104

Rolling financing, 198

Room Cavity Ratio, 63

Specifying light level, 54-55, 57-60
Light Level Worksheet #1, 57

Specifying lumen output, 53, 60-80
Light Level Worksheet #2, 75-80
Lumen method, 61-62

Survey, 37-49

Survey Worksheet, 45-49

System comparisons, rapid start fluorescent, 27-31

Task lighting, 26, 86

Thermal protection in ballasts, 10

Third-party financing, 193-198

Unit power density (UPD), 100, 103

Utility demand side management (DSM) programs, 164-193
Rebates, 170-171, 112
List of rebates, 171-191
Direct assistance, 192
Low-interest loans, 192
Getting involved, 192-193

Very High Output (VHO) fluorescent lamps, 14